SPECIFIC INTERACTIONS and the MISCIBILITY of POLYMER BLENDS

MICHAEL M. COLEMAN
JOHN F. GRAF
PAUL C. PAINTER
The Pennsylvania State University

SPECIFIC INTERACTIONS and the MISCIBILITY of POLYMER BLENDS

PRACTICAL GUIDES FOR PREDICTING & DESIGNING MISCIBLE POLYMER MIXTURES

TECHNOMIC
PUBLISHING CO., INC.
LANCASTER · BASEL

Specific Interactions and the Miscibility of Polymer Blends

a **TECHNOMIC** publication

Published in the Western Hemisphere by
Technomic Publishing Company, Inc.
851 New Holland Avenue
Box 3535
Lancaster, Pennsylvania 17604 U.S.A.

Distributed in the Rest of the World by
Technomic Publishing AG

Printed in the United States of America
10 9 8 7 6 5 4 3 2 1

Main entry under title:
  Specific Interactions and the Miscibility of Polymer Blends: Practical Guides
    for Predicting & Designing Miscible Polymer Mixtures

A Technomic Publishing Company book
Bibliography: p.
Includes index p. 491

Library of Congress Card No. 91-65261
ISBN No. 87762-823-8

To our friend and mentor
Professor Jack L. Koenig
(*he didn't teach us much,*
but what he did teach us was bloody important)

and

To our wives
Mary Jane, Christine, and Cathy
(*for their patience, encouragement and understanding*)[*]

---

[*] We had to say this because they threatened to beat us up.

TABLE OF CONTENTS

Preface, Apologies and Acknowledgments xiii

Glossary of Common Symbols xvii

1. The Thermodynamics of Mixing   **1**

   A. Introduction to Theories of Mixing 1
   B. Intermolecular Interactions 2
     Physical Interactions 6
     "Chemical" Interactions 8
   C. Regular Solutions 10
   D. Polymer Solutions and Blends 14
   E. Phase Separation in Polymer Solution and Blends.. 19
     Conditions for Phase Separation 19
     The Predictions of Simple Models 22
     Closed Loop Phase Diagrams and Lower
     Critical Solution Temperatures 26
     Hydrogen Bonding and Free Volume in
     Polymer Blends and Solutions 35
   F. Simple Models for the Description of the Phase
     Behavior of Polymer Blends 36
   G. References 45

2. A Practical Guide to Polymer Miscibility   **49**

   A. Introduction 49
   B. The Estimation of χ From Solubility
     Parameters 54

*Determination of Group Molar Attraction
and Molar Volume Constants* 56
*The Calculation of Solubility Parameters
and the Errors Involved* 60
*Estimation of the Solubility Parameters of
Strongly Associated Polymers* 64
C. Simple Rules Government Miscibility in
Polymer Blends 70
D. Very Weak or Nonexistent Favorable
Intermolecular Interactions 76
Dispersive Forces 76
Summary 83
E. Relatively Weak Favorable Intermolecular
Interactions 84
Polar Forces 85
Weak Specific Interactions 93
Weak to Moderate Forces 102
Moderate Specific Interactions 106
Summary 126
F. Relatively Strong Favorable Intermolecular
Interactions 127
Moderate Specific Interactions 127
Moderate to Strong Specific Interactions 129
Strong Specific Interactions 136
Summary 151
G. Final Words 152
H. References 153

3. **The Nature of the Hydrogen Bond**   **157**

A. Introduction 157
B. What is a Hydrogen Bond? 159
C. Experimental Chacrcterization of
Hydrogen Bonds 162
D. Hydrogen Bonding in Polymers 164
E. References 169

4. **Equilibrium Constants and the Stoichiometry
of Hydrogen Bonding**   **171**

A. Introduction 171
B. Association Models and
      Equilibrium Constant Definitions 176
C. The Stoichiometry of Hydrogen Bonding and
      the Relationship to Infrared Spectroscopic
      Measurements 180
    Simple Self-Association 180
    Competing Equilibria 185
    Temperature Dependence of the Equilibrium
    Constants 190
    Two Equilibrium Constant Model for
    Self-Association 191
    Carboxylic Acids and Urazoles –
    Formation of Cyclic Species 196
D. Hydrogen Bonding in Polymers 199
E. The Flory Lattice Model for Mixing
      Heterogeneous Polymers 202
    Lattice Model for Simple Self-Association 202
    Lattice Model for Competing Equilibria 212
F. "Transferable" Equilibrium Constants 214
G. References 220

5. **Vibrational Spectroscopy and the**
    Hydrogen Bond   **221**

    A. Introduction 221
    B. Origin of the Vibrational Spectrum 222
        Theoretical Background 222
        Normal Modes of Vibration and
        Normal Coordinate Analysis 223
        Conditions for Infrared Absorption 230
    C. The Practice of Infrared Spectroscopy 232
        Introduction 232
        Instrumentation 233
        Sample Preparation 236
        Quantitative Analysis and the Use
        of Curve Resolving 237
    D. The Vibrational Spectrum of Hydrogen
        Bonded Systems 252

Introduction 252
Potential Energy Diagrams for the
A-H - - B System 253
Bond Lengths, the Enthalpy of Hydrogen Bond
Formation and Frequency Shifts 263
Band Widths and the Temperature Dependence of
Absorption Coefficients 266
E. Quantitative Infrared Spectroscopic
Measurements of the Fraction of
Hydrogen Bonded Groups 272
Self-Association Equilibrium Constants 273
Measurement of Self-Association Constants by
Solution Studies of Small Molecules 279
Inter-Association Equilibrium Constants 285
Copolymer Blends and the Transferability of
Spectroscopically Determined Equilibrium
Constants 293
F. Mapping Phase Diagrams Using Infrared
Spectroscopy 297
G. References 305

6. **Association Models and the Thermodynamics of Mixing Molecules with Strong
Specific Interactions**   **309**

A. Introduction—Why Use Association Models? 309
B. Hydrogen Bonding in Small Molecules—
Open Chain Association in Mixtures where
One Component Self-Associates 314
C. Thermodynamic Properties of Associated
Solutions 320
D. A Simplified Expression for the Free Energy 325
E. Hydrogen Bonding in Polymer Mixtures—
Linear Association 328
Hydrogen Bonding in Polymers –
The Distribution in the Mixture 330
The Free Energy of Mixing–
A Choice of Reference States 333
F. Chemical Potentials 338

G. Dependence of Equilibrium Constants on
  Molecular Weight and Chain Stiffness 340
H. Phase Behavior 343
  Spinodal Equations 343
  Temperature Dependence 347
  Sample Calculations of Spinodals 348
  *Melting Point Depression in Polymer Blends
  where One Component Crystallizes –
  A Demonstration of the Composition
  Dependence of the ΔG_H Term* 354
I. Self-Association and the Enthalpy and
  Entropy of Mixing 360
J. Free Volume Effects in the Mixing of Polymers
  that Hydrogen Bond 368
K. References 374
Appendix 1: Self-Association Through the
Formation of Cyclic Dimers 377
Appendix 2: Linear Self-Association Described
by Two Equilibrium Constants 378

7. **The Calculation of Phase Diagrams of
  Strongly Interacting Polymers**   **381**

  A. Introduction 381
  B. Homopolymer (Self-Associated)–Homopolymer
    (Non Self-Associated) Systems 382
    Polyurethane and Polyamide Blends 382
    Poly(vinyl phenol) Blends 404
    Poly(methacrylic acid) Blends 429
  C. Homopolymer (Self-Associated)–Copolymer
    (Non Self-Associated) Systems 435
    Poly(vinyl phenol) Blends 436
    Amorphous Polyurethane Blends 449
    Poly(methacrylic acid) Blends 452
  D. Homopolymer (Non Self-Associated)–
    Copolymer (Self-Associated) Systems 453
    Preamble 453
    Vinyl Phenol Copolymer Blends 455
    Polyamide "Copolymer" Blends 466

   Methacrylic Acid Copolymer Blends 469
  E. Copolymer (Non Self-Associated)–Copolymer
   (Self-Associated) Systems 470
   Preamble 470
   Vinyl Phenol Copolymer Blends 472
   Methacrylic Acid Copolymer Blends 487
  F. References 489

Index 491

PREFACE, APOLOGIES & ACKNOWLEDGMENTS

*A theory is more impressive the greater the simplicity of its premises is,
the more different kinds of things it relates,
and the more extended is its area of applicability.*
–Albert Einstein

In "Principles of Polymer Chemistry" Paul Flory[1] wrote:

> The critical value of the interaction free energy is so small for any pair of polymers of high molecular weight that it is permissible to state as a principle of broad generality that *two high polymers are mutually compatible with one another only if their free energy of interaction is favorable, i.e., negative.* Since the mixing of a pair of polymers, like the mixing of simple liquids, in the great majority of cases is endothermic, incompatibility of chemically dissimilar polymers is observed to be the rule and compatibility is the exception. The principal exceptions occur among pairs possessing polar substituents which interact favorably with one another.

Indeed, very few miscible blends of non polar (or weakly polar) homopolymers were identified during a thirty year period following the publication of this book. During the 1980's, however, a steadily increasing number of compatible systems have been reported; but, in line with Flory's arguments, many of these can be regarded as "polar," particularly those that involve the formation of hydrogen bonds between the components of the mixture. We believe that miscible mixtures of this type will prove to be fairly common and the purpose of this book is to explore the circumstances in which single phase materials can be obtained. We will also describe a model for the phase behavior of these mixtures which we believe to have predictive value; or, in a simpler form, can at least be used as a practical guide to polymer miscibility.

In describing the phase behavior of "small" molecules it has been common to distinguish between what Fowler and Guggenheim[2] described as normal and associated liquids, the most common examples of the latter corresponding to molecules such as water, alcohols, amines, etc., that form hydrogen bonds. It was recognized at an early stage that the assumptions of simple models, such as regular solution theory, were not valid when applied to these types of mixtures and their unusual or anomalous properties could only be accounted for by recognizing that the molecules were associated in a specific manner (see Hildebrand[3] or Prigogine[4], for example). Indeed, Prigogine[4] proposed that the formation of a complex be treated by using the assumption of a chemical equilibrium between the monomolecules of the associated species, and this approach has formed the basis for the use of so-called association models.

Association models have until recently been largely ignored in treating hydrogen bonding in polymer mixtures. They have most frequently been applied to mixtures of alcohols with simple hydrocarbons, where the equilibrium constants used to describe association have most frequently been determined by a fit to thermodynamic data (e.g., vapor pressures, heat of mixing). In our work we have sought to do two things; first, adapt this approach to a description of the phase behavior of polymer mixtures; second, develop spectroscopic methods that provide an independent measurement of the equilibrium constants.

Our purpose in this book is to explore and describe this approach and attempt to illustrate its broad utility. In putting this work together, however, we found that our aims became decidedly schizophrenic, reflecting the character and interests of the authors and their propensity to yell at one another. We wish to address what we believe to be two overlapping yet different audiences. One would be primarily interested in the broad nature of this approach and the practical applications of a simple model. The second would be more interested in the nitty gritty of the derivations of the equations and some of the fundamental aspects of the spectroscopy of these systems. Accordingly, we have decided to essentially structure this book in the form of a sandwich. We begin with a brief introduction to theories of mixing and the phase behavior of polymer mixtures, followed by a practical guide to polymer miscibility based on a simple balance of forces approach (i.e., between favorable and unfavorable interactions). This chapter also has a more fundamental purpose, however, in that it serves to identify the types of systems in which, by copolymerization or other means, one might introduce the appropriate hydrogen bonding functional groups and obtain a miscible system. Those readers who do not wish, at least initially, to work their way through the derivations can then jump to Chapter 7 where the results of applying a more rigorous model are presented.

The "meat" of the book is the stuff in-between, Chapters 3 through 6, where fundamental aspects of hydrogen bonds, spectroscopy and the application of association models are described. Some derivations are presented in what will appear to the knowledgeable reader as excessive detail. For this we have no apology, in that this book is also intended as a teaching aid for those students new to this field. To continue our analogy, we also offer a "side-dish" with this menu, consisting of computer programs that calculate and display many of the important quantities described in this book (e.g., the stoichiometry of hydrogen bonding and its relationship to infrared measurements, phase behavior, etc.). In our view one can obtain a good feel for the miscibility of many systems by simply "playing with" these programs.

This is in many ways a personal book, reflecting the views, work and idiosyncrasies of the authors. It is not intended as a comprehensive review of the literature on hydrogen bonding and other specific interactions in polymers. It will thus largely be based on our own work and will often ignore important contributions from other authors and groups. For this we do apologize and beg the indulgence of our friends and colleagues in this field. (We feel particularly guilty about our neglect of some of the major contributions made by Eli Pearce, T. K. Kwei, Frank Karasz, Bill McKnight, Don Paul, Virgil Percec and Reimond Stadler).

Finally, much of the work described here was accomplished on the backs of a number of outstanding graduate and post-doctoral students and we would like to acknowledge the significant contributions of (in alphabetical order) Dorab Bhagwagar, Marty High, Stephen Howe, Jiangbin Hu, Joon Lee, Andrew Lichkus, Eric Moskala, Yung Park, Carl Serman, Suresh Shenoy, Dan Skrovanek, Wei-Long Tang Barry Thomson, Dan Varnell, Yun Xu and Jim Zarian. Our particular thanks go to the faculty of the Polymer Science Program at Penn State University for numerous stimulating, but useless, discussions; to Judy Bell and Linda Decker for "bashing our unintelligible scribblings into the computer" and to Yun Xu and Marty High for "volunteering" for the mind numbing, torpor producing task of proofreading this text. They will get their reward in heaven. And we would be remiss if we did not acknowledge those who have supported us financially through the past ten years or so; the Division of Materials Research (Polymers Program) of National Science Foundation; the Department of Energy; the Petroleum Research Fund administered by the American Chemical Society; the E. I. du Pont de Nemours & Company; the Shell Foundation; the ARCO Chemical Company; and Allied-Signal Inc.

REFERENCES

1. Flory, P. J., *Principles of Polymer Chemistry*, Cornell University Press, 1953.
2. Fowler, R. H. and Guggenheim, E. A., *Statistical Thermodynamics*, Cambridge University Press, Cambridge, 1939.
3. Hildebrand, J. H. and Scott, R. L., *The Solubility of Nonelectrolytes*, Third Edition, American Chemical Society Monograph Series, 1950.
4. Prigogine, I. and Defay, R., *Chemical Thermodynamics* (English Translation, Everett, D. H.), Longmars Green and Co., London, 1954.

GLOSSARY OF COMMON SYMBOLS

A	Chemical repeat unit of non self-associating molecule.
B	Chemical repeat unit of self-associating molecule.
G	Gibbs free energy.
G_m	Gibbs free energy of mixing.
G_H	Component of Gibbs free energy associated with hydrogen bonding.
h	h - mer
$\bar{h}^\circ$	Number average hydrogen bonded chain length of self-associated B units in pure B.
h_A	Enthalpy of formation of a hydrogen bond between B and A units.
h_B	Enthalpy of formation of a hydrogen bond for linear chain-like self-association between B units.
h_2	Enthalpy of formation of a hydrogen bond for dimer type self-association between B units.
K_A	Association equilibrium constant for formation of a hydrogen bond between B and A units.
K_B	Equilibrium constant for linear chain-like self-association between B units.
K_2	Equilibrium constant for dimer type self-association between B units.
M_A	Degree of polymerization of A covalent polymer molecules.
M_B	Degree of polymerization of B covalent polymer molecules.
n	n-mer

N_A	Number (moles) of polymer molecules of type A
N_B	Number (moles) of polymer molecules of type B
n_A	Number of A segments $= N_A M_A$
n_B	Number of B segments $= N_B M_B$
r	Ratio of molar volumes of A and B chemical repeat units, V_A / V_B.
R	Gas constant.
T	Temperature.
V_A	Molar volume of chemical repeat unit A.
V_B	Molar volume of chemical repeat unit B.
x_A	Mole fraction of A units.
x_B	Mole fraction of B units.
z	Lattice coordination number.
δ	Solubility parameter
$\Delta\delta$	Solubility parameter difference
$(\Delta\delta)_{Crit}$	Critical value of the solubility parameter difference.
σ	Symmetry number.
Φ	Volume fraction.
Φ_{B_h}	Volume fraction of a hydrogen bonded chain of h B units.
Φ_{B_hA}	Volume fraction of a hydrogen bonded chain of h B units and one A unit.
χ	Flory - Huggins interaction parameter.
χ_{Crit}	Critical Value of χ
μ	Chemical potential.

SUBSCRIPTS AND SUPERSCRIPTS

i	Interacting unit.
°	Property of pure component.
A, B, etc.	as defined above.

## CHAPTER 1

The Thermodynamics of Mixing

An Incomplete and Biased Overview

A. INTRODUCTION TO THEORIES OF MIXING

Guggenheim[1] classified mixtures primarily into two types, "those in which molecular orientation is unimportant and those in which is it all important." In this latter class are, of course, molecules that interact through the formation of strong specific interactions, particularly hydrogen bonds, but the majority of both theoretical and experimental studies of mixing have been aimed at obtaining an understanding of the behavior of mixtures of the first type and it is these simpler systems we will consider first. For binary mixtures of small (i.e., non-polymeric) molecules a particularly simple expression for the Gibbs free energy of mixing is obtained if the energy of interaction between the unlike components is the same as the energy of interaction between like molecules (i.e., the interchange energy is zero):

$$\frac{\Delta G_m}{RT} = n_A \ln x_A + n_B \ln x_B \qquad (1.1)$$

where n_A and n_B are the number of moles of components A and B and x_A, x_B are the corresponding mole fractions. As we will see, this equation is as simple as this subject gets and can be obtained by counting the number of arrangements of A and B molecules on a lattice of $n_A + n_B$ sites:

$$\Omega = \frac{(n_A + n_B)!}{n_A! \; n_B!} \qquad (1.2)$$

[For an athermal system, the natural logarithm of this equation gives the Helmholtz free energy, ΔF_m, but for solids and liquids not too close to the critical point, and at ordinary pressures, terms in PV or VdP are negligible, so that G is practically indistinguishable from F.]

There are a number of critical assumptions that form the basis for this "ideal mixture" treatment, amongst the most crucial of which are:

1. As already mentioned, the forces acting between like molecules and unlike molecules are identical so that mixing is random.

2. The molecules A and B are roughly the same size and shape.

3. "Free volume" is neglected.

These complications conspire to make a precise treatment of the problem of mixing an extremely difficult one, for which tractable solutions have only been obtained through the use of various simplifying assumptions. Our concern here is not a general discussion of these problems, however, but the particular set of assumptions that are necessary for what we believe to be a successful treatment of polymer mixtures, particularly those where there are strong, specific interactions. Accordingly, we will briefly, and more or less qualitatively, consider only those aspects of the subject of mixtures that provide the foundation for the model we will employ – hence the somewhat facetious sub-title of this chapter. The interested reader is referred to other texts for a more complete discussion[2-6]. We start by considering the origin and effect of a non-zero energy of mixing.

B.   INTERMOLECULAR INTERACTIONS

For "small," more-or-less spherical molecules (e.g., CCl_4), it is usual to consider interactions between molecules treated as a whole, but for molecules that consist of a number of chemical units linked together in some fashion (e.g., the n-alkanes, any polymer) it is far more useful to consider interactions between *segments*, sometimes defined in terms of identifiable chemical units, but which can also be defined in terms of some reference volume that may, for example, include part of a polymer chemical repeat unit, or a number of such units.

The interaction energy between molecules or segments can be considered to consist of two components, arising from repulsive and attractive intermolecular forces, respectively. Repulsive forces become significant at short distances and it is convenient to represent the repulsive potential by a term of the form $(\sigma/d)^{12}$, where d is the intermolecular distance. An exponential form [Aexp(-Bd)] has also been used, but there appears to be no fundamental theoretical justification for either choice. Attractive forces are better understood and we will confine our discussion to the energies of these interactions.

Table 1.1
Frequently Encountered Forces

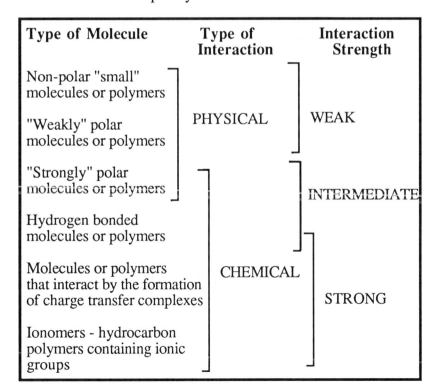

Type of Molecule	Type of Interaction	Interaction Strength
Non-polar "small" molecules or polymers		
"Weakly" polar molecules or polymers	PHYSICAL	WEAK
"Strongly" polar molecules or polymers		INTERMEDIATE
Hydrogen bonded molecules or polymers		
Molecules or polymers that interact by the formation of charge transfer complexes	CHEMICAL	STRONG
Ionomers - hydrocarbon polymers containing ionic groups		

We will somewhat arbitrarily categorize the most frequently encountered forces between molecules or polymer segments as illustrated in table 1.1.

We have employed the usual criterion of "interaction strength," where interactions between non-polar molecules are considered to be "weak" relative to those that are more polar. We will briefly discuss the nature of these interactions below, but it will prove useful to first place our definition of "weak," "intermediate" and "strong" interactions on an energy scale that is relative to RT at ambient temperatures, as this will ultimately lead us to consider separate and distinct ways of describing weak and strong interactions.

Consider a pair of molecules or segments of molecules located adjacent to one another and oriented in such a fashion that attractive forces between them are maximized and for some (perhaps extremely small) interval of time they can be considered to be associated.

Figure 1.1 shows the results of a simple calculation (on a molar basis) using Boltzmann's distribution to determine the fraction of molecules (or, for our case, segments) that have energies greater or equal to an "energy of dissociation," ΔE, at a given temperature. Above interaction strengths of about 3 kcal. mole^{-1} less than 1% of the interacting units would have sufficient energy to dissociate at any particular instant; this is the realm of the (relatively) strong hydrogen bonding interactions typical of the polyamides, polyphenols, polyacids, etc., and also those molecules that interact through the formation of charge transfer complexes or ionic forces. Here, we cannot assume random contacts of the functional groups involved in these interactions, as they are (usually) directional and specific. We can, however, treat the contribution to the free energy of mixing with association models, as we will demonstrate later. Conversely, below say 1 kcal. mole^{-1}, there is a substantial fraction of interacting units that have energies $\geq \Delta E$ and it is more appropriate to assume random mixing and a mean field. Between these two poles, i.e., ΔE values of between approximately 1-3 kcal. mole^{-1}, we have an intermediate case where things are not so clear cut. Following a practice that has been common in describing interactions in low molecular weight species[2,6], we will describe relatively weak interactions as "physical," while those interactions that result in the formation of identifiable species (e.g., dimers, other complexes) as "chemical" interactions. Polymers that interact in this latter manner can usually be clearly recognized on the basis of spectroscopic measurements and in this book we will be largely concerned with a particular type of "chemical" interaction – hydrogen bonds. This is not to say that we will neglect "physical" interactions, but we will describe these in the usual fashion, using solubility parameters or a Flory-Huggins χ.

Unfortunately, it is not clear how we should describe polymers with certain types of strongly polar functional groups (e.g., poly(vinyl chloride), polyacrylonitrile)). There is evidence for association in mixtures involving these polymers, but it is not definitive and at the present time materials with "intermediate" strength interactions fall through the cracks in our quantitative treatment.

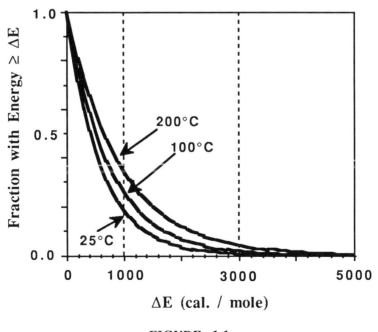

FIGURE 1.1

Physical Interactions

We will define physical interactions as those involving London dispersion forces and weak dipolar interactions. Dispersion interactions have their origin in instantaneous non-isotropic charge distributions within molecules. The time average of these fluctuations is zero, but at some moment an instantaneous dipole can induce a corresponding dipole in a neighboring molecule or segment, so that there is a net attractive force between them. The perturbation of the electronic motion of one molecule by another is related to its perturbation by light as a function of energy (frequency). This can be expressed in terms of the variation of refractive index with frequency, or the *dispersion* of light (hence, dispersion forces). The attractive component of the potential energy, E, for spherical molecules that interact in this manner can (to a first approximation) be expressed in terms of the ionization energy of the molecules (I) and their polarizabilities (α), which for interactions between molecules 1 and 2 is given by:

$$E(d) = -\frac{3}{2}\frac{\alpha_1\,\alpha_2}{d^6}\frac{I_1\,I_2}{(I_1 + I_2)} \qquad (1.3)$$

This is a short range interaction, varying as the inverse sixth power of the distance between molecules. Accordingly, in some models and calculations it is often assumed that only nearest neighbor interactions need to be considered. Furthermore, equation 1.3 was originally formulated to describe the interactions between spherical, symmetrical molecules, so that its application to dispersion forces in asymmetric polyatomic molecules involves some problems (see reference 2).

In the most commonly employed treatments of the thermodynamics of mixing, it is assumed that interactions of this type are pairwise additive. The potential energy of the liquid can then be found by using, for example, the pair correlation function method described by Hildebrand and Scott[2]. The energy term is usually identified with the energy of vaporization, E^v, so that at least for the pure components of a mixture, the energy of interaction between like molecules can be obtained from an experimentally determinable quantity. This energy of interaction can be expressed in terms of the *cohesive energy density*, C, of a substance, equal to $\Delta E^v/V$, where V is the molar volume, or a *solubility parameter*, δ, equal to $(\Delta E^v/V)^{0.5}$. Interactions between the unlike components of a mixture must then be obtained by assuming a relationship to the pure component interaction energies. There will be more on this later, but clearly a number of assumptions have already been piled up (spherical molecules, the additivity of pair interaction potentials, the neglect of higher order terms omitted from equation 1.3, etc.) and we haven't considered polar molecules yet. Nevertheless, the above description provides a useful basis for the definition of interaction parameters, as we will see.

Relatively weak polar forces do not present too much of an additional problem, as dispersion forces are also a major component of the attractive potential for many polar molecules (see reference 6 and citations therein). For molecules where the permanent dipole is not too large, the attractive potential energy of these interactions also has a d^{-6} dependence and can be written[2,6]:

$$E(d) = -\frac{C'_{12}}{d^6} \qquad (1.4)$$

where:

$$C'_{12} = \frac{2}{3}\frac{\alpha_1\alpha_2 I_1 I_2}{(I_1+I_2)} + (\alpha_1\mu_1^2 + \alpha_2\mu_2^2) + \frac{2\mu_1^2\mu_2^2}{3kT} \qquad (1.5)$$

and C'_{12} can be related to the cohesive energy density. The first term represents dispersion interactions, the second dipole/induced-dipole interactions and the last term represents dipole/dipole interactions where the orientations of the molecules are disrupted by thermal motion and an average has been obtained by applying Boltzmann statistics (hence, the d^{-6} dependence, rather than the d^{-3} dependence characteristic of the interaction between two dipoles in a specific orientation relative to one another). Again, higher terms in the London dispersion energy term are neglected, as are multipole interactions (which have terms in d^{-8}, d^{-10}, etc.). Using equation 1.5 it is possible to separate the cohesive energy and hence, solubility parameters into non-polar and (weak) polar contributions. We will not make this distinction. In any event, Hildebrand and Scott[2] have pointed out that for molecules which have buried dipoles, the contribution of polar interactions is small. For molecules with strong dipole moments there is a significant effect, however, and such interactions must be accounted for in some manner.

"Chemical" Interactions

Chemical interactions are usually "strong," specific, and orientation dependent. In polymer blends we are most often concerned with the following intermolecular or inter-segment forces:

a) strong dipoles
b) hydrogen bonds
c) charge transfer complexes
d) ionic interactions in ionomers.

Although a few studies of miscibility in blends where polymer segments interact through charge transfer complexes and ionic forces have appeared, by far the most common and important systems involving strong interactions are (at the time of writing) those involving hydrogen bonds and/or strong dipole interactions. Hildebrand and Scott (reference 2, page 162) considered that molecules with "dipoles capable of forming hydrogen bonds or bridges - - - - are so exceptional in their behavior as to require separate consideration." We can do no less and the bulk of this book is concerned with miscibility in polymer mixtures that is a result of hydrogen bond formation. Accordingly, we will omit any further discussion of charge transfer complexes and ionomers. In this section, we will simply consider strong dipolar forces, as these will be important in several systems we wish to consider, and a separate (but brief) discussion of the nature of the hydrogen bond is presented in Chapter 3. There is one general point that we wish to make before proceeding, however. The methods we will use to obtain the free energy of mixing molecules that hydrogen bond should also, in principle, apply to those that interact through strong dipole forces and the formation of charge transfer complexes. This is because they have a common important feature; they result in the formation of associated species that have a definite stoichiometry (dimers, trimers, etc.). Furthermore, the functional groups involved in such interactions are often localized in a specific part of the molecule, so that chemical interactions are often superimposed upon the physical interactions discussed above. In ionomers, however, there are ionic groups located on hydrocarbon polymer chains and these species form phase separated domains or clusters of various size, a problem that needs to be handled in a different manner.

We complete this section by noting that strong dipole interactions occur between polar molecules and the attractive component of the potential energy is given by:

$$E(d) = -\frac{\mu_1^2 \mu_2^2}{d^3} f(\theta, \phi) \qquad (1.6)$$

where $f(\theta,\phi)$ is a geometric function that depends upon the relative orientation of the dipoles and is equal to 1 for dipoles arranged in a parallel fashion (see references 2 and 6 for a more complete discussion). For $-E(d) \gg kT$, dimers and other associated species will persist for "significant" lifetimes. In liquids the relative orientation of the dipoles will fluctuate with thermal motion, but the degree of orientation will depend not only on the strength of the interaction relative to kT, but also upon the shape of a molecule and its rotational entropy. Clearly, for polymers that have functional groups containing strong dipoles, this will depend upon chain flexibility, which will be a significant factor. A parameter g, defined originally by Kirkwood[7], has been used to account for such hindered rotations. Although values of this parameter can be determined from dielectric measurements, it has not yet been employed in a model that provides useful interaction parameters, but does provide a measure of the significance of polar interactions in a given mixture. For many molecules with strong dipoles the Kirkwood g factor has values close to unity, indicating an absence of significant degrees of association, but molecules that hydrogen bond are set apart as a class in which rotation is strongly hindered. As we noted above, hydrogen bonds are a central concern of this book and we will describe them separately later.

C. REGULAR SOLUTIONS

We now need to consider the assumptions that are necessary in order to obtain useful expressions for the free energy of mixing. If we begin with the fundamental equation:

$$F = -kT \ln Q \qquad (1.7)$$

then the problem of mixing is the construction of the partition function Q for the system to a useful degree of approximation. A fundamental initial assumption is that we can separate the degrees of freedom describing the positions and motions of the centers of mass of the molecules, the translational partition function Q_{tr}, from all other degrees of freedom described by the internal partition function, Q_{int}:

$$Q = Q_{tr} \, Q_{int} \qquad (1.8)$$

The internal partition function includes rotations and this assumption therefore implies that molecules have the same degree of rotational freedom in the mixture as in the pure state. Hydrogen bonds (and other specific interactions) severely restrict rotational mobility and also affect Q_{int} by perturbing the internal energy levels of the interacting molecules. This is a problem that we will return to later. For simple, non-polar roughly spherical molecules, however, this assumption appears reasonable and the free energy of mixing will then be determined by Q_{tr}, since contributions from Q_{int} will cancel when the free energy of the mixture is considered relative to the pure components. The quantity Q_{tr} depends upon the potential energy of the system. This, in turn, is related to the relative positions of the molecules in the mixture (see preceding section), which in the lattice model is obtained from the energy of the system when each molecule is at rest in its equilibrium position in its lattice cell (Q_{latt}). There is also a contribution from the motions of the molecules within their cells (Q_{vib}), which in simple models is assumed independent of composition. Thermal expansion of the lattice is also neglected. These last two factors can be significant and result in so-called free volume effects, which we will also consider later. Even when these factors are neglected, there remains the serious difficulty of evaluating the simplified, yet still largely intractable, expression for the resulting partition function:

$$Q = \sum \Omega (E) \, e^{-E/kT} \qquad (1.9)$$

where $\Omega(E)$ is the number of non-degenerate configurations having the same value of the energy E. The evaluation of $\Omega(E)$ is an extremely difficult problem, but we do know that for molecules of equal size where free volume is neglected:

$$\sum_E \Omega (E) = \frac{(n_A + n_B)!}{n_A! \, n_B!} \qquad (1.10)$$

There are various approximations that have been used to obtain solutions and a discussion of these can be found in Guggenheim's book[4], but the simplest approach, and one that is the basis of regular solution theory and equivalent polymer mixing theories, is to assume that E has the value obtained by averaging over all configurations and is essentially that which would occur in a random mixture (Bragg-Williams assumption), so that the molecules move in a potential field that is unaffected by local variations in composition. As a result, the simplest form of the free energy has separate and additive terms for the entropy and enthalpy of mixing, which can be written:

$$\frac{\Delta G_m}{RT} = n_A \ln x_A + n_B \ln x_B + (n_A + n_B) \, x_A \, x_B \frac{zw}{RT} \qquad (1.11)$$

where w is the exchange energy (per mole) defined as:

$$w = \frac{1}{2}(\varepsilon_{AA} + \varepsilon_{BB}) - \varepsilon_{AB} \qquad (1.12)$$

and ε_{AA}, ε_{BB} are the contact energies between the "like" A, B molecules, respectively, ε_{AB} is the contact energy between a molecule of A and a molecule of B, and z is the lattice coordination number. If a geometric mean assumption is made:

$$\varepsilon_{AB} = (\varepsilon_{AA} \varepsilon_{BB})^{0.5} \qquad (1.13)$$

then the free energy of mixing can be written in terms of composition and the properties of the pure components as:

$$\frac{\Delta G_m}{RT} = n_A \ln x_A + n_B \ln x_B + (n_A + n_B) x_A x_B \frac{z(\varepsilon_{AA}^{0.5} - \varepsilon_{BB}^{0.5})^2}{RT}$$

$$(1.14)$$

In general, for non-polar (and some moderately polar) molecules, the geometric mean assumption appears to work well and the heat of mixing these materials should be positive and unfavorable to mixing.

It is not usual to leave the interaction term in equation 1.14 in a form that explicitly depends upon the lattice parameter z. Hildebrand[2] introduced solubility parameters, which as we discussed above are defined in terms of the cohesive energy of a mole of material or a mixture. In pure component A, for example, if the potential energy is E_A then the cohesive energy density is defined as:

$$C_{AA} = -\frac{E_A}{V_A} \qquad (1.15)$$

where V_A is the molar volume of a A. The solubility parameter is equal to $(C_{AA})^{0.5}$ and for pure low molecular weight materials, is determined from the energy of vaporization (ΔE^v) per unit volume:

$$\delta_A = (C_{AA})^{0.5} = \left(\frac{\Delta E_A^v}{V_A}\right)^{0.5} \tag{1.16}$$

If in a mixture of A and B, the geometric mean assumption is now applied to the cohesive energy densities:

$$C_{AB} = (C_{AA}\, C_{BB})^{0.5} \tag{1.17}$$

then equation 1.14 can be rewritten to obtain a free energy per mole given by:

$$\frac{\Delta G_m}{RT} = x_A \ln x_A + x_B \ln x_B + \frac{V_m}{RT}(\delta_A - \delta_B)^2 \Phi_A \Phi_B \tag{1.18}$$

where V_m is the molar volume of the solution and Φ_A, Φ_B are volume fractions. This equation is the essence of regular solution theory. A detailed description of the application of these ideas (and the contributions made by many great scientists to this field) is presented in Hildebrand's books on solubility, which we have already referred to extensively. The third edition (1950), written in collaboration with Scott (2), is by far the most useful for our purposes, even taking into account the awful typeset used in producing this classic text.

The solubility parameter approach works reasonably well when applied to non-polar or weakly polar molecules in situations where the assumption of random mixing and the neglect of free volume are good approximations. Even here, however, there are problems associated with errors in measuring or calculating solubility parameters, the assumption of central forces (i.e., the molecules are assumed to be spherical and the potential energy depends upon the distance between their centers) and the effect of polar substituents. We will consider the effect of errors in Chapter 2, where we discuss the calculation of solubility parameters for polymers from group contributions. Various efforts to correct for other factors are discussed by Marcus[6]. For our purposes, it is sufficient to consider a very simple approach. The final term in equation 1.18 can be rewritten:

$$\frac{V_m}{RT} \, \Phi_A \, \Phi_B \left[(\delta_A - \delta_B)^2 + 2 \, k'_{AB} \, \delta_A \, \delta_B \right]$$

where k'_{AB} has been used as a correction for non-central forces[6]. Here, we will use k'_{AB} as purely an empirical term that accounts for the breakdown of the geometric mean assumption and other approximations involved in the use of solubility parameters. Alternatively, we could simply use an "adjustable" parameter, χ, in place of the terms in square brackets, which is the usual approach taken in applying equivalent simple models to polymer solutions and blends, as we will discuss in the following section. We will find the use of the above correction term helpful in discussing the "repulsion model", however, which has been widely used in studies of polymer blends, and will return to this topic later.

D. POLYMER SOLUTIONS AND BLENDS

Flory[8] and Huggins[9] employed most of the crucial assumptions of regular solution theory and independently obtained a simple form for the free energy of mixing a monodisperse polymer and solvent:

$$\frac{\Delta G_m}{RT} = n_s \ln \Phi_s + n_p \ln \Phi_p + n_s \, \Phi_p \, \chi_s \qquad (1.19)$$

where χ_s is put equal to the term zw/RT in equation 1.11. There are a few things concerning this equation that we need to consider, as it will form the basis for our attempts to deal with hydrogen bonding in polymer mixtures. To begin with, the first two terms, representing the combinatorial entropy of mixing, were originally obtained using a lattice model where the solvent molecule was used to define the lattice cell size. The polymer was therefore considered to be a flexible chain of connected segments that are each equal in size to a solvent molecule, as illustrated schematically in figure 1.2.

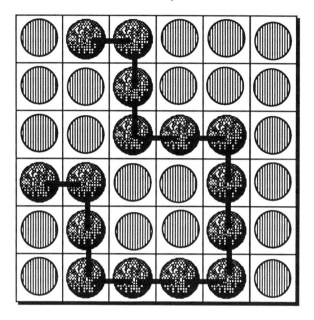

FIGURE 1.2

This has consequences in terms of the definition of χ, which we will return to in a while.

More crucially, the enumeration of the number of configurations available to a chain when placed on a lattice is a complicated problem, even when random mixing is assumed. The original results obtained by Flory and Huggins represent different levels of approximation. Equation 1.19 is the simpler Flory result, but apparently differs from that given by Huggins by a negligible amount, relative to other approximations inherent in this general approach[5]. [Guggenheim's book on mixtures[6] assesses the problems associated with determining the number of configurations available to chains on a lattice in considerable detail and is essential reading for anyone interested in this subject.] We will use Flory's approximations and assumptions to determine the number of configurations available to hydrogen bonded chains and for those not familiar with the procedure, we have reproduced the methodology, as applied to heterogeneous polymers, in Chapter 4, as we will use these equations in our discussion of equilibrium constants.

This brings us to the second point we wish to make. Flory uses a reference state where the molecules are initially separate and oriented and with respect to this reference state, the entropy of mixing, ΔS_m^*, is given by:

$$\frac{\Delta S_m^*}{R} = -\left[n_s \ln \Phi_s + n_p \ln\left(\frac{\Phi_p}{M_p}\right) \right] + (M_p - 1) n_p \ln\left[(z-1)/e\right]$$

(1.20)

where M_p is the "degree of polymerization" of the polymer, defined as:

$$M_p = \frac{V_p}{V_s}$$

(1.21)

and V_p, V_s are the molar volumes of the polymer and solvent, respectively. Equation 1.20 has two components, the combinatorial part that was given in equation 1.19, and an entropy of disorientation given by:

$$\frac{\Delta S_{dis}}{R} = n_p\left[\ln M_p + (M_p - 1) \ln\left[(z-1)/e\right] \right]$$

(1.22)

Upon converting to the usual reference state of the pure, disordered polymer, the ΔS_{dis} term is subtracted from equation 1.20. This separability of the intra and intermolecular terms leads to the conclusion that *in general* the thermodynamic properties of polymer solutions are independent of chain flexibility. In footnotes included in a later review[10], Flory mentioned two exceptions to this rule. The first involves stiff rods, where high concentrations cannot be achieved without imposing a degree of order or alignment on the rods. This does not concern us here. Flory also noted that "specific local interactions with neighbor molecules may, of course, affect the configurations ---"[10]. This will concern us later in this book.

Finally, we return now to equation 1.19 and consider the general situation of mixing molecules of type A and B, where both can be polymeric. If we use molecule B to define the lattice cell size, then χ can be very large, naturally reflecting the size of the molecules. A more useful measure is the interaction energy per segment, which can be defined in terms of the molar volume of the chemical repeat unit of either A or B (V_A, V_B) for example, or some arbitrarily defined reference volume, V_r. It should be kept in mind that the magnitude of χ depends upon the value of V_r employed.

We can now define a free energy per mole of lattice sites by multiplying equation 1.19 by V_r / V, where V is the total volume of the system, to obtain:

$$\frac{\Delta G'_m}{RT} = \left(\frac{\Delta G_m}{RT}\right)\left[\frac{V_r}{V}\right]$$

$$= \frac{\Phi_A}{M_A}\ln \Phi_A + \frac{\Phi_B}{M_B}\ln \Phi_B + \Phi_A \Phi_B \chi_r \tag{1.23}$$

where in the original theory, χ_r was assumed to be purely energetic and for largely non-polar molecules where the geometric mean assumption holds could be related to Hildebrand's solubility parameters by:

$$\chi_r = \frac{V_r}{RT}\left(\delta_A - \delta_B\right)^2 \tag{1.24}$$

For polymer solutions, it was found necessary to add a "fudge factor" $\beta \cong 0.3$ to equation 1.24, to obtain agreement with experiment. This factor is probably related to the relatively large free volume differences between polymers and solvents[11] and for polymer / polymer mixtures we will not consider it further. A more crucial point is that although equation 1.24 is qualitatively useful, deviations between experiment and theory have been observed in various systems. As a result, it is now usual to treat χ as an empirical parameter with both entropic and enthalpic components:

$$\chi = \chi_H + \chi_S \tag{1.25}$$

$$\chi_H = -T\left(\frac{\partial \chi}{\partial T}\right) \qquad (1.26)$$

$$\chi_S = -\frac{\partial(\chi T)}{\partial T} \qquad (1.27)$$

so that on this basis χ should vary as $a + b / T$. In blends with relatively weak interactions this appears to be so, as the neutron scattering studies of Han et al.[12] on polystyrene - poly(vinyl methyl ether) and Russell et al.[13] on blends of polystyrene - poly(methyl methacrylate) indicate. Furthermore, χ can have a strong composition dependence which is particularly pronounced in mixtures with strong, specific interactions, as we will see later.

For polymer solutions, resolution of experimentally determined values of χ into its enthalpic and entropic components usually showed χ_s to be a major contributor, thus setting the stage for so-called free volume theories. We will consider these after a general discussion of phase behavior. We conclude this section with a prescient statement made by Guggenheim in 1953[1] concerning the utility of these simple theories:

> I think that the most important features of the formulae outlined above are the following. It is generally possible to a useful degree of approximation to treat the configurational (athermal) terms in the free energy and the interaction terms separately and additively. It is, however, wrong to associate the former with the entropy, and the latter with the total energy (or the enthalpy). I believe that these features will persist in the most useful future contributions to the subject of solutions.

More recent studies indicate that the mean-field approximation provides a reasonable description of many polymer mixtures, but is not appropriate for polymer solutions near the critical point[14,15] or, of course, for dilute polymer solutions[3].

E. PHASE SEPARATION IN POLYMER SOLUTIONS AND BLENDS

Conditions for Phase Separation

One obvious condition for miscibility is that the free energy of mixing must be negative. Indeed, if for a binary mixture the plot of ΔG_m is concave upward throughout the composition range, as illustrated in figure 1.3a, then the two components are miscible in all proportions. This is because any point on this curve (e.g., Q) has a lower free energy than any two phase system of the same overall composition. For example, if we consider the hypothetical phase separated mixture represented by points P_1 and P_2, the sum of their composition weighted free energies is given by the point Q^*, which has a higher free energy than a miscible mixture Q.

If the free energy curve appears as in figure 1.3b, however, then it is immediately apparent that at any point along the curve between B_1 and B_2 the free energy is not at a minimum. A lower free energy is obtained by phase separation into mixtures of composition B_1, B_2, the points of contact of the double tangent to the free energy curve. Because the chemical potentials are the first derivatives of the free energy with respect to composition and thus the intercepts of this tangent extrapolated to the composition limits (see figure 1.3b), then for the coexisting phases:

$$\Delta\mu_A^1 = \Delta\mu_A^2$$
$$\Delta\mu_B^1 = \Delta\mu_B^2 \tag{1.28}$$

It is important to observe, however, that the portions of the free energy curve between B_1 and the point of inflection S_1, and similarly between B_2 and S_2, are still concave upward and thus mixtures that have compositions between these point are stable against separation into phases consisting of close neighbor compositions, but not stable against separation into phases of composition B_1 and B_2.

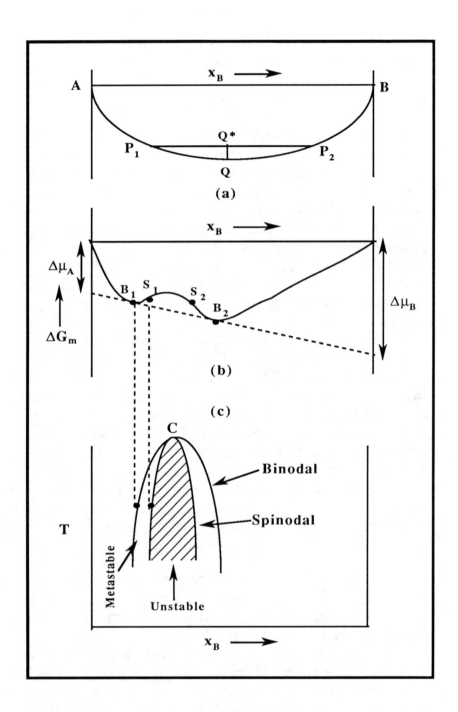

FIGURE 1.3

These *metastable mixtures* are characterized by a nucleation and growth mechanism of phase separation, with a finite cooling into the metastable region required for nucleation.

In contrast, the portions of the curve between S_1 and S_2 are concave downwards. They are unstable and phase separation proceeds spontaneously from small amplitude composition fluctuations, which statistically promote continuous and rapid growth.

The points of inflection S_1, S_2 are characterized by the condition that the second derivative of the free energy with composition is equal to zero,

$$\frac{\partial^2 \Delta G_m}{\partial x_A^2} = 0 \; ; \; \text{or} \; \frac{\partial^2 \Delta G_m}{\partial \Phi_A^2} = 0 \; ; \text{etc.} \qquad (1.29)$$

and we can choose the most convenient concentration scale as our variable. The locus of such points, taken as a function of temperature, define the *spinodal*, while the locus of points defined by equation 1.28, is the *binodal* – as shown in figure 1.3c. These lines coincide at a critical point, C, where in addition to the above criteria:

$$\frac{\partial^3 \Delta G_m}{\partial x_A^3} = 0 \; ; \; \frac{\partial^3 \Delta G_m}{\partial \Phi_A^3} = 0 \; ; \; \text{etc.} \qquad (1.30)$$

The point C is a critical temperature and as we will see shortly both "upper" and "lower" critical solution temperatures can occur. We can summarize the criteria for stability of a binary mixture by stating that the free energy curve should be negative and also concave upward over the composition range, or:

$$\frac{\partial^2 \Delta G_m}{\partial x_A^2} > 0 \qquad (1.31)$$

The Predictions of Simple Models

To evaluate the predictions of theoretical models we require experimental measurements that can, if possible, be related directly to the calculated binodal and spinodal. The former is often equated to cloud point measurements, the temperature at which an initially homogeneous film becomes cloudy or a solution becomes turbid. The spinodal can be determined from scattering studies through the inverse relationship of scattering intensity (extrapolated to zero scattering angle) to the second derivative of the free energy with respect to composition.

For mixtures of low molecular weight materials the coexistence curves are predicted (by the simple models described above) to appear as in figure 1.4a, while the highly asymmetric curves shown in figure 1.4b are predicted for polymer solutions, a result of the large size of a polymer relative to a solvent molecule.

These phase diagrams reflect the intuitive knowledge of everyday life, that two unlike materials tend to become "more miscible" as the temperature is raised. The critical points C are called upper critical solution temperatures (UCST's), since they lie at the top of a two phase region. Applying the criteria described in the preceding section to the Flory-Huggins equation it can be shown that at the critical point in a polymer solution:

$$\Phi_{BC} = \frac{1}{(1 + M_B^{0.5})} \tag{1.32}$$

$$\chi_C = \frac{1}{2}\left(1 + \frac{1}{M_B^{0.5}}\right)^2 \tag{1.33}$$

where we arbitrarily let the subscript B represent the polymer. For $M_B \gg 1$, $\chi_c \cong 0.5$. In general, the Flory Huggins theory leads to a reasonable prediction of χ_c, but there are significant deviations between theoretical predictions and experimental measurements of other quantities.

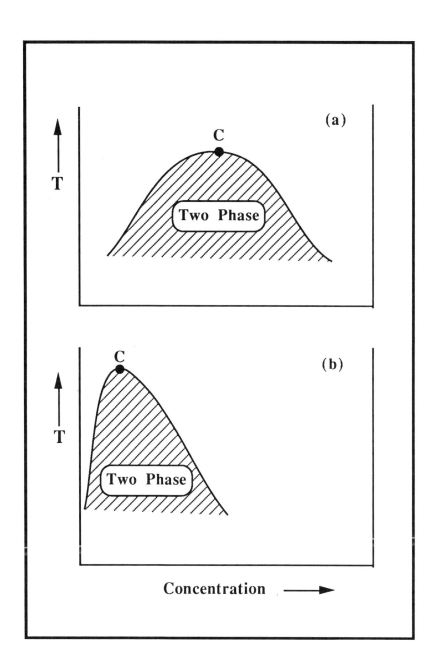

FIGURE 1.4

For example, Schultz and Flory[16] compared calculated binodal curves (from the conditions expressed in equation 1.28) to experimental measurements of turbidity for a set of polystyrene fractions dissolved in cyclohexane. The results are shown in figure 1.5. Qualitatively, the theory predicts the shape of the curves, but the calculated critical concentrations Φ_{B_c} are too low and there are significant deviations at higher polymer concentrations. In some ways, however, this comparison does the Flory-Huggins theory an injustice. Implicit in the equations above is the assumption that the polymer is monodisperse. Stockmayer[17] demonstrated that the spinodal depends only on

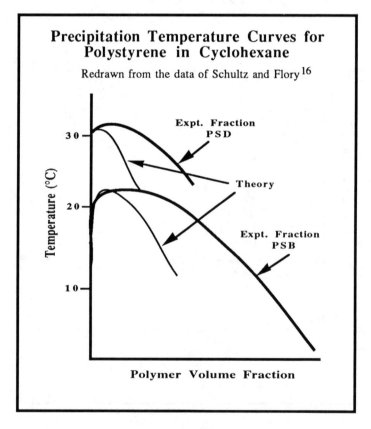

FIGURE 1.5

the weight average molecular weight (and at the critical point depends also upon the z-average).

The binodal, however, depends upon the molecular weight distribution in a far more complex manner[18-20] and its effect has to be determined numerically[21]. In general, the predicted cloud point curves are broadened and flattened. This alone cannot account for the discrepancy between calculated and observed phase behavior, however, without the introduction of a dependence of χ on concentration.

Unlike polymer solutions, the observation of UCST's in the experimentally accessible temperature range of blends of high polymers (i.e., above the T_g and below the degradation point) is rare. For non-polar polymers (where χ is usually positive) this immediately follows from an examination of the equation describing the critical value of χ:

$$\chi_C = \frac{1}{2}\left(\frac{1}{M_A^{0.5}} + \frac{1}{M_B^{0.5}}\right) \qquad (1.34)$$

which shows that for mixtures of high molecular weight (non-polar) materials χ must be exceedingly small in order to obtain a miscible blend. In order to overcome this problem, blends of fairly low molecular weight polystyrene with polybutadiene were studied by Roe and Zin[22], who obtained the cloud points shown in figure 1.6. The curves could only be fitted to the calculated binodal by assuming a dependence of χ on concentration (these authors also studied copolymers, and we will consider these shortly).

If the only discrepancy between observed and calculated phase behavior lay in the deviations of binodals from observed cloud points, then the Flory-Huggins theory could be assumed to be more or less satisfactory. The observation of lower critical solution temperatures (LCST's) in polymer solutions and blends posed a much more dramatic difficulty, however, and we now turn to a discussion of this phenomenon.

Cloud Points for Relatively Low Molecular Weight Polybutadiene - Polystyrene Blends

Redrawn from the data of Roe and Zin[22]

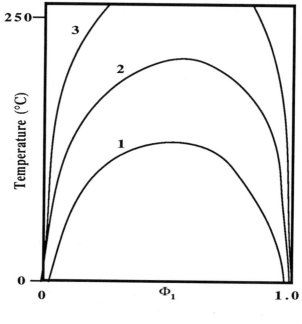

FIGURE 1.6

Closed Loop Phase Diagrams and Lower Critical Solution Temperatures

In most discussions of the phase behavior of polymer solutions and blends it is usual at this point to plunge into a discussion of free volume. We will defer such pleasures for the moment and consider instead the observation of LCST's and closed loops in low molecular weight systems where the components hydrogen bond. The phase diagrams appear very similar to those due to free volume differences in non-polar polymers, but have an entirely different molecular (but not thermodynamic) origin. This will ultimately, in a later part of this book, lead us to a consideration of the balance of forces in polymer blends where the components hydrogen bond (do free volume differences dominate, or are LCST's more dependent upon the entropic changes due to association?).

In 1923 Basil McEwen of His Exalted Highness The Nizam's College, Hyderabad, India, reported the existence of what we now call immiscibility loops in the phase behavior of mixtures that hydrogen bond, as illustrated in figure 1.7 for m-toluidine and glycerol[23]. At temperatures above about 120°C solutions of these materials are single phase. Below this upper critical solution temperature the system phase separates. At still lower temperatures, however, a lower critical solution temperature is encountered and the system again becomes miscible.

m-Toluidine - Glycerol Mixtures

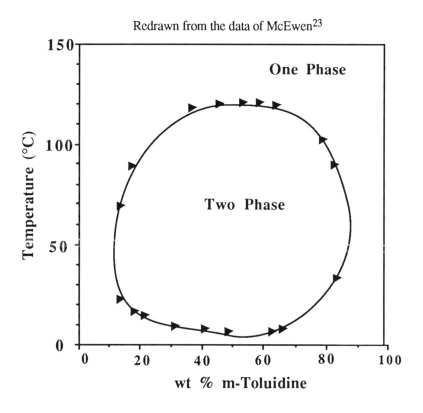

Redrawn from the data of McEwen[23]

FIGURE 1.7

Mixtures of butyl alcohol in water are even more unusual, showing a second upper critical solution temperature at still lower temperatures (ref. 24 and citations therein). Certain aqueous solutions of polar polymers also have closed loop coexistence curves[25,26], as illustrated in figure 1.8 for polyethylene and polypropylene glycol solutions[25].

As early as 1937 Hirschfelder et al.[27] recognized that this type of behavior is characteristic of systems where what was then called hydrogen bridges are present.

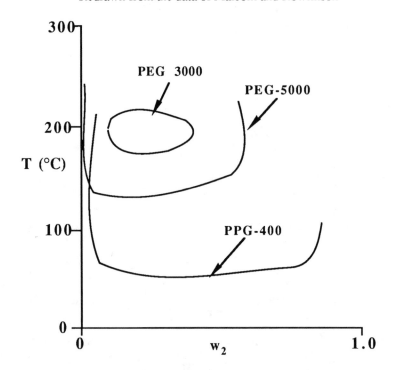

Phase Diagrams for
Aqueous Solutions of Polyethylene (PEG)
and Polypropylene (PPG) Gylcols

Redrawn from the data of Malcom and Rowlinson [25]

FIGURE 1.8

Prigogine and Defay[5] and Copp and Everett[28] presented a thermodynamic analysis of the LCST in associated solutions, demonstrating that this transition is related to a large negative excess entropy in systems where the components can form complexes, for example by hydrogen bonding.

Not all such systems will display LCST's, however, as a second requirement is that the excess enthalpy should be small. In many systems this quantity is large and the situation depends on the number and strength of the various hydrogen bonds that are broken and formed as a function of temperature and composition, and also, although this is not always recognized, the balance between hydrogen bonding and dispersion force contributions.

Phase Diagrams for Polystyrene
(Low Molecular Weight Fractions)
in Acetone

Redrawn from the data of Siow et al.[30]

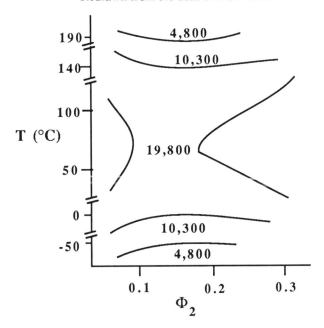

FIGURE   1.9

Nevertheless, we can summarize by saying that until about 1960 most of the mixtures that were known to have LCST behavior were polar, often interacting through the formation of hydrogen bonds, and in these systems it was recognized that the transition was related to entropy changes that are a result of association.

The observation by Freeman and Rowlinson[29] of LCST's in solutions of polyisobutylene with various nonpolar solvents was a turning point in studies of polymer solutions. These authors did not present phase diagrams in their paper, so we reproduce here cloud points obtained from the polystyrene - acetone system by Siow et al.[30], shown in figure 1.9. For the lower molecular weight samples used in this study both a UCST and LCST are observed, but as the molecular weight of the polystyrene is increased these curves shift towards one another and eventually coalesce to form the hour-glass shaped coexistence curves characteristic of a phase separated system. A number of polymer blends also exhibit LCST behavior, and as an example the free energy curve and spinodal of polystyrene - poly(vinyl methyl ether) mixtures, determined by the neutron scattering studies of Han et al.[12], are shown in figure 1.10.

Attempts to quantify the effect of free volume have proceeded along two main lines. The first uses a cell model, described initially by Prigogine[31] and later in a modified from by Flory and co-workers[32-34], which allows the lattice site to vary in size so as that each site, or cell, can contain a portion of the systems free volume, as illustrated in figure 1.11a. In contrast, a hole model has been adopted by Sanchez and Lacombe[35-38], where free volume is represented as vacant lattice sites, as illustrated in figure 1.11b. We will use this latter model later (Chapter 6), but at this point will not consider the various equations developed by these authors. More pertinent to our present qualitative discussion are the observations of Patterson and Robard[39]. Using Flory's model they approximated χ in the original Flory-Huggins theory to the sum of two terms, one describing contact or interactional energies and the second consisting of so-called free volume terms.

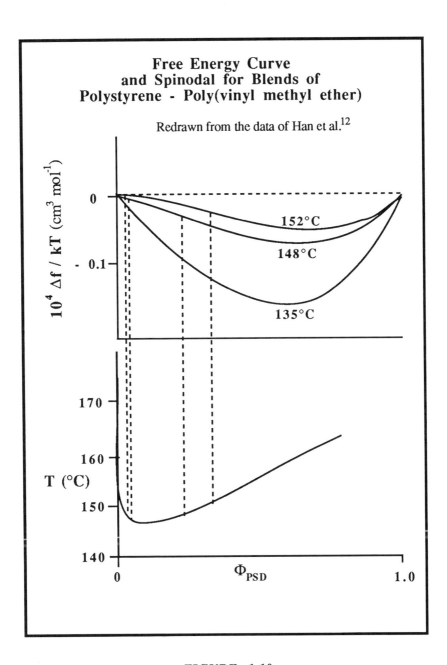

FIGURE 1.10

(a)

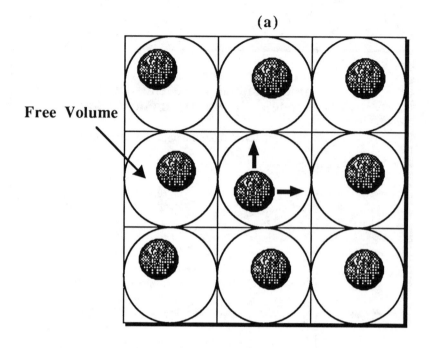

Free Volume

(b)

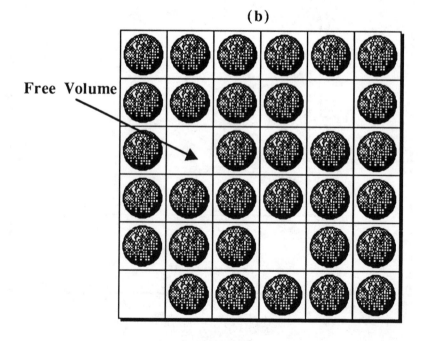

Free Volume

FIGURE 1.11

The former decreases with temperature through the usual $1/T$ dependence, while the latter increases, as shown schematically in figure 1.12 for a polymer solution. Here the interactional terms is assumed positive and the critical value of χ (χ_c) is still defined by equation 1.33. Its value is represented by the horizontal line in this figure. The sum of the interactional and free volume terms can be such that χ crosses this line twice, giving both UCST and LCST behavior with a region of miscibility in-between.

For two polymers, of course, the U shaped curve defining χ would usually lie above the critical value for any appreciable positive value of the interactional term. If this latter quantity is even slightly negative, however, χ can vary with temperature in the manner shown in figure 1.13, where the intersection of χ at the critical value (~ zero) gives an LCST.

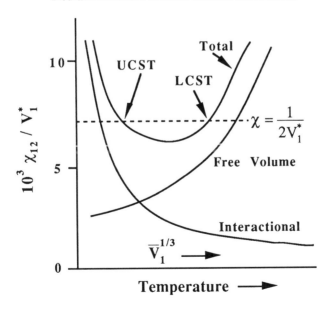

Schematic Representation of a
Polymer - Solvent System

Redrawn from the data of Patterson and Robard[39]

FIGURE   1.12

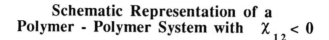

Schematic Representation of a
Polymer - Polymer System with $\chi_{12} < 0$

Redrawn from the data of Patterson and Robard[39]

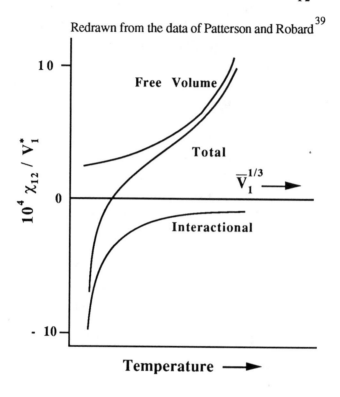

FIGURE 1.13

To summarize LCST (and closed loop) behavior can occur in mixtures in two well-known circumstances. The first is in mixtures of polar, usually hydrogen bonding materials, where entropy changes due to association appear to be the critical factor. The second circumstance involves entropy changes due to free volume differences, observed in solutions and blends of some non-polar polymers.   If we now consider blends or solutions of polymers that hydrogen bond it is clear that both factors will contribute, but will they do so equally or will one dominate over the other?

Hydrogen Bonding and Free Volume in Polymer Blends and Solutions

That LCST behavior can be observed in blends of polymers that hydrogen bond is demonstrated in figure 1.14, which shows (crude) cloud point measurements made on blends of poly(vinyl phenol) with poly(butyl methacrylate)[40].

We will quantitatively demonstrate later in this book that the major factor involved in this phase separation is the entropy change associated with changes in hydrogen bonding. Here, for completeness, we will present qualitative arguments that in the following chapter will allow us to consider a simple model for predicting the miscibility of polymer blends.

Patterson and Robards[39] pointed out that in general free volume differences between polymers are small, while the differences between polymers and most solvents are large.

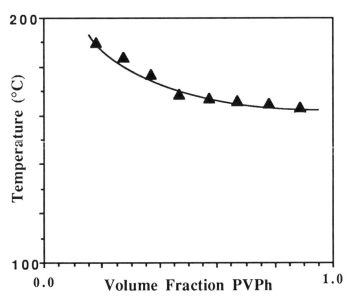

Cloud Point Curve for Poly(Vinyl Phenol) - Poly(n-Butyl Methacrylate) Blends

FIGURE 1.14

Accordingly free volume factors are important in determining the phase behavior of non-polar polymer solutions. In blends characterized by small negative values of χ, poly(styrene) - poly(vinyl methyl ether) for example, they would also be crucial, as we are dealing with the balance between two relatively small forces. In such systems we would therefore also expect to see significant shifts in LCST's as a result of fairly minor changes in other parameters (e.g., molecular weight). In contrast, in systems that hydrogen bond we can have a large negative contribution to the free energy of mixing from hydrogen bonding.

Here we must be aware that these interactions are not confined to unlike A-B contacts, but in many systems of interest one of the components self-associates (e.g., amides, urethanes, etc.). We thus have a balance between the enthalpic and entropic factors associated with *forming* hydrogen bonds between different species at the expense of hydrogen bonds between *like* species. Furthermore, there is often a sizeable unfavorable (to mixing) contribution from dispersion forces in these mixtures. Accordingly, at certain temperatures and compositions the balance between these competing forces can result in phase separation, but it is a balance between large terms that vary sharply with temperature. As we will see, small free volume differences will contribute in only a relatively minor fashion in these systems. Clearly, in *solutions*, this situation could be different, as free volume differences are now more significant. Our concern here, however, is blends of high polymers, and we will start with the assumption that to a first approximation we can neglect free volume effects.

F. SIMPLE MODELS FOR THE DESCRIPTION OF THE PHASE BEHAVIOR OF POLYMER BLENDS

In this section, and the following chapter, we will consider the extent to which we can obtain a qualitative or semi-quantitative understanding of miscibility in polymer blends through the use of simple models, where factors such as free volume are neglected. Because the critical value of χ for the mixing of two polymers is extremely small (see equation 1.34), for many years it was thought that miscibility in most homopolymer systems is a consequence of specific interactions, which can be described in terms of a negative χ parameter. During the last few years, however, there has been considerable interest in blends where at least one component is a copolymer[41-43], usually random, and all the binary χ parameters are positive. This leads to the possibility of obtaining a window of miscibility through a "repulsion" effect between the units of the copolymer. Paul and co-workers[43] have proposed that this concept can be logically extended to certain types of homopolymer blends by assuming that the repeating unit of at least one of the constituents can be subdivided into units that "repel" and this approach was subsequently applied by Ellis to blends of aromatic - aliphatic polyamides[44-46]. The repulsion model has also been successfully applied to various systems where hydrogen bonding interactions are known to occur[43-49] Because there are usually at least three adjustable parameters that are used to fit the observed phase behavior of these systems, however, this simply demonstrates that the model can be made consistent with the data, but is not proof of its validity. In our view the role of specific interactions in systems with polar groups should not be neglected. This is not to say that some sort of "repulsion" effect does not occur, but as we will see this approach is equivalent to the older concept of cosolvency, and we will use this in the general approach that we will apply in the following chapter. Essentially, we will separate interactions into favorable (largely polar) and unfavorable (dispersion and weak polar forces) components, and use a carefully constructed set of solubility parameter group contributions to measure the latter. Miscibility is then determined by the balance between these forces and we will argue that the search for miscible blends should start with those systems where solubility parameter differences are within certain limits, which, in turn, will depend upon the type of polar functional groups that are present.

In describing the repulsion model, we will consider a simple copolymer - homopolymer blend where we let the subscript 1 represent the copolymer, consisting of units A and B, and the subscript 2 represent the homopolymer consisting of units C. The Flory-Huggins equation can then be written[41-43]:

$$\frac{\Delta G_m}{RT} = \frac{\Phi_1}{M_1} \ln \Phi_1 + \frac{\Phi_2}{M_2} \ln \Phi_2 + \Phi_1 \Phi_2 \chi_{blend} \qquad (1.35)$$

where χ_{blend} can be expressed in terms of the binary interaction parameters as:

$$\chi_{blend} = f_A \chi_{AC} + f_B \chi_{BC} - f_A f_B \chi_{AB} \qquad (1.36)$$

where f_A, f_B are the volume fractions of the A and B units present in the copolymer.

This equation follows directly from an analysis in terms of the cohesive energy per mole of the system, which for the blend can be written:

$$-\frac{E_{ABC}}{(x_1 V_1 + x_2 V_2)} = \Phi_A^2 C_{AA} + \Phi_B^2 C_{BB} + \Phi_C^2 C_{CC}$$
$$+ 2\Phi_A \Phi_B C_{AB} + 2\Phi_A \Phi_C C_{AC} + 2\Phi_B \Phi_C C_{BC} \qquad (1.37)$$

where x_i, V_i are the mole fractions and molar volumes of the ith component, the cohesive energy densities of the pure components are given the symbols C_{AA}, C_{BB} and C_{CC}, while those between unlike components are C_{AB}, C_{AC} and C_{BC}. The quantities Φ_A, Φ_B, Φ_C are the volume fractions of segments A, B, C, which are assumed to be randomly mixed. These fractions are related to the ones given above by:

$$\Phi_A = f_A \Phi_1$$
$$\Phi_B = f_B \Phi_1$$
$$\Phi_C = \Phi_2 \qquad (1.38)$$

We are interested in the energy change upon mixing components 1 and 2, so for the pure copolymer and homopolymer we write:

$$-\frac{E_{AB}}{(x_A V_A + x_B V_B)} = f_A^2 C_{AA} + f_B^2 C_{BB} + 2 f_A f_B C_{AB} \quad (1.39)$$

$$-\frac{E_C}{V_C} = C_{CC} \quad (1.40)$$

so that:

$$\frac{\Delta E_m}{V} = E_{ABC} - x_1 E_{AB} - x_2 E_C \quad (1.41)$$

Substituting and rearranging terms we can obtain:

$$\frac{\Delta E_m}{V} = \left[f_A (2C_{AC} - C_{AA} - C_{CC}) + f_B (2C_{BC} - C_{BB} - C_{CC}) \right.$$
$$\left. - f_A f_B (2C_{AB} - C_{AA} - C_{BB}) \right] \Phi_1 \Phi_2 \quad (1.42)$$

This is equivalent to equation 1.36 with $\chi_{AC} = (2C_{AC} - C_{AA} - C_{CC}) / RT$, etc. The negative ($\chi_{AB}$) terms in equations 1.36 and 1.42 are a result of replacing copolymer AB contacts with AC, BC contacts. This is the repulsion effect and it has been argued that if χ_{AB} is large enough then an overall negative value of χ_{blend} can be obtained, even though all binary interaction parameters are positive. This says nothing about the conditions under which this is likely to occur, however, and to explore this we first consider some work on ternary systems by Scott[50,51], published forty years ago, but still relevant to contemporary studies of blends. If a polymer, say C, is dissolved in a mixture of two solvents, A and B, then the cohesive energy can be written:

$$\frac{\Delta E_m}{V} = \Phi_A \Phi_C (2C_{AC} - C_{AA} - C_{CC}) + \Phi_B \Phi_C (2C_{BC} - C_{BB} - C_{CC})$$
$$+ \Phi_A \Phi_B (2C_{AB} - C_{AA} - C_{BB}) \quad (1.43)$$

Note that the last term in equation 1.43 is positive, while the corresponding term in equation 1.42 is negative. This is because of a difference in reference states. For the ternary solution considered here, the reference states are the separate A, B, C components. For the blend, the copolymer units A and B are initially in the same phase in the reference state of pure copolymer.

Scott[50] pointed out that if the geometric mean assumption is made, $C_{AB} = (C_{AA} C_{BB})^{0.5}$, etc. then equation 1.43 can be expressed in terms of solubility parameters, but, more importantly, can be rearranged into a form that demonstrates that the energy of mixing polymer C with solvents A and B is equivalent to mixing C with a hypothetical new solvent whose solubility parameter is the volume fraction average of B and C. This has been called *cosolvency*. If a particular polymer (high molecular weight) is characterized by a solubility parameter of say 10 (cal. cm^{-3})$^{0.5}$, then (in the absence of specific interactions) it would not be separately miscible with solvents A or B if $\delta_A = 8$ and $\delta_B = 12$ (cal. cm^{-3})$^{0.5}$. Appropriate mixtures of these solvents have a volume fraction average value of the solubility parameter δ_{AB} close to 10 (cal. cm^{-3})$^{0.5}$, however, and these would therefore be miscible.

Scott[51] extended this analysis to random copolymers and demonstrated that for the copolymer - homopolymer system described above the equivalent result is obtained. i.e.:

$$\frac{\Delta E_m}{V} = \Phi_A \Phi_B \left[\overline{\delta}_{AB} - \delta_C \right]^2 \qquad (1.44)$$

where $\overline{\delta}_{AB}$ is the volume fraction average of the solubility parameters describing the A, B copolymer segments. If the geometric mean assumption is correct, then χ_{blend} is always positive. Nevertheless, χ_{blend} could become vanishingly small over a certain composition range and this can be attributed to a "repulsion" effect. It is entirely equivalent to the phenomenon of cosolvency, however, and requires that the segments of the copolymer be characterized by solubility parameters that bound on "each side" the value for the homopolymer. The window of miscibility obtained in this way is, of course, much narrower than in the polymer C, solvent A, solvent B system because of the smaller entropy of mixing, but would still be there and would depend upon the molecular weights of the components through the critical condition:

$$(\chi_{\text{blend}})_{\text{Crit}} = \frac{1}{2}\left[\frac{1}{M_1^{0.5}} + \frac{1}{M_2^{0.5}}\right]^2 \qquad (1.45)$$

The solubility parameter description of polymer blends was discussed by Paul and Barlow[43] who then described the effect of deviations from the geometric mean assumption by defining parameters k_{ij} as follows:

$$C_{ii} = (1 - k_{ii})(C_{ii}C_{jj})^{0.5} \qquad (1.46)$$

where $k_{ij} = k'_{ij}$, defined in section C. Accordingly:

$$\frac{RT\chi_{\text{blend}}}{V_{\text{ref}}} = \left[\bar{\delta}_{AB} - \delta_C\right]^2 + 2k_{AC}f_A\delta_A\delta_C$$

$$+ 2k_{BC}f_B\delta_B\delta_C - 2k_{AB}f_Af_B\delta_A\delta_B \qquad (1.47)$$

These authors then considered an example where for the copolymer $\delta_A = 8$ and $\delta_B = 11$ (cal. cm^{-3})$^{0.5}$, while the solubility parameter for the homopolymer is given by $\delta_C = 9$ (cal. cm^{-3})$^{0.5}$. They noted that for $k_{AC} = k_{BC} = 0$, very small values of k_{AB} (>1/176) would be required to obtain a negative χ and a window of miscibility. For the system chosen, however, χ would still go to zero for a certain value of copolymer composition because of the "cosolvent" effect and a window of miscibility, albeit narrower, would still be obtained.

One would certainly expect some deviations from the geometric mean assumption, but the question which has not been addressed in any detail is the extent of these deviations necessary to obtain a negative χ or a repulsion effect where "cosolvency" does not occur. Furthermore, it seems to us unreasonable to assume that the energies of A, B contacts deviate while those of A, C and B, C contacts do not. (i.e. k_{AC}, $k_{BC} \neq 0$).

We have therefore calculated values of $RT\chi_{\text{blend}}/V_{\text{ref}}$ as a function of copolymer composition for a system similar to that described by Paul and Barlow[43], with $\delta_A = 8$, $\delta_B = 11.0$ and δ_C =10 (cal. cm^{-3})$^{0.5}$. Lacombe and Sanchez[52] calculated deviations from the geometric mean assumption that were usually of the order of 2%, but for one or two systems were as large as 10%. We will assume that the copolymer constituents deviate by this larger figure (i.e., $k_{AB} = 0.1$) while initially the deviations of the AC, BC contact energies are only of the order of 1% (i.e., $k_{AC} = k_{BC} = 0.01$). We then obtain the miscibility window shown in figure 1.15, where χ_{blend} is negative over a wide range of copolymer composition. If the deviations of the copolymer/ homopolymer energies from the geometric mean are just 2% instead of 1%, however, this window is narrowed considerably, and if $k_{AC} = k_{BC} = 0.025$ then χ_{blend} is positive throughout the composition range.

Similarly, it is also a requirement that the homopolymer have a solubility parameter within the range or close to that encompassed by the copolymer components. Figure 1.16 shows the original conditions described above (figure 1.15) and, in addition, plots of $RT\chi_{blend}/V_{ref}$ versus composition for $\delta_C = 11.0$ and $\delta_C = 11.5$ (cal. $cm^{-3})^{0.5}$. Again, the "window of miscibility" diminishes and disappears in this series, even though very large deviations in AB contacts ($k_{AB} = 0.1$) relative to AC, BC contacts were allowed ($k_{AC} = k_{BC} = 0.01$).

To summarize, the possibility of exothermic interactions (i.e., a negative value of χ_{blend}) through the repulsion effect requires relatively large deviations of the copolymer contact energies from the geometric mean assumption and, simultaneously, relatively small deviations of the copolymer - homopolymer contact energies. Furthermore, it also requires a "cosolvent effect", unless extraordinarily large values of k_{AB} are postulated. This *combination* of requirements is quite stringent. For example, one could envisage a copolymer consisting of a statistical distribution of polar (A) and non-polar groups (B), such that the "repulsion" between these segments would be significant and the deviation from the geometric mean perhaps unusually large. If mixed with a non-polar homopolymer (C), however, one would also anticipate equally large deviations of AC interactions from the geometric mean. In other words, the structural factors that lead to deviations for the copolymer pair

DEVIATIONS FROM THE
GEOMETRIC MEAN ASSUMPTION

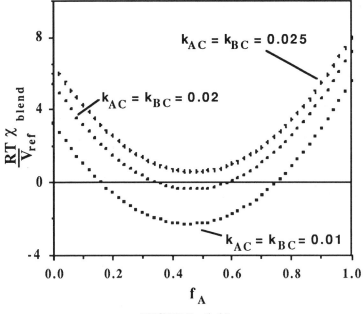

FIGURE 1.15

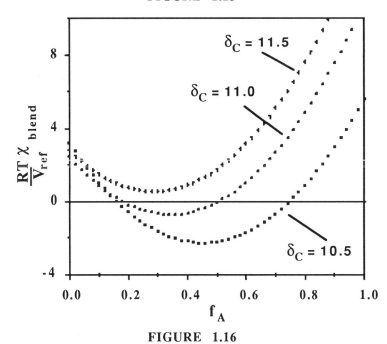

FIGURE 1.16

interactions will also come into play, in one form or another, for interactions between the copolymer and homopolymer segments.

In our view a more reasonable explanation for observed "miscibility windows" in most copolymer - homopolymer blends is a combination of a cosolvency or repulsion effect, where a χ describing dispersive interactions becomes small over a certain range of compositions, while superimposed upon this is a specific interaction that occurs between the blend components, giving a negative contribution to the free energy of mixing and thus "broadening" the composition limits of the window.

The separation of terms describing favorable and unfavorable interaction parameters was mentioned in section C and was discussed in more detail many years ago by Hildebrand and Scott[2]. There are a number of ways that specific interactions can be handled, but in our view a particularly useful approach is the use of association models that account for both non-random contacts of strongly interacting functional groups and "self-association" of the pure components by defining equilibrium constants that can be *independently* determined by spectroscopic measurements. The associated species are allowed random contacts with one another, however, so that unfavorable to mixing London dispersion (and weak polar) forces are separately accounted for by the usual Flory Huggins $\Phi_A \Phi_B \chi$ term and we can write the following equation for the free energy:

$$\frac{\Delta G_m}{RT} = \frac{\Phi_A}{M_A} \ln \Phi_A + \frac{\Phi_B}{M_B} \ln \Phi_B + \Phi_A \Phi_B \chi + \frac{\Delta G_H}{RT}$$

$$(1.49)$$

where the ΔG_H term accounts for the *free energy* changes that are a result of specific interactions. Although it appears that we have simply added a term to the Flory-Huggins expression, equation 1.49 can be obtained by the formal application of a Flory type lattice model, as we will discuss this and (briefly) other approaches in Chapter 6. The parameters in the ΔG_H term can be measured spectroscopically in many, but not all, systems and if we obtain χ from solubility parameters then we can *predict* phase behavior (from the derivatives of equation 1.49 with respect to composition). Before proceeding to describe the details of this approach, however, we will spend some time in the next chapter showing how equation 1.49 can be used in a qualitative or semi-quantitative fashion to obtain a "feel" for the miscibility of polymer systems and to identify potentially miscible partners (and / or eliminate others from consideration). This approach is not rigorous, but in our view, is of considerable practical utility and, moreover, works remarkably well.

G. REFERENCES

1. Guggenheim, E. A., *Discussions of the Farad. Soc.*, 1953, **15**, 24.
2. Hildebrand, J. H. and Scott, R. L.,*The Solubility of Nonelectrolytes*, Third Edition, American Chemical Society Monograph Series, 1950.
3. Flory, P. J., *Principles of Polymer Chemistry*, Cornell University Press, 1953.
4. Guggenheim, E. A., *Mixtures*, Clarendon Press, Oxford, 1952.
5. Prigogine, I. and Defay, R., *Chemical Thermodynamics* (English Translation, Everett, D. H.), Longmars Green and Co., London, 1954.
6. Marcus, Y., *Introduction to Liquid State Chemistry*, John Wiley and Sons, London, 1977.
7. Kirkwood, J. G., *J. Chem. Phys.*, 1939, **7**, 911, ibid *Trans. Farad. Soc.*, 1946, **42A**, 7.
8. Flory, P. J., *J. Chem. Phys.*, 1941, **9**, 660.
9. Huggins, M. L., *J. Chem. Phys.*, 1941, **9**, 440.
10. Flory, P. J., *Discussions of the Farad. Soc.*, 1970, **49**, 7.
11. Patterson, D., *Rubber Chemistry and Technology* ,1967, **40**, 1.
12. Han, C. C., Bauer, B. J., Clark, J. C., Muroga, Y., Matsushita, Y., Okada, M., Tran-cong, Q., Chang, T. and Sanchez, I. C., *Polymer*, 1988, **29**, 2002.
13. Russell, T. P., Hjelmjr, R. P., Seeger, P. A., *Macromolecules*, 1990, **23**, 890.
14. Joanny, J. F. *C. R.Acad. Sc. Paris Ser B*, 1978, **286**, 89.
15. deGennes, P. G., *J. Physique Lett.*, 1977, **38**, L441.
16. Shultz, A. R. and Flory, P. J., *J. Am. Chem. Soc.*, 1952, **74**, 4760.
17. Stockmayer, W. H., *J. Chem. Phys.*, 1949, **17**, 588.

18. Tompa, H., *Polymer Solutions*, Butterworths, London, 1956.
19. Koningsveld, R. and Staverman, A. J., *J. Polym. Sci.*, A2, 1968, **6**, 305; 325; 349.
20. Solc, K., *Macromolecules*, 1970, **3**, 665.
21. Roe, R. J. and Lu, L., *J. Polym. Sci. Pol. Phys. Ed.*, 1985, **23**, 917.
22. Roe, R. J. and Zin, W. C., *Macromolecules*, 1980, **13**, 1221.
23. McEwen, B., *J. Chem. Soc.*, 1923, **123**, 2279, and 2284, *ibid.*,. 1924, **124**, 1484.
24. Moriyoski, T., Kaneshina, S., Aihara, K. and Yabunota, K., *J. Chem. Thermodyn.*, 1975, **7**, 537.
25. Malcom, G. N. and Rowlinson, J. S., *Trans. Farad. Soc.*, 1957, **53**, 921.
26. Nord, F. F., Bier, M. and Timasheff, N., *J. Am. Chem. Soc.*, 1951, **73**, 289.
27. Hirschfelder, J., Stevenson, D., and Eyring, H., *J. Chem. Phys.*, 1937, **5**, 896.
28. Copp, J. L. and Everett, *Discussions of the Farad. Soc.*, 1953, **15**, 174.
29. Freeman, P. I.. and Rowlinson, J. S., *Polymer*, 1959, **1**, 20.
30. Siow, K. S., Delmas, G. and Patterson, D., *Macromolecules*, 1972, **5**, 29.
31. I. Prigogine, *The Molecular Theory of Solutions*, Interscience Publishers, Inc., New York, N.Y., 1957.
32. P. J. Flory, R. A. Orwoll, and A. Vrij, *J. Am. Chem. Soc.*, 1964, **86**, 3507.
33. *ibid*, 1964, **86**, 3515.
34. P. J. Flory, *J. Am. Chem. Soc.*, 1965, **87**, 1833.
35. I. C. Sanchez and R. H. Lacombe, *J. Phys. Chem.*, 1976, **80**, 2352.
36. R. H. Lacombe and I. C. Sanchez, *J. Phys. Chem.*, 1976, **80**, 2568.
37. I. C. Sanchez and R. H. Lacombe, *Polym. Letters Ed.*, 1977, **15**, 71.
38. I. C. Sanchez and R. H. Lacombe, *Macromolecules*, 1978 11, 1145.
39. Patterson, D. and Robard, A., *Macromolecules*, 1978, **11**, 690.
40. Serman, C. J., Painter, P. C. and Coleman, M. M., *Polymer*, in press.
41. Kambour, R. P., Bendler, J. T. and Bopp, R. C., *Macromolecules*, 1983, **16**, 753.
42. ten Brinke, G., Karasz, F. E. and MacKnight, W. J., *Macromolecules*, 1983, **16**, 1927.
43. Paul, D. R. and Barlow, J. W., *Polymer*, 1984, **25**, 487.

44. Ellis, T. S., *Polymer*, 1988, **29**, 2015.

45. Ellis, T. S., *Macromolecules*, 1989, **22**, 742.

46. Ellis, T. S., *Polymer*, 1990, 31, 1058.

47. Brannock, G. R., Barlow, J. W. and Paul, D.R., *J. Pol. Sci., Poly. Phys.*, 1990, **28**, 871.

48. Jo, W. H., Lee, S. C., *Macromolecules*, 1990, **23**, 2261.

49. Zhu, K. J., Chen, S. F., Tai, H., Pearce, E. M. and Kwei, T. K., *Macromolecules*, 1990, **23**, 150.

50. Scott, R. L., *J. Chem. Phys.*, 1949, **17**, 268.

51. Scott, R. L., *J. Polym. Sci.*, 1952, **9**, 423.

52. Lacombe, R. H. and Sanchez, I. C., *J. Phys. Chem.*, 1976, **80**, 2568.

## CHAPTER 2

A Practical Guide to Polymer Miscibility

*Practical men know where they are, but not always whither they are going;
thinkers know whither we are going, but not always where we are.*
—*Bernard Shaw*

A. INTRODUCTION

In this chapter we will present a qualitative "guide" to miscibility that is based on the assertion that the free energy of mixing can be written in the following form:

$$\frac{\Delta G_m}{RT} = \frac{\Phi_A}{M_A} \ln \Phi_A + \frac{\Phi_B}{M_B} \ln \Phi_B + \Phi_A \Phi_B \chi + \frac{\Delta G_H}{RT}$$
(2.1)

The segmental interaction parameter χ is assumed to represent "physical" forces only, while the ΔG_H term reflects free energy changes corresponding to specific interactions, most commonly, but not necessarily, hydrogen bonds. We have discussed the rationale for separating these contributions and neglecting free volume in chapter 1 and we will further justify this approach by directly deriving equation 2.1 using a lattice model in chapter 6. In essence, the formation of hydrogen bonds alters the number of configurations available to the chains and we can account for this by determining the probability that a set of non hydrogen bonded chains would spontaneously occur in a configuration equivalent to that found in a hydrogen bonded system. We then arrive at equation 2.1, with the factor ΔG_H expressed in terms of various concentration terms and equilibrium constants that describe the distribution of hydrogen bonded species present at a particular concentration

49

and temperature. This is no advance on present theories, of course, unless we can determine these parameters by independent experimental measurement. This we can do (for many, but not all, systems) using infrared spectroscopy (see chapter 5). Accordingly, given a method of estimating the value of χ we could, in principle, predict whether or not a particular blend should be miscible. If it were not, we could also determine the most appropriate candidates for copolymerization to make it miscible (i.e., co-monomers that would reduce the average value of χ for the chain). In this chapter we will elaborate on these ideas and discuss the use of solubility parameters in the calculation of χ. We will then show that by making informed guesses concerning the magnitude of the ΔG_H specific interaction term we can obtain a good qualitative understanding of the miscibility of most polymer blend systems.

We commence our discussion by restating the equation describing the critical value of χ in a non-polar system.

$$\chi_{Crit} = \frac{1}{2}\left[\frac{1}{M_A^{0.5}} + \frac{1}{M_B^{0.5}}\right]^2 \tag{2.2}$$

Figure 2.1 illustrates graphically the dramatic effect of the degree of polymerization on χ_{crit} for the special case of a blend where $M_A = M_B$. Consider, for example, two polymers having $M_A = M_B = 1000$ (roughly a molecular weight of 100,000 in most cases); in order to achieve molecular mixing the value of χ needs to be < 0.002 (and this is before other unfavorable effects, such as "free volume", are taken into account).

If we are dealing with weak interactions (principally dispersion forces) only we can, to a first approximation, relate χ to the Hildebrand solubility parameters of the two polymers using the equation:

$$\chi = \frac{V_r}{RT}\left[\delta_A - \delta_B\right]^2 \tag{2.3}$$

Assuming a reference volume, $V_r = 100$ cm^3 mole^{-1}, a critical value of the solubility parameter difference, $\Delta\delta_{crit}$, may be calculated (see figure 2.2). Again, for the case of a polymer blend containing two polymers having $M_A = M_B = 1000$ the difference in the solubility parameters must be less than about 0.1 (cal. cm^{-3})$^{0.5}$ to ensure molecular mixing. Accordingly, *for polymers that weakly interact*, it is necessary to measure or calculate the solubility parameters of polymers to an accuracy of better than ± 0.05 (cal. cm^{-3})$^{0.5}$ in order to obtain any reasonable prediction of miscibility in these types of polymer blends. As we will see later, this requirement is beyond the capabilities of any known method of determining the solubility parameters of polymers.

If, in addition to weak dispersion forces, there are favorable, usually specific interactions, then the possibilities for obtaining miscible mixtures becomes more numerous. In chapter 1 we arbitrarily categorized interactions into those that are weak, intermediate or strong, according to the fraction of molecules that have energies greater or equal to an "energy of dissociation", ΔE, at a given temperature and for reference we reproduce this plot in figure 2.3.

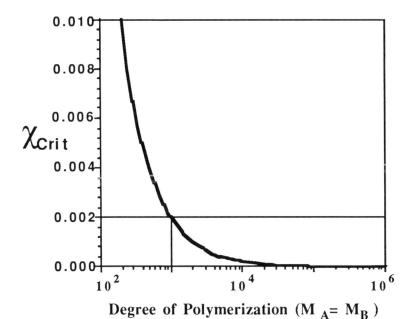

FIGURE 2.1

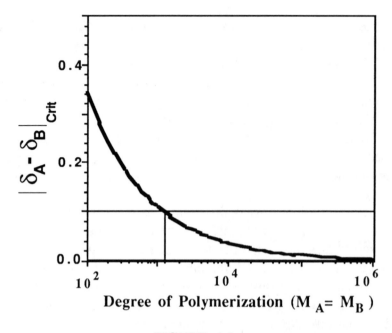

FIGURE 2.2

Essentially, we describe "weak" interactions as those whose energy is less than 1 kcal. mole^{-1}, as it can be seen from figure 2.3 that for such systems there is a substantial fraction of units that have energies $\geq \Delta E$. Strong interactions ($\Delta E > 3$ kcal. mole^{-1}) result in significant degrees of association, while intermediate or moderate interactions lie between these poles. We will define χ so that it reflects weak "physical" forces only, according to equations 2.1 and 2.3, and we conceptually separate out the favorable (negative) contributions in the $\Delta G_H/RT$ term; this model is therefore most appropriately applied to those systems where favorable interactions can be unambiguously identified with specific localized parts or functional groups in the molecules. Simply put, the presence of such favorable intermolecular interactions then effectively increases the magnitude of χ_{crit} and permits the toleration of a greater difference in the non-hydrogen bonding solubility parameters of the two polymers, $\Delta \delta$. This is illustrated schematically in figure 2.4, which shows a plot of $\Delta \delta$ versus χ

calculated from equation 2.3 assuming a reference volume of 100 cm^3 mole^{-1} at a temperature of 298 K.

As noted above, for very weak or "repulsive" interactions χ values of < 10^{-3} are necessary for miscible blend systems, which corresponds to $\Delta\delta$ values of < 0.1 (cal. cm^{-3})$^{0.5}$. However, if certain functional groups contribute relatively weak to intermediate strength favorable interactions (between about 1 and 3 kcal. mole^{-1} - figure 2) χ values of up to about 0.1 may be tolerated, corresponding to an upper $\Delta\delta$ value of close to 1 (cal. cm^{-3})$^{0.5}$. To complete the picture, when much stronger interactions are involved (>3 kcal. mole^{-1}) χ values of greater than 0.1 can be tolerated and miscible blends are possible even when $\Delta\delta$ exceeds 1 (cal. cm^{-3})$^{0.5}$. Remember that χ, as defined here represents "physical" forces only.

This is the key to our simple guide to polymer miscibility and as we mentioned above centers on the questions: how do we estimate the value of χ and what potentially favorable intermolecular interactions exist in the system ? It is here where we now focus our attention.

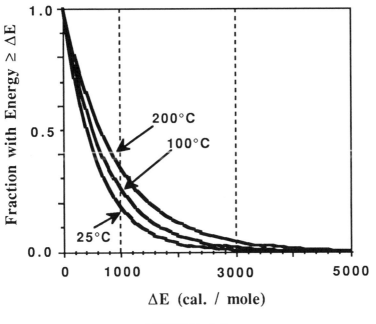

FIGURE 2.3

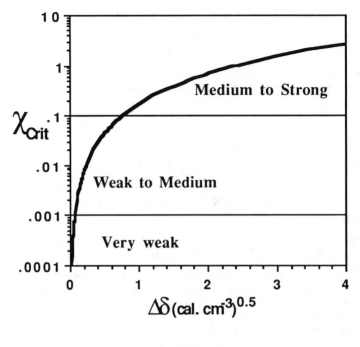

FIGURE 2.4

B. THE ESTIMATION OF χ FROM SOLUBILITY PARAMETERS

The uses and abuses of solubility parameters are legendary. On the one hand, they are extensively applied and highly valued by the polymer chemist confronted by specific practical problems, such as those found in the surface coatings industry; on the other, they are disdained by many theoretically inclined polymer scientists on the compelling grounds of their many experimentally demonstrable inadequacies. Introduced by Hildebrand[1], solubility parameters in their original form are only applicable to molecules that have physical interactions (see Chapter 1). Many attempts have been made to extend the solubility parameter concept into systems that involve hydrogen bonding and other polar intermolecular interactions[2]. While these approaches have been useful as guides for polymer solubility in low molecular weight solvents, they have not been very useful in the prediction of polymer - polymer miscibility.

In contrast to low molecular weight compounds, where solubility parameters can be directly calculated from heat of vaporization measurements (i.e. from $\delta = [\Delta E^v / V]^{0.5}$, where ΔE^v is the energy of vaporization to a gas at zero pressure), polymer solubility parameters have to be determined *indirectly* , either by experiment (e.g. swelling measurements of a lightly crosslinked polymer in a series of solvents of known solubility parameters) or by calculation from group molar attraction constants. The errors inherent in the indirect experimental methods used to determine polymer solubility parameters are too large to be useful in the prediction of polymer miscibility (a cursory glance at the wide range of reported values for typical polymers listed in table 8.3 of van Krevelen's book[3] should be sufficient confirmation of this assertion). The most common method used to estimate polymer solubility parameters therefore involves calculation through the use of group molar attraction constants[3-5]. At first glance, it appears a trivial task to calculate the solubility parameter of a polymer. One only has to consider the groups present in the repeat unit of the polymer, refer to tables of molar attraction constants (F) attributed to either Small[5], Hoy[4] or van Krevelen[3], and use the relationship:

$$\delta = \frac{\sum F_i}{V} \qquad (2.4)$$

Table 2.1 shows the results of such a calculation for two polymers, poly(methyl methacrylate) (PMMA) and poly(ethylene oxide) (PEO). All three methods give very similar results for the solubility parameter of PMMA ($\delta = 9.1 \pm 0.2$ (cal. cm^{-3})$^{0.5}$) and the choice of molar volume makes little difference (i.e. V_{expt} from density measurements of the polymer, or V_r and V_g which are the rubbery and glassy molar volumes, respectively, calculated from a correlation determined separately by van Krevelen - table 4.6 , reference 6). This appears fine, but now consider the calculated solubility parameters for PEO.

Table 2.1
Calculated Solubility Parameters

Polymer	Molar volume cm^3 mole^{-1}	Calculated solubility parameter (cal. cm^{-3})$^{0.5}$		
		Small[8]	Hoy[7]	van Krevelen[6]
PMMA	V_{expt} = 85.6	9.1	9.2	9.3
	V_r = 87.8	8.9	9.0	9.1
	V_g = 86.5	9.0	9.1	9.2
PEO	V_{expt} = 38.9	8.6	9.7	10.3
	V_r = 41.4	8.1	9.1	9.6

The values range from 8.1 to 10.3 (cal. cm^{-3})$^{0.5}$, depending upon which set of molar attraction constants is employed and, *of equal importance*, which value of the molar volume is used.

Discrepancies, in general, are particularly acute for polymers containing small repeat units. Incidentally, blends of PMMA and PEO are reported to be miscible[6] and although they are both polar polymers it is reasonable to assume that any favorable interactions between them are relatively weak. Accordingly, if one were trying to "prove" the necessity of closely matched solubility parameters for the miscibility of this particular blend, a jaundiced selection of molar attraction constants and molar volumes can be easily found to substantiate this hypothesis. This is hardly a satisfactory situation, however, and we decided to review the underlying assumptions made and the errors involved in determining solubility parameters from group molar attraction constants[7].

Determination of Group Molar Attraction and Molar Volume Constants

We have obtained data pertaining to the liquid molar volumes (V) and solubility parameters (δ) of selected organic compounds that were deemed particularly appropriate as models for polymeric materials, from the data bank compiled by Daubert and Danner at Penn State University[8]. After several preliminary studies a final data set of 210 compounds was chosen such that its members exhibit little if any predilection towards self-association; this set includes 54 linear and branched aliphatic hydrocarbons; 33 mono-, di- and tri-alkyl substituted aromatic hydrocarbons; 51 linear and branched unsaturated, unconjugated, aliphatic hydrocarbons; 35 aliphatic and aromatic esters; 16 ethers; 11 ketones; 6 mixed ether / ester / ketones and 4 tertiary amines. In addition, after it was determined that inclusion of organic compounds that are known or suspected to be weakly self-associated did not materially affect the final results, 12 primary and secondary chlorides; 10 nitriles and 23 primary and secondary amines were added to the data set. It is important to emphasize that no compounds that contain groups known to strongly self-associate, such as alcohols and substances containing carboxylic acid, amide, urethane or similar groups were included. Details of the specific compounds chosen and a summary of their properties are included in the thesis of Serman[9].

A 255 x 18 matrix, M, was prepared using the Mac II - MATLAB program (The MathWorks Inc., Sherborn, MA 01770) for use on a Macintosh II computer. The 18 columns contained the number of the following chemical groups in each of the model compounds: $-CH_3$, $-CH_2-$, $>CH-$, $>C<$, C_6H_3, C_6H_4, C_6H_5, $CH_2=$, $-CH=$, $>C=$, $-COO-$, $-CO-$, $-O-$, $-Cl$, $-CN$, $-NH_2$, $>NH$ and $>N-$. We settled upon this particular group subdivision, which is almost identical to that employed by Small[5] in 1952, after the results of many preliminary calculations had convinced us that the errors inherent in the correlation method employed by Hoy[4] did not warrant further reduction. There are simply not enough data on model compounds with the appropriate structural subtleties to determine their group contributions with sufficient confidence.

Two additional 255 x 1 matrices were formed from the experimental values of the molar volumes (V) and the product of molar volumes and solubility parameters (F). Orthogonal (or "QR") factorization was performed to solve the overdetermined set of linear equations. A best solution, in the least squares sense, was computed by matrix division[10]:

$$V^* = M \backslash V \text{ and } F^* = M \backslash F \qquad (2.5)$$

Accordingly, V^* and F^* each represent 1 x 18 solution vectors that contain the molar volume and molar attraction constants, respectively, for the chemical groups mentioned above. The results are displayed in tables 2.2A and 2.2B together with the F contributions that were determined previously by Small[5]. The correspondence between our values of the molar attraction constants and those determined previously by Small is striking. Only the values for the ether group differ appreciably, which does have significance in our subsequent calculations of the solubility parameters of polyethers.

Although it is gratifying to have independently confirmed Small's values, this would frankly be trivial were it not for the recognition of a fundamental assumption inherent in the determination of group molar attraction constants that has important ramifications. The molar attraction constants are determined from experimental values of the solubility parameters of the model liquids *multiplied by their respective molar volumes* (i.e, $\Sigma F = \delta . V$). Accordingly, we contend that subsequent calculations of the solubility parameter of an unknown organic liquid or polymer *must also use group contributions to molar volumes based upon a correlation using the same set of experimental data.* In other words, using our molar attraction constants (or those of Small, Hoy etc.) with some arbitrary experimental or calculated molar volume to determine the solubility parameter of a polymer is specious. It is necessary to have both V^* and F^* from the same set of model compounds to be consistent in calculating the solubility parameter of a polymer. The data listed in tables 2.2A and 2.2B allowed us to determine just how accurately we can do this[7].

Table 2.2A
Unassociated Groups

GROUP	V*	F* (This work)	F (Small)
	$(cm^3\ mole^{-1})$	$((cal.\ cm^3)^{0.5}\ mole^{-1})$	
$-CH_3$	31.8	218	214
$-CH_2-$	16.5	132	133
$>CH-$	1.9	23	28
$>C<$	-14.8	-97	-93
C_6H_3	41.4	562	-
C_6H_4	58.8	652	658
C_6H_5	75.5	735	735
$CH_2=$	29.7	203	190
$-CH=$	13.7	113	111
$>C=$	-2.4	18	19
$-OCO-$	19.6	298	310
$-CO-$	10.7	262	275
$-O-$	5.1	95	70
$>N-$	-5.0	-3	-

Table 2.2B
Weakly Associated Groups

GROUP	V*	F* (This work)	F (Small)
	$(cm^3\ mole^{-1})$	$((cal.\ cm^3)^{0.5}\ mole^{-1})$	
$-Cl$	23.9	264	260
$-CN$	23.6	426	410
$-NH_2$	18.6	275	-
$>NH$	8.5	143	-

The Calculation of Solubility Parameters and the Errors Involved

Multiplication of the matrix M with V* (or F*) yields a set of calculated values of the molar volumes (or total F) for the model organic liquids based upon the respective group contributions (Tables 2.2A and 2.2B). Figure 2.5 shows a graph of the calculated molar volumes plotted against the experimental values for the 255 organic liquids employed in the original data set. A least squares fit of the data reveals an excellent correlation (a perfect correlation, of course, would be an intercept of zero and a slope and correlation coefficient of unity). The standard error of estimate was determined to be 1.9 cm^3 mole^{-1}.

A similar result, albeit slightly less perfect, is seen in the graph of the calculated versus experimental values of total F (figure 2.6). Here the corresponding standard error of estimate was calculated to be 24 ((cal. cm^{-3})$^{0.5}$ mole^{-1}).

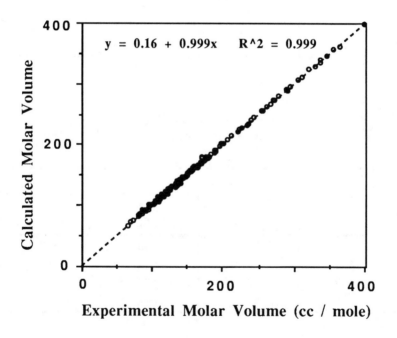

FIGURE 2.5

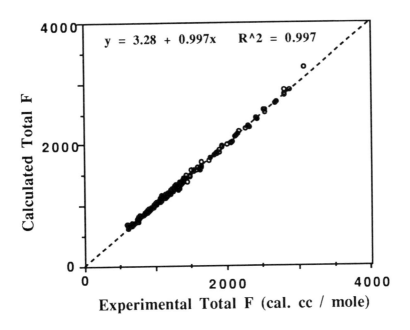

FIGURE 2.6

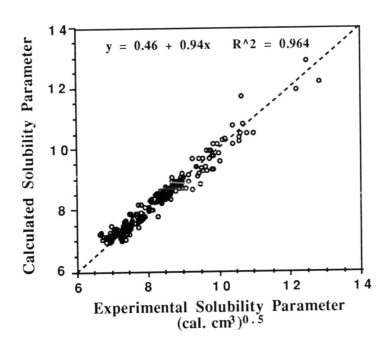

FIGURE 2.7

Of primary relevance to our studies of polymer blends, however, is the comparison of the experimental and calculated solubility parameters, which is presented graphically in figure 2.7. There is considerable scatter in the results as $\delta = \Sigma F / \Sigma V$ and the errors in molar volume and F compound. The standard error of estimate was determined at 0.21 (cal. cm^{-3})$^{0.5}$. This is a very important result. We have alluded to the fact that for high molecular weight polymer blends, *in the absence of favorable intermolecular interactions*, miscibility is only feasible when the solubility parameters of the two polymers are within about 0.1 (cal. cm^{-3})$^{0.5}$ of one another.

If the error in calculating solubility parameters is approximately ±4 times this figure (at the 2σ, 95% confidence level), it is little wonder that solubility parameters have been poor predictors of the phase behavior of polymer blends *that have only physical interactions*.

Finally, it is instructive to calculate the molar volumes of polymer repeat units and compare them to the experimental densities of *amorphous* polymers (or the amorphous component of a semicrystalline polymer) reported at 25°C. Obviously, we have to take into account the glass transition temperatures of the amorphous polymers. van Krevelen, for example, has reported different correlations for glassy and rubbery molar volumes of polymers at 25°C (Table 4.6 - ref 3). We again emphasize, however, that if we wish to estimate the solubility parameter of a polymer from a set of molar attraction constants we must be consistent and employ a correlation based upon the same set of molar volumes from which it was derived.

We should add that for our purposes we are really only concerned with polymers in blends that are "liquid-like", as our model for the free energy of mixing of polymers necessitates equilibrium conditions. Since V* was determined from organic liquids it is in closer accord to polymers in the "rubbery" state. We thus appear to be restricted to a comparison between amorphous polymers that have T_g's below room temperature. It is possible, however, to estimate a rubbery molar volume from the experimental densities of glassy amorphous polymers at 25°C using the empirical relationships $\alpha_l T_g \sim 0.16$[12] and $(\alpha_l -$

α_g) $T_g \sim 0.115$[13] where α_l and α_g are the liquid and glass volume coefficients of thermal expansion, respectively.

Figure 2.8 shows a comparison between the calculated (using Tables 2.2A and 2.2B) and the experimental molar volumes at 25°C of 98 polymers of widely different chemistries obtained from density measurements reported in numerous sources in the literature[9]. About one third of the polymers have T_g's ≥ 25°C and their glassy molar volumes were "corrected" to give rubbery molar volumes, using the relationships mentioned above. (For example, polystyrene has a T_g of 100°C and a density of 1.065 g cm^{-3} at 25°C. This corresponds to a glassy molar volume of 97.8 cm^3 mole^{-1} which when "corrected", yields a rubbery equivalent of 95.6 cm^3 mole^{-1}.) The agreement between the calculated and experimental polymer molar volumes appears quite reasonable, as shown by the fine least squares fit of the data. Only eight examples were found to differ by ≥ 5 %. A standard error of estimate of 2.8 cm^3 mole^{-1} was determined - twice that determined for the analogous organic liquids.

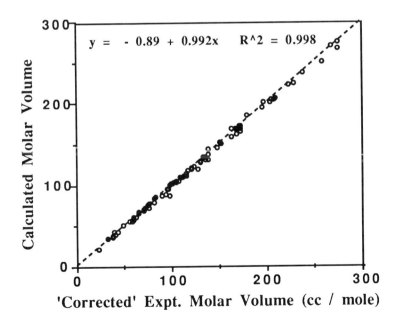

$$y = -0.89 + 0.992x \qquad R^2 = 0.998$$

FIGURE 2.8

Table 2.3 shows a comparison of the experimental[3] and our calculated solubility parameters of some representative weakly self-associated and essentially non-hydrogen bonding polymers. We contend that as we have been consistent and have used molar volume and attraction constants derived from the same set of model compound data, that this is the best we can do and we have to accept a potential error of at least ± 0.4 (cal. cm^{-3})$^{0.5}$.

Estimation of the Solubility Parameters of Strongly Associated Polymers

We have separated the molar attraction and molar volume contributions into two parts (tables 2.2A and 2.2B) to emphasize a point. Table 2.2A contains the values for groups that we can confidently assume do not self-associate to any appreciable extent. Those group contributions listed in table 2.2B, however, were derived from model organic compounds that are known to weakly self-associate. This may not be too significant, but it is a matter of degree and when we consider strongly self-associating groups, such as alcoholic and phenolic hydroxyls, amide, urethane and carboxylic acid groups, the problem becomes acute. Model organic compounds containing such groups were not included in our calculations of molar attraction constants[7]. Why is this?

We can again use figure 2.2 to illustrate the problem. Experimental solubility parameters of model organic compounds are calculated primarily from vapor pressure data (i. e. $\delta = [\rho . \Delta E_{vap} / M]^{0.5}$). If the energy of self-association is less than say 1 kcal. mole^{-1}, a large fraction of molecules have energies equal to or greater than the "dissociation" energy. Accordingly, at the boiling point of typical low molecular weight compounds of this type, individual molecules are the predominant species and M is simply the molecular weight of a 'monomer'. In contrast, for energies in excess of say 3 kcal. mole^{-1}, a significant fraction of the molecules remain associated at the boiling point and may be thought in terms of a distribution of 'h-mers'.

Table 2.3
Comparison Between Experimental and Calculated
Polymer Solubility Parameters

| POLYMER | SOLUBILITY PARAMETER $(cal.\ cm^{-3})^{0.5}$ | |
	$\delta_{Expt}{}^6$	$\delta_{Cal'd}$
Polyisobutene	7.8 - 8.1	7.2
1,4-Polybutadiene	8.1 - 8.6	8.1
1,4-Polyisoprene	7.9 - 10.0	8.1
Polystyrene	8.5 - 9.3	9.5
Poly(vinyl chloride)	9.4 - 10.8	9.9
Poly(vinyl acetate)	9.4 - 11.1	9.6
Poly(methyl acrylate)	9.7 - 10.4	9.6
Poly(ethyl acrylate)	9.25 - 9.4	9.3
Poly(propyl acrylate)	9.05	9.1
Poly(butyl acrylate)	8.8 - 9.1	8.9
Poly(isobutyl acrylate)	8.7 - 11.0	8.7
Poly(methyl methacrylate)	9.1 - 12.8	9.0
Poly(ethyl methacrylate)	8.9 - 9.2	8.9
Poly(butyl methacrylate)	8.7 - 9.0	8.7
Poly(isobutyl methacrylate)	8.2 - 10.5	8.5
Poly(benzyl methacrylate)	9.8 - 10.0	9.8
Polyacrylonitrile	12.5 - 15.4	13.8
Polymethacrylonitrile	(10.7)	11.9
Poly(methylene oxide)	10.2 - 11.0	10.5
Poly(tetramethylene oxide)	8.6	8.8
Poly(propylene oxide)	7.5 - 9.9	8.5
Poly(ethylene terephthalate)	9.7 - 10.7	11.5

In any event, the determination of the solubility parameter is equivocal, as M is no longer just the molecular weight of the monomer, but some larger value reflecting an average molecular weight of the associated species. This is exemplified in the hydrogen bonded dimer formation of carboxylic acids, as was recognized by Hoy[4].

However, since we use a model that separates the repulsive "physical" forces, embodied in the interaction parameter χ , from the favorable "attractive" forces, contained in the $\Delta G_H/RT$ term of equation 1, what we really require is an estimation of the *non-hydrogen bonded solubility parameters* of strongly self-associated polymers, such as poly(4-vinyl phenol) (PVPh), phenoxy, polyamides etc. In other words, we need the solubility parameters of these polymers that would be obtained if they did not strongly self-associate through hydrogen bonding. This is not a trivial problem. One approach has been to calculate the solubility parameter of a *homomorph*, a closely related but non-hydrogen bonding polymer. For example, substitution of a methyl group for the labile proton in a polyamide, polyurethane or polyphenol etc. has been attempted, but the methyl group itself causes a significant perturbation, especially if the repeat unit is relatively small, and this leads to a serious underestimation of the value of the solubility parameter. We prefer to employ the following procedure.

Table 2.4 lists estimated non-hydrogen bonded solubility parameters for a series of polyamides: aliphatic polycaprolactams (denoted nylon 6 etc.) and aromatic polyisophthalamides (denoted nylon n-I).

Nylon n-I

There are two sets of results. The first are calculated using a combination of individual -CO- and >N-H group contributions

Table 2.4
Calculated Non-Hydrogen Bonded
Solubility Parameters for Polyamides

POLYAMIDE	SOLUBILITY PARAMETER $(cal. \ cm^{-3})^{0.5}$	
	[CO + NH]	[CO + N]
Nylon 3	12.8	13.5
Nylon 4	11.7	11.9
Nylon 5	11.0	11.0
Nylon 6	10.5	10.4
Nylon 7	10.1	10.0
Nylon 8	9.9	9.8
Nylon 9	9.7	9.6
Nylon 10	9.5	9.4
Nylon 11	9.4	9.3
Nylon 12	9.3	9.1
Nylon 2-I	13.3	13.9
Nylon 3-I	12.7	13.1
Nylon 4-I	12.1	12.5
Nylon 5-I	11.8	12.0
Nylon 6-I	11.5	11.6
Nylon 7-I	11.2	11.3
Nylon 8-I	11.0	11.0
Nylon 9-I	10.8	10.8
Nylon 10-I	10.6	10.6
Nylon 12-I	10.3	10.3

given previously (F* and V* - tables 2.2A and 2.2B) to construct a contribution for the amide (-CO-NH-) group given in table 2.5. The rationale here is that these individual -CO- and >N-H group contributions were determined from compounds that are polar, but not strongly self-associated, and as such represent the non-hydrogen bonded contributions of the amide group.

The second set are calculated for an analogous hypothetical polyamide molecule, but without the N-H proton. In effect we calculate the solubility parameter using a combination of the individual -CO- and >N- group contributions for the amide group. This result is also given in table 2.5. In this case the rationale is that the errors involved in eliminating the proton are reasonably small, especially when the repeat unit is relatively large and, again, both the -CO- and >N- group contributions were derived from essentially unassociated model compounds. It is pleasing to see that the two sets of results are in reasonably close agreement.

Table 2.5
Groups Determined by Addition or Subtraction

GROUP	V (cm^3 mole^{-1})	F* ((cal. cm^3)$^{0.5}$ mole^{-1})
-C$_6$H$_2$-	34.1	475
-C$_6$H$_{10}$-	81.2	687
-C$_6$H$_{11}$	94.5	806
-C$_5$H$_4$N-	66.5	766
-CO-O-CO-	30.3	560
-O-CO-O-	24.7	393
-NH-CO-	19.2	405
-N-CO-	5.7	259
-NH-CO-O-	28.1	441
-N-CO-O-	14.6	295

Table 2.6
Non - Hydrogen Bonded Solubility Parameters

POLYMER	SOLUBILITY PARAMETER $(\text{cal. cm}^{-3})^{0.5}$
Poly(4-vinyl phenol)	10.6 - 11.0
Poly(vinyl alcohol)	10.2 - 10.6
Phenoxy	10.2
Amorphous Polyurethane[6]	11.2

Using the same principles the non-hydrogen bonded solubility parameter of an amorphous polyurethane (APU) employed in one of our previous studies[11] was calculated. Comparable values of 11.2 and 11.3 $(\text{cal. cm}^{-3})^{0.5}$ were determined for APU using the [-OCO- + >N-H] and [-OCO- + >N-] group contributions for the urethane group, respectively (see table 2.5).

In contrast to the polyamide and polyurethane polymers, for hydroxyl containing polymers (PVPh, phenoxy and PVOH etc.) we do not have the luxury of being able to dissect the O-H group into two group contributions that are derived from essentially unassociated molecules.

Nevertheless, we can calculate the solubility parameter for hypothetical analogues that are missing the hydroxyl proton by employing the ether group contributions, both F* and V*, for the hydroxyl group. Table 2.6 summarizes the results of such an exercise. To be candid, estimating the non-hydrogen bonded solubility parameters in this manner is rather crude and subject to considerable error, especially for relatively small polymer repeat units. We have reason to be confident that this approach has merit, however. This is based upon the argument that we have successfully predicted the gross phase behavior of a wide variety of PVPh, polyamide and polyurethane blends using values of the non-hydrogen bonded solubility parameters close to those given in tables 2.5 & 2.6. This work will be reviewed in chapter 7.

C. SIMPLE RULES GOVERNING MISCIBILITY IN POLYMER BLENDS - *A Question of Balance*[#]

In the following discussion it will be convenient to consider several different categories of polymer blends, as suggested in figure 2.4; those involving progressively stronger intermolecular forces from the very weak to the strong. However, *in general*, if the reader accepts the arguments presented so far and wishes to search for new miscible polymer blend systems, then the first rule of thumb is to minimize χ which, in turn, necessitates looking for closely matched *non-hydrogen bonded* solubility parameters. This is hardly a new concept, except that we have extended it into the realm of polymer mixtures involving strong favorable intermolecular interactions by separating out the physical contributions into an interaction parameter, χ , which is estimated from the *non-hydrogen bonded* solubility parameters. In a nutshell, if χ is close to zero, the favorable intermolecular interactions present will drive the system towards miscibility. Just how close the non-hydrogen bonded solubility parameters have to be to one another depends upon the *type, number and relative strength* of the specific favorable intermolecular interactions that are present.

Borrowing a concept (cosolvency) used extensively for gauging polymer solubility in the surface coatings industry, consider the case of blending a homopolymer having a solubility parameter, δ_A, with a random copolymer composed of two repeat units that individually solubility parameters, δ_B and δ_C , respectively, that span δ_A. As we discussed in chapter 1, a random copolymer of a given composition can be considered in terms of an average solubility parameter, $\bar{\delta}_{BC}$, which if matched to δ_A would, in principle, result in a miscible blend mixture. The "window of miscibility", i.e. the range of copolymer compositions that are miscible with the homopolymer about this point will then depend upon the *type, number and relative strength* of the favorable intermolecular interactions that may be present, as illustrated in figure 2.9.

[#] With apologies to the Moody Blues, for those who remember the sixties.

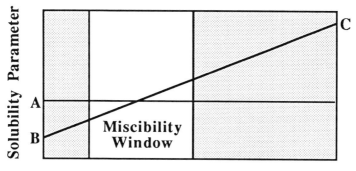

FIGURE 2.9

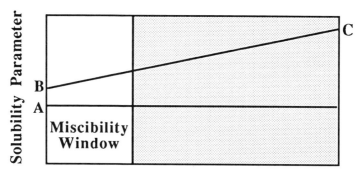

FIGURE 2.10

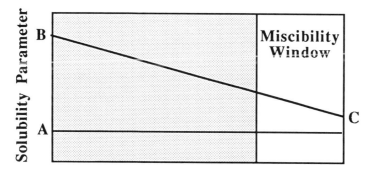

FIGURE 2.11

It is important to recognize, however, that in our scheme miscibility windows are not just restricted to intermediate copolymer composition ranges, but also can exist at either extreme of the copolymer composition range. If the non-hydrogen bonded solubility parameter of polymer A lies outside the range of the solubility parameters embraced by the B-co-C copolymer then we can obtain miscibility windows shown schematically in figures 2.10 and 2.11, providing that there are specific interactions between appropriate polymer units, in these examples the A and B segments. In the example shown in figure 2.10, polymer A is miscible with polymer B, because the favorable intermolecular interactions overwhelm the unfavorable contribution arising from the difference in the non-hydrogen bonding parameters. But as B is diluted through copolymerization with C, a point is reached where the combination of increased solubility parameter difference and decreased number of favorable interaction sites act in concert to tip the balance in favor of phase separation. In the other case (figure 2.11), polymer A is immiscible with polymer B, because the contributions from favorable intermolecular interactions are not enough to overwhelm the unfavorable contribution arising from the difference in the non-hydrogen bonding solubility parameters. But, as B is diluted through copolymerization with C the solubility parameter difference between polymer A and the B-co-C copolymer decreases, and although there is a parallel decrease in the number of favorable interaction sites, the balance tips in favor of miscibility.

Then if the solubility parameters of polymer A and polymer C are close enough, a miscibility window such as that illustrated in figure 2.11 is possible. In order to arrive at a simple predictive scheme there are therefore two major factors that we must consider as a function of copolymer composition: (i) the variation of the non-hydrogen bonded solubility parameter difference and (ii) the *type, number and relative strength* of the intermolecular interactions that may be present. To illustrate further, consider the case of a copolymer in which one comonomer interacts favorably with the homopolymer, but the other does not. A good example is the styrene-co-vinyl phenol

(STVPh) copolymer blends with poly(methyl methacrylate) (PMMA). In this case styrene may be viewed simply as *diluting* the vinyl phenol segment, thereby reducing the concentration of sites available to interact (hydrogen bond) with the MMA segment. In other words, the contribution from the $\Delta G_H/RT$ term of equation 2.1 varies with copolymer composition which, in turn, means that the *critical* values of the interaction parameter, χ_{Crit}, or solubility parameter difference, $(\Delta\delta)_{Crit}$, also changes in some fashion with copolymer composition. If we are able to even qualitatively obtain an indication of the miscibility of copolymer blends we need to obtain an idea of the *form* of this variation. As we will discuss in Chapter 6, the magnitude of the $\Delta G_H/RT$ term can be quantitatively determined for hydrogen bonded polymer blend systems such as STVPh - PMMA if one has a knowledge of the molar volumes of the individual segments, the relevant enthalpies of hydrogen bond formation and equilibrium constants describing self- and what we call inter-association (favorable interactions between different types of functional groups). Fortunately, these parameters are readily obtained from group contributions and infrared spectroscopic studies. Accordingly, in this case the *critical* values χ_{Crit} and $(\Delta\delta)_{Crit}$ may be calculated directly from equations 2.6 and 2.7 (which are analogous to equations 2.2 and 2.3 cited for the non polar case).

and:

$$\chi_{Crit} = \frac{1}{2}\left[\frac{1}{M_A^{0.5}} + \frac{1}{M_B^{0.5}}\right]^2 + f(\Delta G_H) \qquad (2.6)$$

$$(\Delta\delta)_{Crit} = \left[\frac{RT\,\chi_{Crit}}{V_B}\right]^{0.5} \qquad (2.7)$$

where $f(\Delta G_H)$ is a function of the second and third derivatives of the ΔG_H term with respect to composition.

Figure 2.12 shows graphically the results of theoretical calculations, using equations that we will describe later in this book, of $(\Delta\delta)_{Crit}$ as a function of copolymer composition for hypothetical blends of a homopolymer A with a copolymer B-co-C where C can be considered an inert diluent. All segments were assumed to have molar volumes of 100 cm^3/mole and the temperature was taken to be 25°C. Equilibrium constants describing the self-association of B, K_B, and the inter-association between A and B, K_A, were initially given equal values of 10, 1 and 0.1 as illustrated in figure 2.12.

We have found that all of these curves can be adequately described by the *same* polynomial of the form $y = a + bx + cx^2$ multiplied by some constant that defines the position on the y-axis, i.e. $(\Delta\delta)^\circ_{Crit}$, the *critical value for the two homopolymers* A and B. Additionally, the error introduced by assuming a single polynomial when K_A differs by an order of magnitude from K_B (i.e. when self-association is stronger than inter-association or vice versa) is not large. This is illustrated in figure 2.13 where we show the curves for K_A/K_B ratios of 10, 1 and 0.1 adjusted to a $(\Delta\delta)^\circ_{Crit}$ value of 2.0. The results are contained in a relatively narrow band denoted by the shaded area.

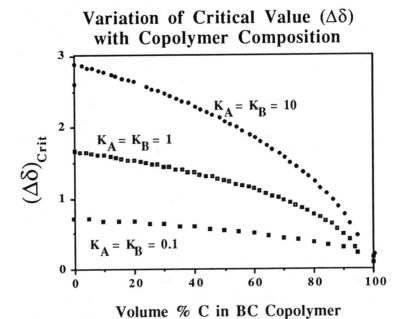

Variation of Critical Value ($\Delta\delta$) with Copolymer Composition

$K_A = K_B = 10$

$K_A = K_B = 1$

$K_A = K_B = 0.1$

$(\Delta\delta)_{Crit}$

Volume % C in BC Copolymer

FIGURE 2.12

Adjusted (Scaled) Values

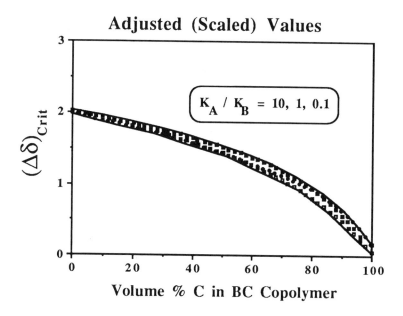

FIGURE 2.13

Although there are obvious limitations in this simple approach, we have performed many simulations and have concluded that $(\Delta\delta)_{Crit}$ as a function of copolymer composition for the majority of polymer blend systems can be reasonably approximated by a single polynomial that is scaled by a constant $(\Delta\delta)^{o}_{Crit}$ value and this is what we employ in our forthcoming qualitative **Miscibility Guide**.

Let us summarize this section by making these final comments. The basis for our qualitative guide to polymer miscibility rests upon our ability to reasonably approximate three major components; (1) the non-hydrogen bonded solubility parameters of homo- and copolymers; (2) the scaling of $(\Delta\delta)^{o}_{Crit}$ as a function of the type, number and relative strength of the specific or favorable molecular interactions present and (3) the variation of $(\Delta\delta)_{Crit}$ as a function of dilution when the concentration of the interacting species decreases. We have discussed in detail the errors involved in calculating solubility parameters and this must always be kept in mind. In other words, there is always a temptation to over-interpret, especially if the result fits one's preconceived notions. We expect that someone, somewhere (not the more perceptive reader, of course!) will calculate the solubility parameters of two polymers, find them to be identical, observe the blend to be immiscible and conclude that our miscibility guide is of no use to man or beast. Perhaps it isn't, but one must take account of the errors involved and emphasize *trends*, not specifics. Additionally, we should be at our best in estimating $(\Delta\delta)^{\circ}_{Crit}$ when the molar volumes of our interacting segments are of the order of 100 cm³/mole. Segments with significantly larger molar volumes are already "diluted" in our terms and the reader may need to consider reducing the value when large repeating units are considered. And finally, the function describing the variation of $(\Delta\delta)_{Crit}$ with dilution is really only applicable to polymers where the equilibrium constants describing self-association and inter-association are within an order of magnitude of one another.

D. VERY WEAK OR NONEXISTENT FAVORABLE INTERMOLECULAR INTERACTIONS

Dispersive Forces $(\Delta\delta)^{\circ}_{Crit} = 0.1$

We do not wish to belabor the point, but it is apparent that the solubility parameters of polymers cannot be determined with sufficient accuracy to adequately predict the phase behavior of polymer blends of this type. Nevertheless, solubility parameters may be employed to discern significant *trends* in the phase behavior of systematically chosen sets of polymer blends - ones in which there are variations in copolymer composition or where a homologous series of polymers is considered.

Butadiene-co-acrylonitrile (BAN) polymers, have solubility parameters that span a wide range from approximately $\delta = 8.1$ to 13.8 (cal. cm⁻³)^0.5. What might we anticipate if we were to blend BAN copolymers of varying copolymer composition with a non-polar homopolymer, such as polystyrene (PS), which has a solubility parameter within this range ($\delta = 9.5$ (cal. cm⁻³)^0.5) ?

This is a convenient place to introduce the "**Miscibility Guide**" computer program that readily calculates such information and comes with this book. We will restrict ourselves to examples derived from the program written for the Macintosh II computer (a reflection of the author's bias), but similar versions for other Apple and IBM compatible computers are available. "Double clicking" (launching the application) on the icon "**Miscibility Guide 020**" produces the window shown in figure 2.14. The window labelled **Groups** contains small icons denoting the groups listed in Tables 2.2A and 2.2B, with the individual values of the group molar volume and molar attraction constants contained within the computer program. (Additions, subtractions, changes etc., to this data base can be made using an **Editor** - see Operating Manual). Let us start by entering in the groups of PS (figure 2.15). "Double clicking" on the icons for CH_2, CH and C_6H_5 groups displays in the **Segment A** window the group icon and boxes containing the number of each group, which may be varied by selecting the box and entering the desired number.

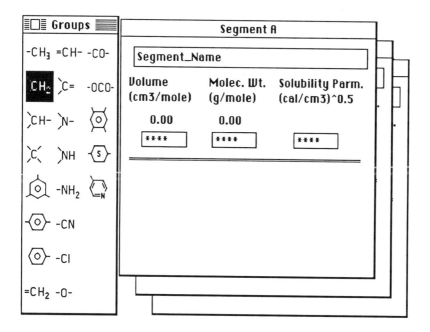

FIGURE 2.14

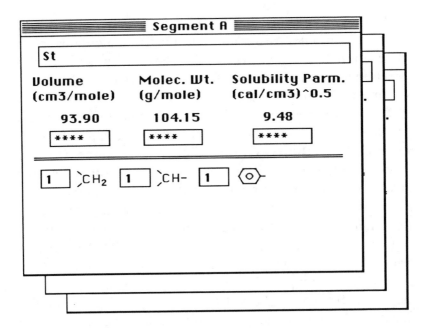

FIGURE 2.15

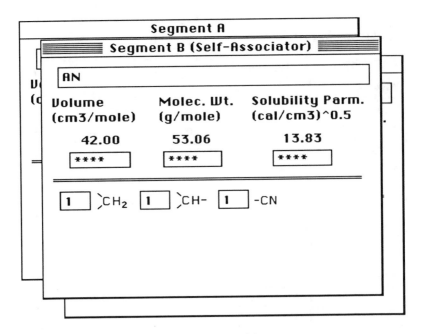

FIGURE 2.16

The calculated molar volume, molecular weight and solubility parameter are displayed automatically in the window (the boxes immediately below these values can be used to enter an alternative value if desired). The segment may be also be labelled, in our case St, and saved.

Selecting successively the **Segment B (Self-Associator)** and **Segment C (Diluent)** windows, we can now repeat the process for the acrylonitrile and butadiene segments, respectively, as illustrated in figures 2.16 and 2.17. Selecting **Run Guide** from the **Guide** window yields the window shown in figure 2.18. In the upper left hand side of figure 2.18 one has the choice of "diluting" either styrene or acrylonitrile to form a copolymer with butadiene - the unselected segment thus becomes the homopolymer considered. The copolymer composition may be selected in one of two ways. First, by selecting the "radio" button **Composition**, in terms of either molar, weight or volume % or fraction.

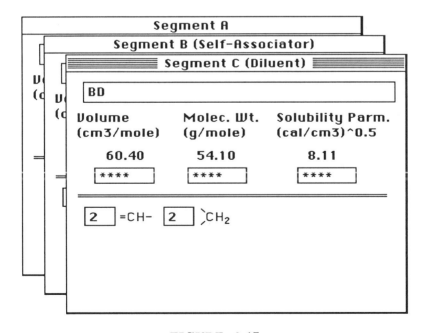

FIGURE 2.17

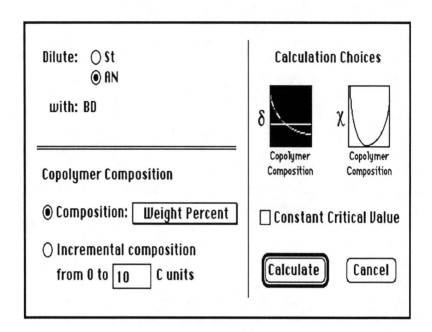

FIGURE 2.18

PS - Poly(AN-co-BD) Blends

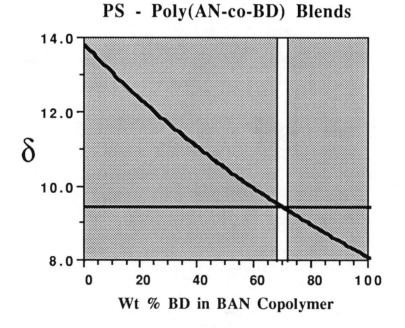

FIGURE 2.19

Alternatively, selecting **Incremental composition** permits one to incrementally add individual units to the **Segment B** (e.g. a homologous series of polyesters containing increasing number of methylene units - see later).

By selecting one of the icons on the right hand side, one may opt for either plots of the solubility parameters versus copolymer composition or the interaction parameter, χ , as a function of copolymer composition. The latter employes a value of 100 cm^3 mole^{-1} for the reference volume, V_r, in the program. For now we will ignore the **Constant Critical Value** box and select the δ **icon** followed by **Calculate** and we should obtain a window like that shown in figure 2.19. Figure 2.19 shows the calculated solubility parameter of BAN as a function of weight % butadiene (BD) in the copolymer, together with that of PS (the line at 9.5 (cal. cm^{-3})$^{0.5}$ running parallel with the x axis).

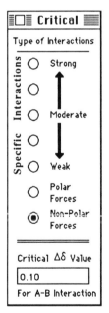

Where these two lines intersect, i.e. when $\delta_{BAN} = \delta_{PS}$ and $\chi = 0$, miscibility is theoretically assured, as $\Delta H_m = 0$. Although the favorable entropy of mixing is very small, it is sufficient to impart miscibility. The breadth of the miscibility window, expressed in terms of the range of copolymer composition over which the blend is predicted to be miscible, denoted by the unshaded portion in the graph, is determined by the critical value of the solubility parameter difference, $\Delta\delta$. The window labelled **Critical** (seen at left) has 7 radio buttons that contain *initial* values of $\Delta\delta$ (i.e. $(\Delta\delta)^{\circ}_{Crit}$)[#] that we have determined (on the basis of detailed calculations) reasonably represent the range of potentially favorable intermolecular inter-actions as we progress

---

[#] In an A-co-B copolymer, for example, where one type of segment has a functional group that can take part in favorable interactions, while the other does not, $\Delta\delta$ will vary with copolymer composition, as shown in figures 2.12 and 2.13. The *initial* value of $(\Delta\delta)_{Crit}$ is that corresponding to a homopolymer of the segments that form specific interactions.

from purely dispersive forces ($\Delta\delta = 0.1$ (cal. cm^{-3})$^{0.5}$) through polar ($\Delta\delta = 0.5$ (cal. cm^{-3})$^{0.5}$) to weak, moderate and strong hydrogen bonding (or other specific) interactions ($\Delta\delta = 1.0$, 1.5, 2.0, 2.5 and 3.0 (cal. cm^{-3})$^{0.5}$). In the case of the BAN - PS blends only weak physical intermolecular forces are involved and a critical value of ≤ 0.1 (cal. cm^{-3})$^{0.5}$ is applicable, resulting in a very narrow range of miscibility. (Acrylonitrile segments are polar, but here they are mixed with non-polar groups.)

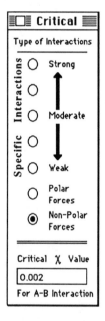

The corresponding estimate of χ for PS - BAN blends is shown in figure 2.20 as a function of copolymer composition and may be displayed by selecting **Run Guide** from the **Guide** window, followed by the χ **icon** (figure 2.18) and **Calculate**. Critical χ values are determined (equation 2.7) from the critical solubility parameter differences using the chosen reference volume. The narrow parabolic shaped variation of χ as a function of BAN copolymer composition has a minimum value of $\chi = 0$ at approximately 30% AN. Recall that for $M_A = M_B = 1000$, in the absence of favorable intermolecular interactions, the critical value of χ is ≤ 0.002. Accordingly, there is a very narrow range of BAN composition ($\approx$ 30-34% AN in this case) over which there is the best chance of finding a miscible blend of BAN and PS. This allowed composition range is actually more restrictive than this, since unfavorable "free volume" effects have not been taken into account and this would further narrow the range. Moreover, recognizing the errors involved in determining individual solubility parameters, the "true" curve may be significantly displaced on either side of the minimum value shown on the x-axis of figure 2.20. The overall shape of the curve would be the same but finding the "window" could, in practice, be difficult !

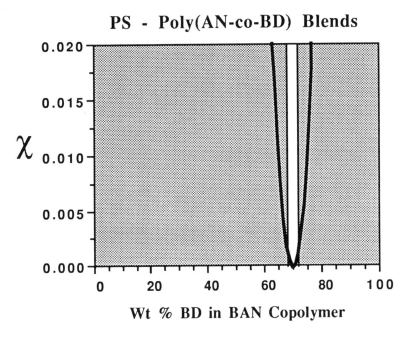

PS - Poly(AN-co-BD) Blends

Wt % BD in BAN Copolymer

FIGURE 2.20

Summary

For the case of polymer blend systems in which the *absence of favorable intermolecular interactions* can be assumed:

1. The probability of finding a miscible blend of two homopolymers of high molecular weight is exceedingly low, since this implies the almost perfect matching of solubility parameters ($\Delta\delta \leq 0.1$ (cal. $cm^{-3})^{0.5}$).

2. As the error in estimating polymer solubility parameters far exceeds the accuracy required to predict miscibility, it makes more sense to use the method to *eliminate* potential candidates. For example, blends of PS and polyisoprene (δ =9.5 and 8.1±0.4 (cal. $cm^{-3})^{0.5}$, respectively) are readily rejected as a potentially miscible pair, as the difference between the solubility parameters, even if one assumes the most favorable values at the

error limits, significantly exceeds 0.1 (cal. cm^{-3})$^{0.5}$. Furthermore, one can assume that it is very unlikely that a non-polar homopolymer such as PS will mix with a copolymer of styrene and any other non-interacting comonomer (e.g. poly(styrene-co-acrylonitrile) (SAN), SBR etc.) since the difference in the solubility parameters becomes increasingly larger the greater the concentration of the comonomer.

3. If the object is to find a miscible blend of a given non-polar homopolymer, a good strategy might be to employ the "cosolvent" approach described above. An experimental approach of trial and error might be necessary to find the rather narrow window defining the limits of miscibility, because of the inherent errors involved in estimating solubility parameters.

E. RELATIVELY WEAK FAVORABLE INTERMOLECULAR INTERACTIONS

Examples of miscible binary polymer blends that fit into the category of relatively weak favorable intermolecular interactions include PMMA - PEO[6]; PVAc - PEO[17]; PS - PVME[18]; PS - poly(phenylene oxide) (PPO)[19] and poly(butylene terephthalate) (PBT) - polyarylate[20]. Keeping in mind the errors involved in estimating polymer solubility parameters, the calculated difference between the solubility parameters of these polymer pairs is $\Delta\delta$ = 0.3, 0.2, 1.0, 0.3 and 0.1 (cal. cm^{-3})$^{0.5}$ mole^{-1}, respectively. Polymers containing vinyl chloride (VC) or acrylonitrile (AN) segments mixed with polar polymers containing ester, acetate or acrylic groups also fit into this category and since there has been a number of systematic studies of these blends we will concentrate our attention on these systems. But first let us consider polymer blends where there exists the potential for relatively weak favorable intermolecular interactions between the components that may be categorized as simple polar forces.

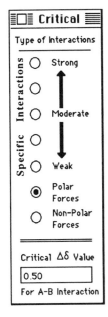

Type of Interactions

Specific Interactions

○ Strong

○

○ Moderate

○

○ Weak

◉ Polar Forces

○ Non-Polar Forces

Critical Δδ Value

0.50

For A-B Interaction

Polar Forces $(\Delta\delta)^{\circ}_{Crit} = 0.5$

We mentioned above that PEO is miscible with PMMA[6]. Certainly, we would anticipate that blends of this type would fall under the category of polar type interactions simply on the basis of the oxygen containing functional groups that are present. Let us see then what our simple guide tells us about such systems. We can do this in one of two ways. First, we can calculate the solubility parameter plots for PEO blends with a homologous series of poly(n-alkyl methacrylates). In terms of the **Miscibility Guide, Segment A** is PEO, **Segment B** PMMA and **Segment C** a methylene group which may be viewed as the diluent.

After selecting **PMMA** as the polymer to be diluted with methylene, **Incremental composition** is chosen for the calculation of up to 10 methylene groups, followed by **Run Guide** (figure 2.18). The result is displayed in figure 2.21, where we employ an upper limit of $(\Delta\delta)^{\circ}_{Crit} = 0.5$ (cal. cm^{-3})$^{0.5}$ - the second button in the **Critical** window. This may seem like an arbitrary choice of $(\Delta\delta)^{\circ}_{Crit}$, and to some extent it is. Nonetheless, as we will see, this assumption works extremely well when applied to polymers that clearly have polar groups, but also do not hydrogen bond at all. The results suggest that PEO should be miscible with PMMA (zero methylenes in the side group), on the edge of miscibility with poly(ethyl methacrylate) and immiscible with the higher homologues.

An alternative way of considering blends of this type is to hold PMMA constant and consider a homologous series of poly(n-alkyl ethers). One returns to **Run Guide** under the **Guide** menu with now a simple ether segment selected as the "polymer" (in effect, a hypothetical polyoxygen !!) to be diluted with methylene. The result is shown in figure 2.22.

Poly(PEO) and Poly(PMMA-co-Methylene) Blend

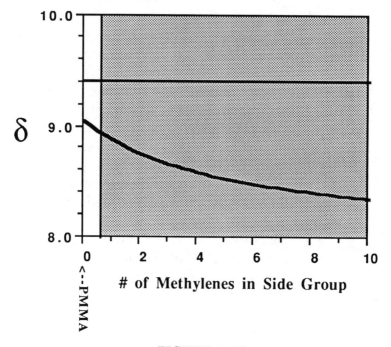

FIGURE 2.21

PEO is once again predicted to be miscible with PMMA using the value of $(\Delta\delta)^{\circ}_{Crit} = 0.5$ (cal. cm^{-3})$^{0.5}$ and so is poly(trimethylene oxide) and poly(tetramethylene oxide) (PTHF). On the other hand, poly(methylene oxide), poly-(pentamethylene oxide) and the higher homologues are predicted to be immiscible.

Another example that fits the category of polar forces is the system of blends involving aromatic polyesters and polyarylate. The chemical repeat of the latter polymer is shown below:

Poly(PMMA) and Poly(Ether-co-Methylene) Blend

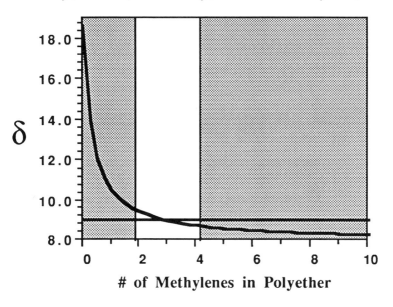

FIGURE 2.22

The repeat that we define for use in our calculations contains only one favorable or polar interaction site, in this case the ester group, and is thus defined as half of the translational repeat unit for this polymer. One simply defines the polyarylate in **Segment A** (figure 2.23) in terms of one ester and one methyl group, 1.5 disubstituted benzene groups and 0.5 tertiary carbon groups. Similarly, **Segment B** is an aromatic ester group made up of one half a disubstituted benzene and one ester group while **Segment C** is simply a methylene group. The **Aromatic ester** is selected as the "polymer" to be diluted with methylene. The result is shown in figure 2.24. As we have defined the segments, one methylene in the polyester chemical repeat corresponds to poly(ethylene terephthalate) (PET), two to poly(butylene terephthalate) (PBT), three to poly(hexamethylene terephthalate (PHMT) etc. Using the critical value of $(\Delta\delta)^{\circ}_{Crit} = 0.5$ (cal. cm^{-3})$^{0.5}$ we predict that PET is on the edge of miscibility; PBT is miscible while PHMT and the higher homologues are immiscible. This appears to be in line with the experimental evidence.

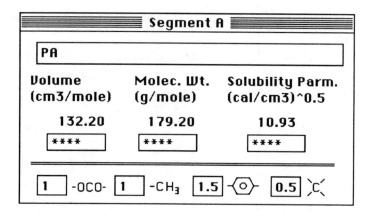

FIGURE 2.23

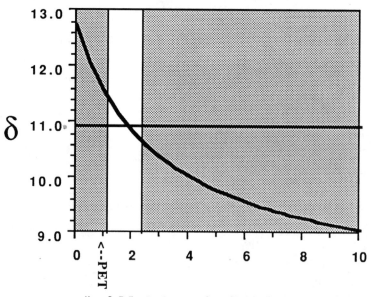

FIGURE 2.24

Now lets get more adventurous and consider blends of *two* copolymers in which the dominant intermolecular forces are polar. Cowie et al.[21] have recently published miscibility maps for a number of polymer blend systems based upon the segments styrene (St), methyl methacrylate (MMA), acrylonitrile (AN) and N-phenyl itaconimide (PII). There are, of course, many blend combinations possible and we will restrict ourselves to systems where it is reasonable to judge the predominant intermolecular interactions involved.

NPII   MMA

With this in mind the simplest system that Cowie and coworkers considered is the blend of St-co-MMA (SMMA) with St-co-PII (SPII). Again, there are no obvious opportunities for intermolecular hydrogen bonding and we can conveniently view this as a polar blend system where each polar component is diluted by the non-polar styrene segment.

Returning to the **Miscibility Guide**, select **Copolymer-Copolymer** from the **Option** menu and a new **Segment D (Diluent)** will be displayed, as illustrated in figure 2.25. The segment information for St is entered into both **Segment D** and **Segment C** (or recalled from a previously saved file). The segment informations for PII and MMA are entered into **Segment A** and **Segment B**, respectively. Now **Run Guide** is selected from the **Guide** menu and a window that resembles figure 2.26 appears. The window is split into two halves corresponding to the two different copolymers. Note that we can choose the diluent segment for **Segments A** and **B** by selecting the appropriate radio button (both styrene in this particular case). The **Composition** of the copolymers may be chosen in terms of incremental, molar, weight or volume fraction or percent by selecting the appropriate box. We will chose volume fraction in order to be consistent with the published experimental data of Cowie et al. Again, for now, we will *ignore* the two **Constant Critical Value** boxes and we are now ready to select **Calculate** and produce a miscibility map. The program calculates the solubility parameters differences for the range of copolymers at 5% intervals and produces a matrix or miscibility map. On the x and y axes are the copolymers expressed in concentration of C in a A-co-C and D in B-co-D copolymers, respectively. The small black dots denote blends calculated to be immiscible, where the difference in the solubility parameters of a particular A-co-C copolymer composition with a particular B-co-D copolymer composition exceeds the critical value, $(\Delta\delta)_{Crit}$. Remember that this critical value varies with copolymer composition and both copolymers have to be considered when adjusting the critical solubility parameter difference, $(\Delta\delta)_{Crit}$, from that of the initial critical value $(\Delta\delta)^o_{Crit}$ displayed in the **Critical** window (employing the quadratic function described previously).

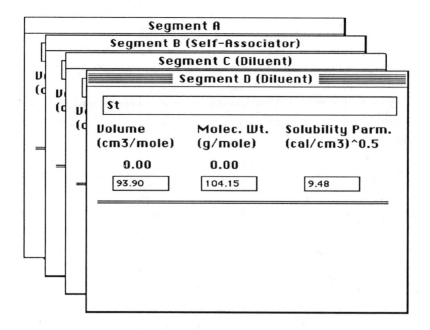

FIGURE 2.25

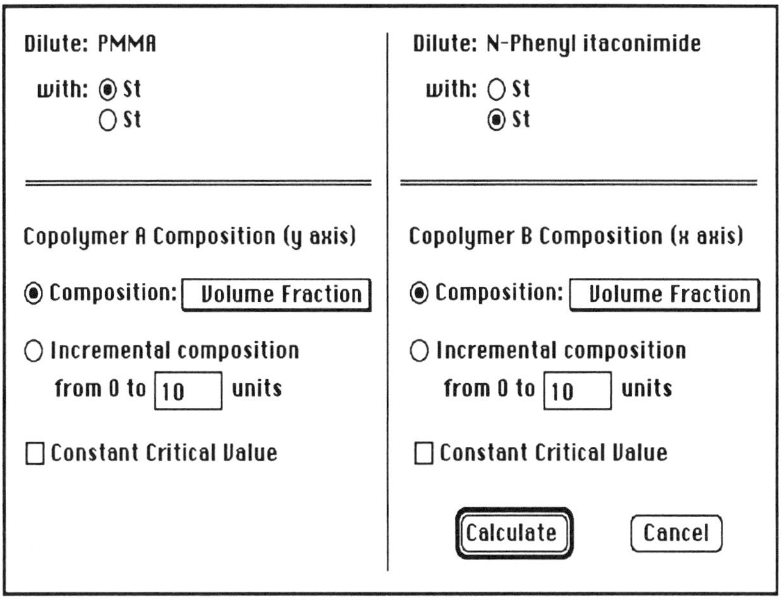

```
┌─────────────────────────────────────────────────────────────────┐
│  Dilute: PMMA              │  Dilute: N-Phenyl itaconimide        │
│    with: ◉ St              │    with: ○ St                        │
│          ○ St              │          ◉ St                        │
│  ═══════════════════════   │  ═══════════════════════════════     │
│                            │                                      │
│  Copolymer A Composition (y axis) │ Copolymer B Composition (x axis) │
│  ◉ Composition: [ Volume Fraction ]│ ◉ Composition: [ Volume Fraction ]│
│  ○ Incremental composition │  ○ Incremental composition           │
│     from 0 to [10]  units  │     from 0 to [10]  units            │
│  □ Constant Critical Value │  □ Constant Critical Value           │
│                            │        [ Calculate ]   [ Cancel ]    │
└─────────────────────────────────────────────────────────────────┘
```

FIGURE 2.26

PS, PMMA and poly(N-phenyl itaconimide) (PPII) have calculated solubility parameters of 9.5, 9.1 and 12.9 (cal. cm^{-3})$^{0.5}$, respectively. We must caution that the latter value is subject to additional error, as no allowance has been made for the fact that the PII segment is cyclic or that the itaconimide group has been approximated from the addition of a tertiary nitrogen and two ketone groups. Nonetheless, we know that PII has a rather high solubility parameter and it is the *trends* that we emphasize. These will prove to be remarkably accurate.

In figure 2.27 a diagram is presented that shows schematically the variation of the solubility parameters of the SMMA and SPII copolymers as a function of the concentration of styrene - the non-polar "diluent". Pure PMMA and PPII are inherently immiscible because the difference in the solubility parameters, ~ 3.8 (cal. cm^{-3})$^{0.5}$, far exceeds the critical value applicable for polar forces, $(\Delta\delta)^{\circ}_{Crit} = 0.5$. However, as we dilute both polymers with styrene the solubility parameters approach one another and at the extreme we have the case of PS mixing with itself (this had better be a miscible "mixture"!).

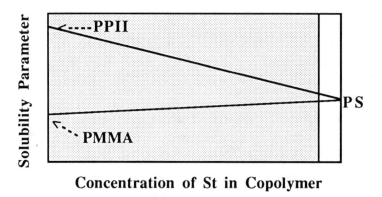

FIGURE 2.27

Remember, as we dilute the system with the non-polar styrene segment, the concentration of polar groups is deminishing, which implies that $0.5 > (\Delta\delta)_{Crit} > 0.1$. The question is just how large of an area are miscible mixtures predicted to occupy on the miscibility map ? The calculated miscibility map is shown in figure 2.28.

It is immediately apparent that the system is almost totally immiscible as only a very small area of predicted miscibility exists at the extreme top right hand side of the map, which corresponds to essentially pure PS. This should not surprize us at this stage since the addition of MMA to the St chain decreases the average solubility parameter of SMMA somewhat, while the opposite occurs by the addition of PII to the St polymer chain and quickly raises the average solubility parameter of SPII copolymer. Therefore the difference in solubility parameters between the two copolymers increases with increasing amounts of MMA or PII. The large black circles in figure 2.28 denote immiscible blends that were reported by Cowie et al.[21] and they are all contained within the predicted immiscible region of the miscibility map. We are obviously pleased with this agreement, but we must not get overly enthusiastic since this is not really a good test of our predictive capabilities, as only a few systems, all immiscible, were reported in this study.

Styrene-co-Methyl Methacrylate Blends with Styrene-co-Phenyl Itaconimide

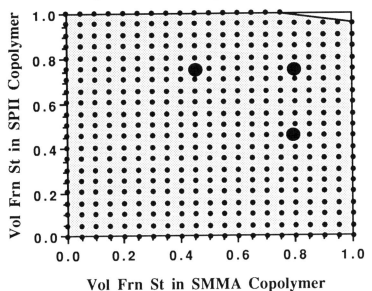

FIGURE 2.28

Weak Specific Interactions $(\Delta\delta)^{\circ}_{Crit} = 1.0$

Poly(vinyl chloride) (PVC) and polyacrylonitrile (PAN) may be classified as moderately self-associated polymers. The precise nature of the intermolecular interactions involved between the chemical moieties of PVC or PAN, be they dipolar, relatively weak hydrogen bonds or a combination of both, is the subject of debate. For our purposes here, however, this is not important.

Let us now consider the PVC - SAN blend system. PVC is neither miscible with PS nor PAN but is miscible with SAN copolymers containing 11.5 to 26 % acrylonitrile[22]. Paul and his coworkers[22] attribute the miscibility of the PVC - SAN blends to the intramolecular repulsion of the styrene and acrylonitrile units within the SAN copolymer chain, using the binary interaction or repulsion model discussed previously.

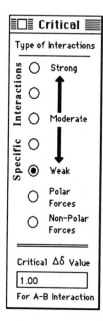

These authors also discuss possible favorable intermolecular interactions occurring between chlorinated compounds and aromatic rings. Interactions involving the acrylonitrile segment are also possible (we have observed, for example, that the carbonyl group in miscible polycaprolactone (PCL) - SAN blends shifts to lower frequency, suggesting that it is involved in a specific interaction with presumably the α-hydrogen of the acrylonitrile unit[23] - see later). The precise nature of the intermolecular interactions is again not important to our arguments here - only the fact that they exist and that they are, on our scale, relatively weak specific interactions.

PVC has a solubility parameter (δ = 9.9 (cal. cm^{-3})$^{0.5}$) that is intermediate between that of PS (δ = 9.5) and PAN (δ = 13.8). In the manner described above, the **Miscibility Guide** can now be employed to calculate the solubility parameter of SAN as a function of styrene content, as displayed in figure 2.29.

Again, it is convenient to view this as diluting a polar polymer, PAN, with an inert diluent, styrene. For blends such as these, which involve weak specific interactions (in our opinion, relatively weak hydrogen bonds) we employ an upper limit of $(\Delta\delta)^{\circ}_{Crit}$ = 1.0 (cal. cm^{-3})$^{0.5}$ - the third button in the **Critical** window. The results suggest a range of miscible SAN copolymers containing from about 3 to 31% AN, which is in acceptable agreement with the experimental data of Paul and his colleagues. Although it is possible to obtain even better agreement by tinkering with the value of $(\Delta\delta)^{\circ}_{Crit}$, within the range of polar to weak specific interactions, i.e. $(\Delta\delta)^{\circ}_{Crit}$ = 0.5 to 1.0 (cal. cm^{-3})$^{0.5}$, we do not believe this is appropriate - the major trends in miscibility have been successfully predicted with the simple value of $(\Delta\delta)^{\circ}_{Crit}$ corresponding on our scale to weak, favorable, specific interactions.

PVC and Poly(AN-co-St) Blend

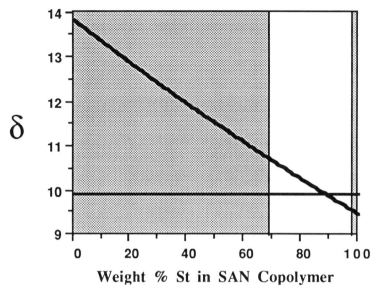

FIGURE 2.29

PVC and Poly(AN-co-St) Blend

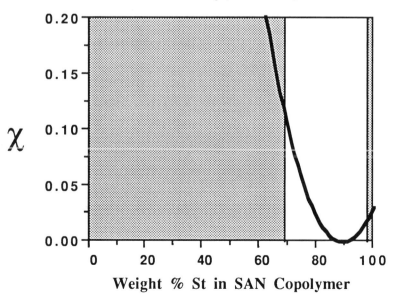

FIGURE 2.30

The corresponding diagram of χ (physical forces only) versus copolymer composition for PVC - SAN blends is shown in figure 2.30. The curve resembles a parabola in shape and goes through a minimum value of zero at a composition of about 12% acrylonitrile. *If their were no favorable intermolecular interactions* (i.e. $\chi_{Crit} < 0.002$) and if the estimated values of the solubility parameters are reasonably accurate, this implies a narrow "miscibility window" for blends of PVC with SAN containing between about 6 and 14% acrylonitrile. However, the presence of relatively weak favorable specific intermolecular interactions serves to "open" the composition range where miscibility can occur. The critical value of χ for weak specific intermolecular interactions corresponding to a $(\Delta\delta)^{\circ}_{Crit} = 1.0$ (cal. cm^{-3})$^{0.5}$ is $\chi_{Crit} \cong 0.17$, (approximately two orders of magnitude greater than the case involving no favorable interactions), and yields a prediction for miscible PVC - SAN blends that is in good agreement with the experimental range of miscibility determined by Paul and coworkers. Also, let us not loose sight of the fact that by simply considering the solubility parameters of PVC and SAN copolymers and recognizing that relatively weak favorable intermolecular interactions exist in this system, we have predicted with reasonable accuracy the most probable range of miscibility without recourse to the postulation of a "repulsion" effect.

To illustrate that the above is not an isolated fortuitous result, let us now consider PVC - BAN blends. Zakrzewski[24] found that BAN copolymers having butadiene contents between 55 and 77% were miscible with PVC over the entire range of *blend* compositions. In terms of the **Miscibility Guide**, the segment information for butadiene is simply substituted for styrene in **Segment C**. Figures 2.31 and 2.32 show, respectively, the calculated solubility parameter and χ plots for PVC - BAN blends as a function of acrylonitrile content. Employing the critical value of $(\Delta\delta)^{\circ}_{Crit} = 1.0$ (cal. cm^{-3})$^{0.5}$ for weak specific intermolecular interactions (identical to that used for the PVC-SAN blends) we predict a miscible range from 45 to 75% butadiene which is in fine agreement with Zakrzewski's data.

PVC and Poly(AN-co-BD) Blend

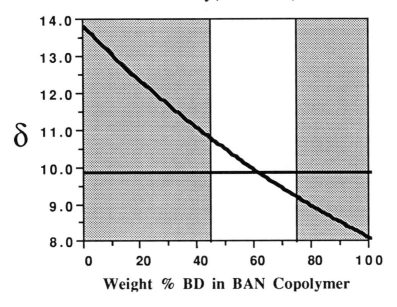

FIGURE 2.31

PVC and Poly(AN-co-BD) Blend

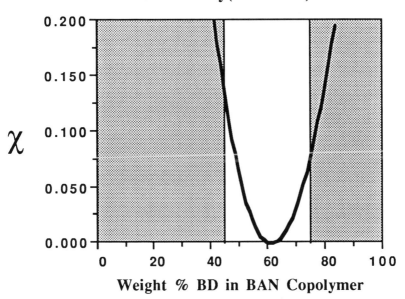

FIGURE 2.32

Again, we do not recommend "tinkering", but cannot resist pointing out that an intermediate value of $(\Delta\delta)^{\circ}_{Crit} = 0.75$ (cal. cm^{-3})$^{0.5}$ almost duplicates the experimental range of miscibility. The curve of χ versus BAN copolymer composition again has a parabolic shape, but this time goes through a minimum value of zero at a composition of about 60% butadiene. *With no favorable intermolecular interactions* (i.e. $\chi_{Crit} < 0.002$) a "miscibility window" for blends of PVC with BAN containing between about 58 and 65% butadiene is suggested (actually the range is even narrower if free volume effects are considered). In any event, the presence of the relatively weak favorable intermolecular interactions, corresponding to a $(\Delta\delta)^{\circ}_{Crit} = 1.0$ (cal. cm^{-3})$^{0.5}$ or alternatively, a χ_{Crit} value of about 0.17, identical to that for the PVC - SAN results mentioned above, leads to the predicted miscibility window for blends of PVC with BAN copolymers containing from about 45 to 75% butadiene. Again, a very good agreement with experimental observation.

Another example that fits in this category of weak specific intermolecular interactions is the SAN and styrene-co-maleic anhydride (SMAH) copolymer - copolymer blend system. There have been extensive experimental studies performed on the miscibility of these copolymers by Hall et al.[25], Kim et al.[26], and Aoki[27]. Here again we have the special case of two copolymers containing a common diluent; styrene.

PS, PAN and MAH have calculated solubility parameters of 9.5, 13.8 and 17.8 (cal. cm^{-3})$^{0.5}$, respectively. Frankly, the latter value is subject to additional error, because no allowance has been made for the fact that the MAH segment is cyclic and the solubility parameter of the anhydride group has been estimated from the simple addition of an ester and a ketone group. Notwithstanding, we know MAH has a relatively high solubility parameter and let's see if we get reasonable predictions using this preliminary estimate. Figure 2.33 shows schematically the variation of the solubility parameter of SAN as a function of AN concentration in the copolymer. Also included in the figure are the calculated solubility parameters of two hypothetical SMAH copolymers containing 20 and 40% MAH.

Note that the solubility parameters of the SMAH copolymers cross at different points on the SAN curve.

This implies, of course, that the range of miscibility for SAN copolymers will vary with SMAH copolymers of different compositions. Returning to the **Miscibility Guide**, we select **Copolymer-Copolymer** from the **Option** menu. The segment information for MAH and AN are entered into **Segment A** and **Segment B**, respectively, while that of the diluent St is entered into both **Segment C** and **Segment D**. After **Run Guide** is selected from the **Guide** menu and appropriate copolymer composition units have been selected (as in figure 2.26) the computer calculates the solubility parameter differences at 5% intervals for the range of copolymer compositions and produces a miscibility map similar to that shown in figure 2.34. On the x and y axes are the copolymers expressed in concentration of St in a MAH-co-St and in a AN-co-St, respectively.

St-co-AN Blends with St-co-MAH Copolymers

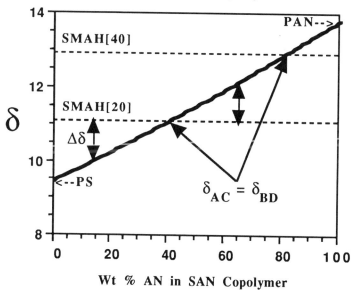

Wt % AN in SAN Copolymer

FIGURE 2.33

Poly(MAH-co-St) and Poly(AN-co-St) Blend

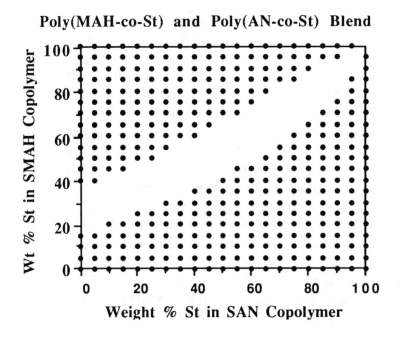

FIGURE 2.34

The small black dots denote blends calculated to be immiscible, where the difference in the solubility parameters of a particular MAH-co-St copolymer composition with a particular AN-co-St copolymer composition exceeds the critical value, $(\Delta\delta)_{Crit}$. Unlike the AN - ester interaction (see later), infrared studies indicate that the intermolecular interactions between AN and MAH segments are surprisingly weak [31] and a value of $(\Delta\delta)^{\circ}_{Crit} = 1.0$ (cal. cm^{-3})$^{0.5}$ has been found to be appropriate. The predicted miscible area for SAN - SMAH blends is the narrow corridor between the forest of black dots illustrated in figure 2.34. How, then, does this stack up to the experimental data? Figure 2.35 shows a comparison of a portion of the calculated miscibility map to data filched from figure 6 in the paper of Kim et al.[26] The large black and white circles denote, respectively, immiscible and miscible blends from Kim's experimental data.

SAN Blends with
Styrene-co-Maleic Anhydride

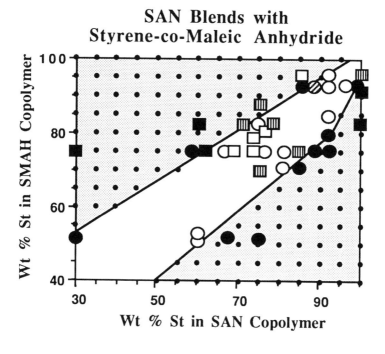

Wt % St in SMAH Copolymer

Wt % St in SAN Copolymer

FIGURE 2.35

The corresponding data of Hall et al.[25] are denoted by the black and white squares. Half black, half white symbols represent systems which exhibit lower critical temperatures. Agreement between prediction and experiment is remarkably good. Furthermore, figure 2.36 shows a corresponding comparison of the experimental data of Aoki (figure 6, reference 27) plotted in units of volume fraction (simply a recalculation after selecting **Run Guide** and units of volume fraction - figure 2.27). It is not perfect, but given the uncertainty of the MAH segment solubility parameter, the calculated region of miscibility is in reasonably agreement with experiment. Certainly, if this **Miscibility Guide** is used in the spirit that we intended, i.e. a "prognosticator of major trends in miscibility", then the simple solubility parameter calculation together with the recognition that weak favorable intermolecular forces are present, has been remarkable successful.

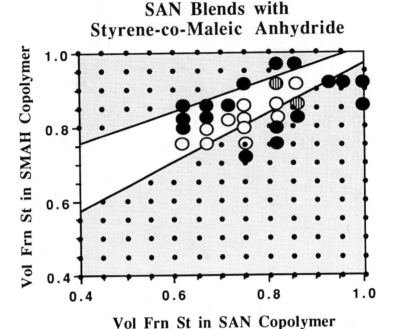

SAN Blends with Styrene-co-Maleic Anhydride

FIGURE 2.36

Weak to Moderate Forces $(\Delta\delta)^{\circ}_{Crit} = 1.5$

Cruz et al.[29] have found that bis-phenol A polycarbonate (PC) is miscible with polyesters having CH_2 / COO ratios of between 2 and 5 (with partial miscibility up to 7). Similarly, Fernandes et al.[30] has determined that tetramethyl substituted analogue, PTMC, is miscible with polyesters having CH_2 / COO ratios of between 5 and 10 (with partial miscibility at 4). Again, while the precise nature of the intermolecular interactions occurring between polycarbonates and polyesters is not crucial to the arguments presented here, it should be noted that a frequency shift (≈ 3 cm^{-1}) is observed in the carbonyl stretching mode of the aliphatic polyester, poly(ε-caprolactone) (PCL), in an amorphous miscible blend with PC. This is roughly half that seen in analogous PCL - PVC [31-33] blends (see below), and suggests an intermolecular interaction that may be somewhat stronger on our scale than the weak intermolecular forces used above.

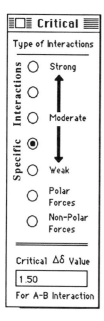

In terms of the **Miscibility Guide**, **Segment B** is PC (or PTMC), **Segment A** simply an ester group -COO and **Segment C** a methylene group. After selecting **Run Guide** and the button for **Incremental** composition the calculation was performed for 10 methylene groups. Figures 2.37 and 2.38 show the calculated solubility parameter of aliphatic linear polyesters as a function of the number of methylene units in the polyester repeat. Included in these figures are lines representing the calculated solubility parameters of PC and PTMC (10.6 and 9.5 (cal. cm^{-3})$^{0.5}$, respectively). [Please note that the values of the group contributions for the carbonate group are also subject to uncertainty since they were estimated from the addition of the values for ester and ether groups and not determined from model organic carbonate data].

The overall form of the two curves is in reasonable accord with the experimental observations. Assuming a critical value of the solubility parameter difference of $(\Delta\delta)^{o}_{Crit} = 1.5$ (cal. cm^{-3})$^{0.5}$, half way between that for weak and moderate forces, (which makes intuitive sense, since the strength of the intermolecular interactions appears to be roughly half that of polyester - PVC blends - see later), the breadth of the miscibility windows are predicted quite well. We could improve the correlation between the calculated predictions and experimental observations by adjusting the solubility parameters (especially those of the polycarbonates) within the bounds of known error, but this would detract from the elementary nature of this guide and again we prefer to emphasize *trends* and not specifics.

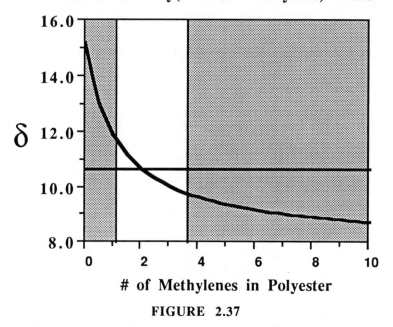

PC and Poly(Ester-co-Methylene) Blend

δ

# of Methylenes in Polyester

FIGURE 2.37

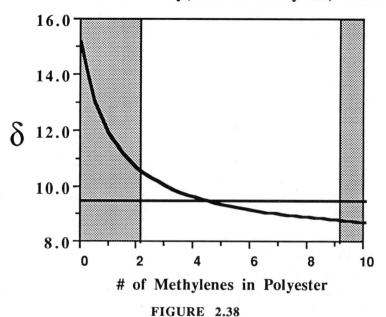

PTMC and Poly(Ester-co-Methylene) Blend

δ

# of Methylenes in Polyester

FIGURE 2.38

At this point we are feeling rather pleased with ourselves and just a little smug, but rather than sitting on our laurels let us examine another copolymer - copolymer system that involves weak to moderate forces. A second system studied by Cowie and his coworkers is the St-co-AN blends with St-co-NPII. Unlike the St-co-MMA blends with St-co-NPII, however, which were discussed above under the category of polar forces (figure 2.28), in this case there is the potential for a weak hydrogen bond between the methine proton of AN and either or both the tertiary nitrogen atom and carbonyl groups of NPII.

NPII   AN

For the **Miscibility Guide**, the segment information for NPII is introduced into **Segment A**, AN into **Segment B**, and St into both **Segments C** and **D**. The resultant miscibility map is shown in figure 2.39. Here on the x and y axes are the copolymers expressed in volume fraction of St in St-co-AN and St-co-NPII, respectively. The predicted miscible region obtained by assuming a value of $(\Delta\delta)^{\circ}_{Crit} = 1.5$ (cal. cm^{-3})$^{0.5}$ resembles a wedge in the miscibility map emanating from the top right hand corner (corresponding to pure PS). This is not surprizing, since the addition of similar concentrations of AN or NPII into the St chain increases the average solubility parameter of St-co-AN and St-co-NPII copolymers by roughly the same magnitude. Therefore the difference in solubility parameters between the two copolymers of similar concentrations of St stays roughly constant.

Styrene-co-Acrylonitrile Blends with Styrene-co-N-Phenyl Itaconimide

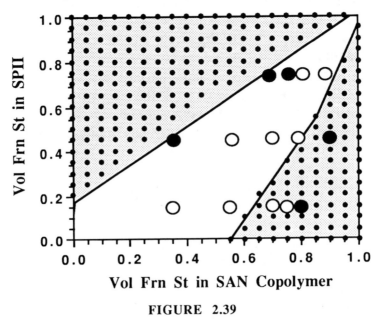

FIGURE 2.39

Figures 2.39 also shows a comparison of our predictions to the experimental results reported by Cowie et al.[21] (black and unfilled filled circles correspond to immiscible and miscible blends, respectively). This is not the best agreement with experiment that we have obtained, but it is important to recognize that we have again been successful in identifying the major trend in miscibility for this blend system using the simple concepts outlined in this chapter. We again resist "tinkering" with the magnitude of the estimated solubility parameter of NPII to obtain a better fit to experiment, as we believe that it will detract from the practical nature of this guide.

Moderate Specific Interactions $(\Delta\delta)^{\circ}_{Crit} = 2.0$

Infrared evidence for the presence of specific interactions involving PVC and the carbonyl group of polymers containing ester, acetate and acrylate groups was reviewed by us in 1984[33].

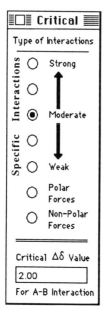

There is still a debate as to whether this interaction should be described as a relatively weak hydrogen bond between the carbonyl group and the methine proton of PVC, as we still favor, or a dipole-induced dipole interaction involving the carbonyl group and the C-Cl bond of PVC, or a Lewis acid - base type of interaction[34]. To reiterate, however, the precise nature of the interaction is not important to the arguments presented here. Nevertheless, these intermolecular interactions are expected to be significantly stronger than those occurring between PVC and the SAN or BAN polymers, and judging from the comparatively minor frequency shifts (< 6 cm^{-1}) in the carbonyl band of miscible polyester (or polyacetate or polyacrylate) blends with PVC[33], we are still in the range of our so-called relatively weak specific interactions.

Woo, Barlow and Paul reviewed the previous literature on PVC blends with aliphatic polyesters in their 1985 paper[35]. Following their studies these authors concluded, in essence, that linear aliphatic polyesters with CH_2 / COO ratios of less than between 3 and 4 were immiscible with PVC. At greater CH_2 / COO ratios, miscible systems are observed but phase separation at elevated temperatures becomes apparent at ratios greater than ten. Judging from the trends in their cloud point curves (figure 7 - ref. 35) and assuming that equilibrium conditions can be achieved, PVC forms miscible blends at ambient temperature with linear aliphatic polyesters having CH_2 / COO ratios of up to about 15. This PVC - polyester "miscibility window" was described by Woo et al. in terms of a binary interaction model and the strong unfavorable interactions between the CH_2 and COO groups of the polyester. However, as we will describe below, our simple concept of a balance between unfavorable physical forces and favorable specific interactions, which is the main theme of this chapter, leads to a similar conclusion.

Let us now see what our **Miscibility Guide** predicts. **Segment A** is now an ester group -COO, **Segment B** PVC and **Segment C**, the diluent, a methylene group. After selecting **Run Guide** and the button for **Incremental** composition the calculation was performed for 10 methylene groups. Figures 2.40 and 2.41 shows the calculated solubility parameter of aliphatic linear polyesters as a function of the number of methylene units in the polyester repeat unit, defined so as to contain one COO group, and an estimate of χ for PVC - polyester blends, respectively. Using a value of $(\Delta\delta)^{\circ}_{Crit} = 2.0$ (cal. cm^{-3})$^{0.5}$ we predict a miscible range extending from 2 to 8 CH$_2$ groups in the aliphatic polyester repeat. The curve of χ versus the number of methylene units is informative, with the familiar shape of a potential energy diagram, and goes through a minimum value of zero at about 3.5 methylene groups.

PVC and Poly(Ester-co-Methylene) Blend

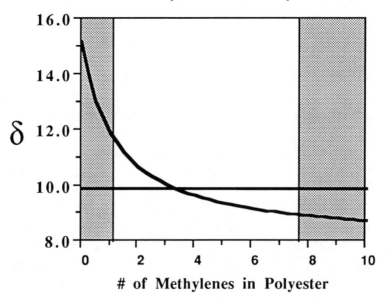

FIGURE 2.40

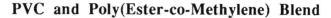

PVC and Poly(Ester-co-Methylene) Blend

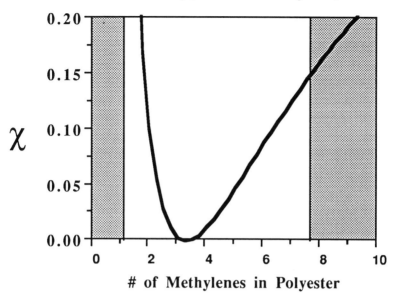

FIGURE 2.41

Below 3 methylene groups χ rises very sharply. In contrast, above 4 methylene groups χ increases, but much more gradually and at a decreasing rate with increasing numbers of methylene groups. This strikes a chord of familiarity and is in accord with the description of the sharp and diffuse divisions between miscibility and immiscibility of these blends at low and high CH_2 / COO ratios, respectively[35]. Once again, the presence of relatively weak specific intermolecular interactions serves to increase the range of miscibility.

As we will see, this critical $(\Delta\delta)^{\circ}_{Crit}$ value of approximately 2.0 (cal. cm^{-3})$^{0.5}$ will be applicable to other PVC blends with polymers of similar chemistry to the polyesters, such as acrylates, methacrylates and acetates. This is at the outer edge of our category of relatively weak favorable (specific) intermolecular interactions. It is worth repeating that while not perfect, we have predicted with reasonable accuracy the most probable range of miscibility for PVC blends with linear aliphatic polyesters by simply considering the solubility parameters of PVC and polyester "copolymers" and recognizing that relatively weak favorable intermolecular interactions exist in this system.

Systematic studies of PVC blends with a series of linear polymethacrylates[36] and polyacrylates[36,37] have been performed by Walsh and coworkers. There is less uncertainty concerning the phase behavior of the former and we will rely on these experimental studies for our purposes here. In summary, Walsh and McKeown found that PVC was miscible with all the linear polymethacrylates up to poly(n-pentyl methacrylate). Poly(n-hexyl methacrylate) appears to be on the edge of the miscibility window, as it phase separated at temperatures below 125°C.

Substituting the segment information of PMMA for the ester group in **Segment A** leads, after recalculation, to figure 2.42 which shows the calculated solubility parameter of the polymethacrylates as a function of the number of methylene groups in the ester side group (i.e. PMMA has no methylene groups in the side group, PEMA one, etc.).

Here, PVC (δ =9.9 (cal. cm^{-3})$^{0.5}$) has a calculated solubility parameter that lies outside the range of both PMMA (δ =9.1) and PE (δ =8.0); the limit when there are an infinite number of methylene units in the repeat. Accordingly, the estimated value of $\Delta\delta$ always exceeds $\Delta\delta_{Crit}$ for the case of a polymer blend involving *no favorable intermolecular interactions*. The smallest value of $\Delta\delta$ corresponds to the blend of PVC and PMMA. Increasing the number of methylene units in the side chain increases the value of $\Delta\delta$.

Using a value of $(\Delta\delta)^{\circ}_{Crit}$ = 2.0 (cal. cm^{-3})$^{0.5}$ we predict a miscible range extending from PMMA up to poly(n-butyl methacrylate) (i.e. 3 CH$_2$ groups in the side group). Incidentally, if we use the observations of Walsh and McKeown to back calculate the value of $(\Delta\delta)^{\circ}_{Crit}$ we arrive at a value of approximately 2.2 (cal. cm^{-3})$^{0.5}$ - a result that yields a prediction precisely in accord with the experimental observations. This is "tinkering", however, and we have cautioned against this, although we are not always capable of resisting temptation!

PVC and Poly(PMMA-co-Methylene) Blend

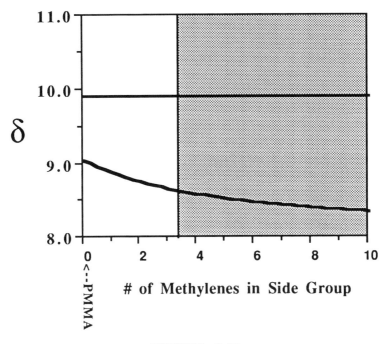

FIGURE 2.42

The last PVC blend example concerns ethylene-co-vinyl acetate (EVA). There are unfortunately many conflicting reports in the literature concerning the miscibility of these blends. Two of the more recent papers that painstakingly document the prior literature are those of Rellick and Runt[38] and Cruz-Ramos and Paul[39]. Sample preparation, a strong $\Delta\chi$ effect, relatively low lower critical solution and degradation temperatures conspire to make the system difficult to study. Nonetheless, the PVC - EVA system has been featured prominently as an example supporting the repulsion effect. Shiomi et al.[40], using samples cast from THF solution, experimentally determined that a wide compositional range of EVA's containing from about 45% to 85% VAc are miscible at ambient temperature. Figure 2.43 shows the calculated solubility parameter of EVA as a function of ethylene content.

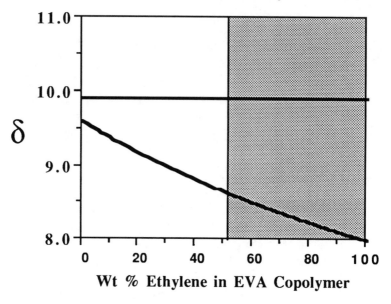

PVC and Poly(VAc-co-Ethylene) Blend

FIGURE 2.43

This is another example where the solubility parameter of PVC lies outside the range of both PVAc (δ =9.6) and PE (δ =8.0 (cal. cm^{-3})$^{0.5}$). The smallest value of $\Delta\delta$ corresponds to the blend of PVC and PVAc.

If we assume that the interactions between PVC and acetate groups are roughly equivalent to those occurring between PVC and ester or acrylate groups considered above, then we should also expect a value of $(\Delta\delta)^{\circ}_{Crit}$ = 2.0 (cal. cm^{-3})$^{0.5}$ to be applicable in this case. Our simple calculation predicts that PVC should be miscible with EVA copolymers containing from about 48 to 100% VAc. This is at variance with the experimental results quoted above and was the impetus that lead us to reexamine the PVC - PVAc blend system. While we confirmed that films cast from THF were invariably multiphased, apparently because of a powerful $\Delta\chi$ effect, a sample that exhibits only one intermediate Tg consistent with a miscible blend can be produced by the slow evaporation of methyl ethyl ketone solution[36]. Accordingly, our simple predictions now appear to be in agreement with experimental observations, although at the time of writing this is still in dispute.

Other systems considered in this section involve polymers containing the acrylonitrile segment blended with methacrylate polymers. PMMA, for example, forms miscible mixtures with SAN copolymers containing from approximately 8 to 35% AN[42-45] and with acrylonitrile-co-p-methyl styrene (pMSAN) that has 2 - 34% AN[46]. In addition, SAN and pMSAN are miscible over certain compositional ranges with poly(ethyl methacrylate) (PEMA), poly(n-propyl methacrylate) (PPMA), poly(cyclohexyl methacrylate) (PCHMA), poly(chloromethyl methacrylate) (PCMMA) and poly(chloroethyl methacrylate) (PCEMA), but immiscible with poly(n-butyl methacrylate) (PBMA). These results are summarized below in Table 2.7. Paul and his coworkers[45] have also recently reviewed the literature pertaining to SAN blends and reported an extensive study of SAN mixed with MMA copolymers containing the comonomers phenyl methacrylate (PhMA), cyclohexyl methacrylate (CHMA) and t-butyl methacrylate (tBMA).

Table 2. 7
Observed Miscibility Windows for SAN and pMSAN Blends

Polymer	Observed Range of Miscibility		
	% AN in SAN	% AN in pMSAN	Reference
PMMA	9 - 35	12 - 34	42, 46
	8 - 30		43
	9 - 39		44
PEMA	4 - 30	10 - 34	43, 46
PPMA	2 - 25		43
PBMA	immiscible	immiscible	43
PCHMA	0 - 20	0 - 22	46
PCMMA	12 - 37	13 - 42	46
PCEMA	12 - 43	18 - 46	46

In common with the vinyl chloride segment, the methine proton of the acrylonitrile (AN) segment is capable of forming a weak to moderate hydrogen bond with ester and ketonic carbonyl groups (as shown schematically below). This intermolecular interaction has been detected spectroscopically from frequency shifts of the infrared carbonyl band[21].

One question that comes immediately to mind is that if PVC and PAN contain similar and comparable sites that are potentially available for interaction with polyesters and the like, why are there not an abundance of miscible PAN homopolymer blends ? And there are not. The simple answer is that there is a very large difference between the solubility parameters of PVC and PAN (9.9 and 13.8 (cal. cm^{-3})$^{0.5}$, respectively). Whereas the former is not far removed from the general range of common aliphatic polyesters, polyacrylates, polymethacrylates etc. (a range encompassing roughly 8.5 to 9.6 (cal. cm^{-3})$^{0.5}$), PAN is not even within spitting distance. In the best case, for example, the difference between the solubility parameters of PAN and poly(methyl acrylate) (PMA: ($\delta = 9.6$)) is 4.2 (cal. cm^{-3})$^{0.5}$, way outside the limits of a value of $(\Delta\delta)^{\circ}_{Crit}$ for weak to moderate hydrogen bonding (or very strong hydrogen bonding for that matter). It is therefore no real surprize that in order to use the weak to moderate hydrogen bonding potential of the AN segment to form miscible polymer blends, we must copolymerize with a comonomer that reduces the overall solubility parameter down to a point where the difference between the solubility parameter of the AN copolymer and, for example, the polyacrylate, becomes within the critical $\Delta\delta$ range appropriate to weak to moderate hydrogen bonding. Copolymerizing AN with St does just that.

Let us first consider the SAN - PMMA blends. Employing the **Miscibility Guide** we leave the segment information of MMA in **Segment A** and introduce AN and St into **Segments B** and **C**, respectively. Figure 2.44 shows the calculated solubility parameter of SAN copolymers as a function of weight % AN, together with that of PMMA. Using the same initial critical value of $\Delta\delta$ as that for the vinyl chloride - ester interaction (i.e. $(\Delta\delta)^{\circ}_{Crit} = 2.0$ (cal. cm^{-3})$^{0.5}$), we predict a range of miscibility from about 3 to 30% AN which is in pretty good agreement with experiment.

Even more encouraging is the trend observed as we consider SAN blends with PEMA, PPMA and PBMA, illustrated in figure 2.45, and see that it compares very favorably to the experimental data summarized in table 2.7.

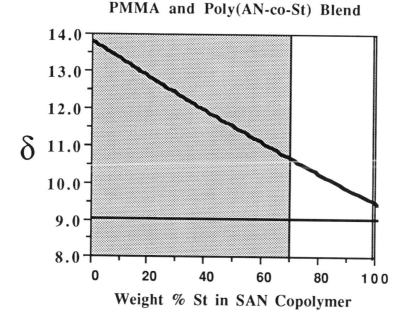

FIGURE 2.44

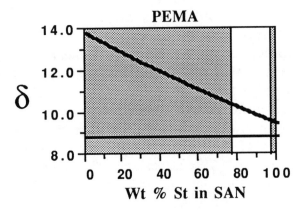

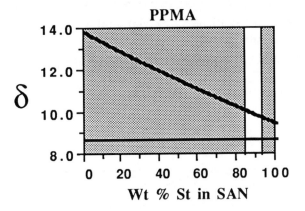

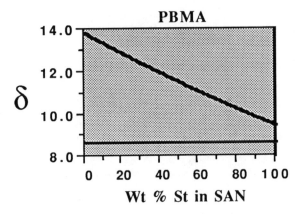

FIGURE 2.45

Employing the $(\Delta\delta)^{o}_{Crit}$ value of 2.0 (cal. cm^{-3})$^{0.5}$, one can see that the predicted window of miscibility narrows appreciably from 3 - 23% AN for blends with PEMA, 7 - 16% AN for PPMA and is totally absent (immiscible) for PBMA. In essence, as we increase the number of methylene groups in the poly(alkyl methacrylates) the solubility parameter difference increases systematically until it exceeds that of the critical value over the entire range of SAN copolymer compositions for mixtures with PBMA.

We will illustrate three more examples (figure 2.46); leaving the reader to explore the copious number of polymer blends that involve copolymers of AN with polymers containing ester type functionalities. Let's look first at PCHMA - SAN blends. The monosubstituted cyclohexyl group is included in the **Groups** window and the solubility parameter of PCHMA is calculated to be 9.2 (cal. cm^{-3})$^{0.5}$ - just slightly above that of PMMA. The predicted window of miscibility is from about 0 to 35% AN which is in the right "ballpark", although somewhat greater than the reported experimental value of 0 to 20% (table 2.7).

How about PMMA blends with pMSAN ? Here we dilute AN with *para*-methyl styrene which has a segment solubility parameter of 9.4 (cal. cm^{-3})$^{0.5}$ - surprisingly close to that of styrene. The results indicate that PMMA should be miscible with pMSAN copolymers containing from about 0 to 32% AN. Experimentally a range of 12 to 34% is found[46]. Not bad, although we tend to overestimate the lower end! Finally, let us consider the PCMMA blends with pMSAN. The chlorine atom in PCMMA raises the solubility parameter to approximately 10.1 (cal. cm^{-3})$^{0.5}$ and the predicted window of miscibility is enlarged from about 3 to 60% AN. This is rather wider than experimentally observed (13 - 42% AN), but the *trend* is still correct and one certainly would not have difficulty in choosing the most likely copolymer composition range for miscibility.

Let us now up the ante and compare our predictions with the experimental miscibility maps for SAN blends with MMA-co-PhMA, MMA-co-CHMA and MMA-co-tBMA copolymers determined by Nishimoto et al.[45] and MMA-co-NPII determined by Cowie et al.[21]

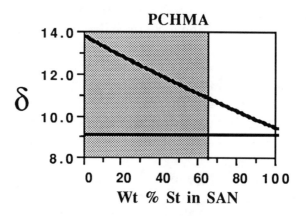

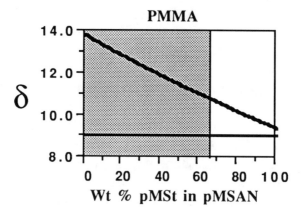

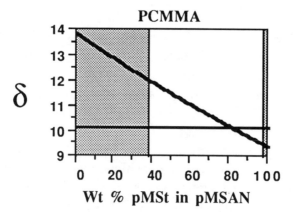

FIGURE 2.46

Figure 2.47 shows a schematic diagram of the variation of the solubility parameter of SAN as a function of AN concentration in the copolymer. Also included in the figure are the calculated solubility parameters for the MMA-co-PhMA, MMA-co-CHMA and MMA-co-tBMA copolymers. Note that the three methacrylate comonomers affect the average solubility parameter of the MMA copolymers in different ways: PhMA raises the average solubility parameter, tBMA lowers it and CHMA makes little difference.

Accordingly, in terms of our simple model, for a particular SAN copolymer of a given composition, the difference in solubility parameter decreases, which implies enhanced miscibility, in the order tBMA > CHMA > PhMA for MMA copolymers containing the same volume fraction of comonomer. Let's run the program and see how we do.

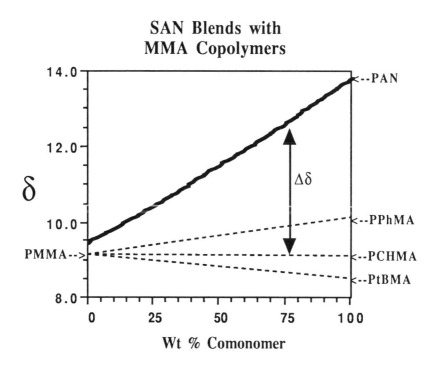

FIGURE 2.47

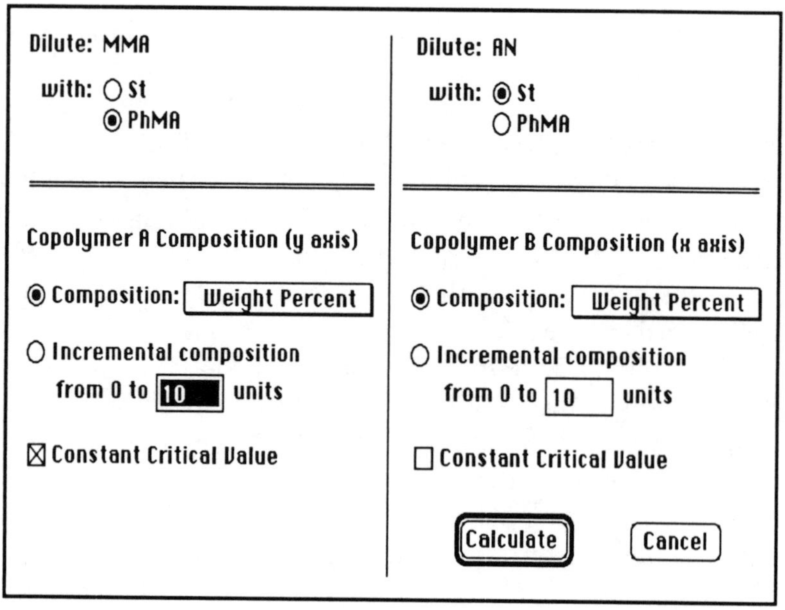

FIGURE 2.48

Returning to the **Miscibility Guide**, select **Copolymer-Copolymer** from the **Option** menu. The segment information for MMA is introduced into **Segment A** , AN into **Segment B**, St into **Segment C** and PhMA into **Segment D.** Now **Run Guide** is selected from the **Guide** menu and the window illustrated in figure 2.48 appears. In the **Copolymer A** section we select MMA to be "diluted" with PhMA, but in this case we need to *activate* the **Constant Critical Value** option by selecting the appropriate box (see figure 2.48). This is because we are not really diluting PMMA with PhMA, but substituting one methacrylate segment for another and the concentration of interaction sites (methacrylate carbonyl groups) does not varying significantly with copolymer composition. In other words, the random methacrylate copolymer may be viewed in our terms as a "homopolymer" containing an average "methacrylate segment". In contrast, the **Copolymer B** segment AN is "diluted" with St and here the **Constant Critical Value** option is *not* activated, since we must take into consideration the effect of the reduction of the number of interaction sites per unit volume with increasing St concentration. The **Composition** of the copolymers is chosen in terms of weight % to correspond to the experimental values of Nishimoto et al.[45]

After selecting **Calculate** the miscibility map shown in figure 2.49 is produced. On the x and y axes are the copolymer compositions expressed in weight % St in SAN and weight % of PhMA in MMA-co-PhMA, respectively. The small black squares again denote blends calculated to be immiscible, where the difference in the solubility parameters of a particular SAN copolymer composition with a particular MMA-co-PhMA copolymer composition exceeds the critical value, $\Delta\delta$. As discussed above, the value of $(\Delta\delta)^{\circ}_{Crit} = 2.0$ (cal. cm^{-3})$^{0.5}$), is appropriate for the AN - methacrylate interactions.

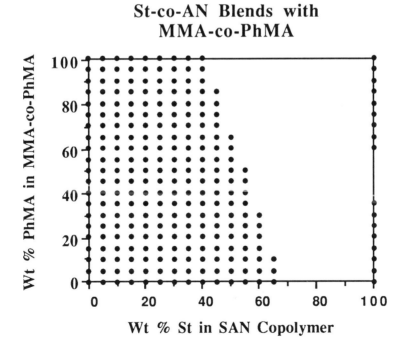

St-co-AN Blends with MMA-co-PhMA

FIGURE 2.49

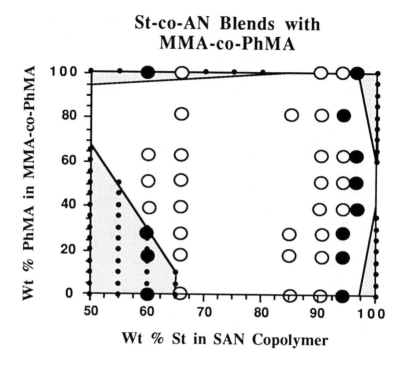

FIGURE 2.50

The predicted miscible region for SAN blends with MMA-co-PhMA blends is the substantial unfilled area at the right hand side of the miscibility map shown in figure 2.49.

In summary, PMMA is predicted to be miscible with SAN polymers containing a few % to 35% AN while MMA copolymers rich in PhMA are predicted to be miscible with SAN copolymers containing a greater amount of AN (up to 60%). In effect, as we increase the solubility parameter of MMA-co-PhMA, a greater concentration of AN in SAN can be tolerated. Figure 2.50 shows a scale expanded version of figure 2.49 and compares the predicted results with the published experimental data of Paul and his colleagues.[45]   Black and white circles denote, respectively, immiscible and miscible blends. Even our critics, (yes, believe it or not, we have some!), must allow that we have predicted with commendable accuracy the major *trends* in the miscibility of these blends.

Returning to the **Miscibility Guide** we can easily substitute the segment information for CHMA (and subsequently, tBMA) into **Segment D** and obtain the scale expanded miscibility maps illustrated in figures 2.51 and 2.52.

On the x axes of figures 2.51 and 2.52 are, respectively, the copolymer compositions expressed in weight % of CHMA in MMA-co-CHMA and tBMA in MMA-co-tBMA. In common are the y axes, which express the weight % St in SAN. The small black squares once again denote blends calculated to be immiscible using a value of $(\Delta\delta)^{\circ}_{Crit} = 2.0$ (cal. cm^{-3})$^{0.5}$). The predicted miscible region for SAN blends with MMA-co-CHMA of all compositions is the unshaded rectangular area in figure 2.51 extending to about 35 % AN in SAN. This is to be expected, since the substitution of CHMA for MMA in the MMA-co-CHMA copolymer does not materially change the average solubility parameter.

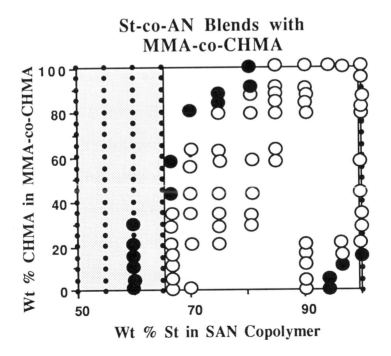

St-co-AN Blends with MMA-co-CHMA

Wt % St in SAN Copolymer

FIGURE 2.51

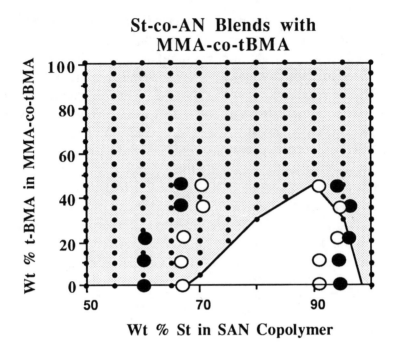

**St-co-AN Blends with
MMA-co-tBMA**

FIGURE 2.52

In contrast to this, the predicted miscible region for SAN blends with MMA-co-tBMA (figure 2.52) is much more restricted, especially for the MMA copolymers rich in tBMA. This reflects the decrease in the solubility parameter of MMA-co-tBMA copolymers with tBMA content (see figure 2.47) so that only a relatively small amount of AN in SAN can be tolerated.

Also included in figures 2.51 and 2.52 are the experimental data of Nishimoto et al. (reference 45, figures 3 and 7). Black and white circles again denote, respectively, immiscible and miscible blends. While the agreement between our predictions and Nishimoto's experimental results is not perfect, we believe that the reader should be impressed by the fact that the major *trends* are yet again correct and that with an embarrassingly simple calculation we have been able to satisfactorily match the miscibility maps of the three copolymer - copolymer blend systems. Dabbling with the values of $(\Delta\delta)^\circ_{Crit}$ and/or the individual solubility parameters within the error limits we established earlier in this chapter can lead to an even better match to the experimental values, but, as we have said before, this we feel detracts from the primary purpose of this guide, which is to predict major trends *without* adjusting parameters.

We have already considered a number of polymer blend systems studied by Cowie et al.[21] involving polar (figure 2.28) or weak hydrogen bond (figure 2.39) forces and the four segments, styrene, methyl methacrylate, acrylonitrile and N-phenyl itaconimide. Our final polymer blend system involving SAN copolymers concerns mixtures with MMA-co-NPII copolymers, which now embodies the stronger hydrogen bonding associated with AN - ester type interactions.

Returning to the **Miscibility Guide**, since the segment information for MMA, AN and St is already in **Segment A**, **Segment B** and **Segment C**, respectively, we only need to input NPII into **Segment D**. As before, we "dilute" AN with St, but make the assumption that substitution of MMA with NPII by copolymerization does not significantly affect the concentration of interacting segments, which means that we again *activate* the **Constant Critical Value** option (figure 2.48). We recognize, of course, that the AN to NPII interaction is weaker than the corresponding AN to MMA interaction, which complicates matters, but in the spirit of this simple guide we will press on with unbounded optimism and see if ignoring this factor seriously effects our predictive capabilities. The calculations are performed in terms of volume fraction in order to correspond to the experimental data of Cowie et al. The result is displayed in figure 2.53. As the principle intermolecular interaction involves that between the AN and MMA segments, the appropriate value of $(\Delta\delta)^\circ_{Crit}$ is 2.0 (cal. cm^{-3})$^{0.5}$.

The predicted miscible region for SAN blends with MMA-co-NPII is the unshaded area in figure 2.53. Black and white circles again denote, respectively, immiscible and miscible blends. Given the approximations inherent in this practical guide, this is not a bad prediction of observed experimental data!

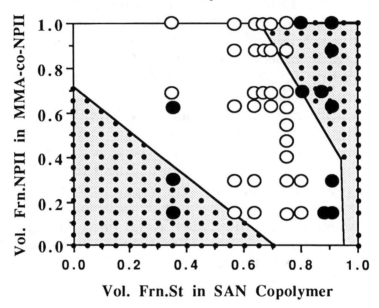

St-co-AN Blends with MMA-co-N-Phenyl Itaconimide

Vol. Frn.St in SAN Copolymer

FIGURE 2.53

Summary

For the case of polymer blend systems in which there are *relatively weak favorable intermolecular interactions* :

1. The presence of polar groups that are capable of relatively weak favorable intermolecular interactions (i.e. those with dissociation energies of between about 1 and 3 kcal. mole[-1]) increases the probability of finding miscible blends, compared to mixtures of non-polar polymers that have only dispersive forces between segments.

2. Simply stated, these favorable interactions serve to counteract the unfavorable contribution to the free energy of mixing that we express express in the χ (physical interactions) parameter. We may conveniently view this in terms of a greater

tolerance for solubility parameter differences between the two polymers (up to about 2.0 (cal. cm^{-3})$^{0.5}$).

3. The first "rule of thumb" for predicting polymer miscibility still remains the matching of the non-hydrogen bonded solubility parameters. We must still recognize the errors involved in calculating these quantities, but now we have some "room to breathe" and are not required to be so stringent.

F. RELATIVELY STRONG FAVORABLE INTERMOLECULAR INTERACTIONS

Examples of miscible binary polymer blends that fit into the category of systems with relatively strong favorable intermolecular interactions include polymers containing amide, urethane, hydroxy and carboxylic acid groups etc., blended with polymers that have ether, acrylate, acetate, ester, oxazoline, pyridine groups etc. present in their (average) chemical repeat. For these systems it is also possible to calculate $\Delta G_H / RT$ directly (equation 2.1), as infrared measurements can be used to obtain a measure of the number and strength of the hydrogen bonds formed in these mixtures. This more rigorous approach is discussed in detail later in this book and here we will keep to our general discussion of the effects of steadily increasing the strength of the intermolecular interactions. We start where we left off, with what we characterize as moderate intermolecular forces.

Moderate Specific Interactions $(\Delta \delta)^{\circ}_{Crit} = 2.0$

We will briefly consider the limited data available on blends involving the aliphatic hydroxyl - ester type interaction. Harris et al.[48] have determined that Phenoxy resin is miscible with aliphatic polyesters containing 3 to 5 methylenes, a relatively narrow range. In addition, we have reported FTIR studies pertaining to the relative strength of the interactions attributed to self-association of Phenoxy and association between the phenoxy and poly(ϵ-caprolactone) (PCL)[49].

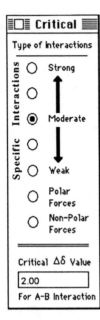

A comparison of the frequency shifts of the poly(ε-caprolactone) (PCL) carbonyl stretching mode in miscible blends with phenoxy and PVPh[33], (approximately 13 and 26 cm^{-1}, respectively), suggests that the relative strength of the intermolecular interactions in the case of the phenoxy blend is roughly half that of the analogous PVPh blend. The *non-hydrogen bonded* solubility parameter for phenoxy has been estimated at 10.2 (cal. cm^{-3})$^{0.5}$ (Table 2.5). Using a moderate hydrogen bonding value of $(\Delta\delta)^{\circ}_{Crit} = 2.0$ (cal. cm^{-3})$^{0.5}$ for the aliphatic hydroxyl - ether interaction, (frankly, at the time of writing, our best "guesstimate"),we obtain the plot shown in figure 2.54.

Phenoxy and Poly(Ester-co-Methylene) Blend

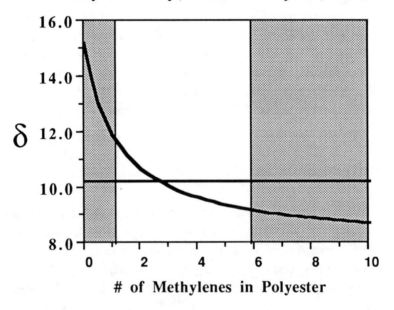

FIGURE 2.54

In the *absence* of any favorable intermolecular interactions between phenoxy and the polyesters, these simple calculations predict that only the linear polyester, $-(CH_2)_3-COO$, is likely to be miscible with phenoxy.

The "window of miscibility" opens with the occurrence of favorable interactions, and a moderate hydrogen bonding interaction corresponding to a $(\Delta\delta)^\circ_{Crit} = 2.0$ (cal. $cm^{-3})^{0.5}$ predicts that polyesters containing about 2 to 6 methylenes should be miscible. Not a bad agreement with that reported[48], but again we prefer to emphasize that the correct *trend* is predicted.

Moderate to Strong Specific Interactions $(\Delta\delta)^\circ_{Crit} = 2.5$

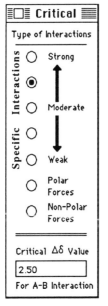

We first consider blends of a series of polyisophthalamides (PIPA or nylon n-I) with two different polyethers, poly(ethylene oxide) (PEO) and poly(vinyl methyl ether) (PVME) that were the subject of recent studies[50], and then briefly mention similar amorphous polyurethane (APU) - polyether blends[11].

In essence, we found experimentally that the nylons 8-I through 10-I (the upper limit of the polymers synthesized) were completely miscible in the amorphous state with PEO, while PIPA polymers containing < 7 methylene groups in the translational repeat became progressively less miscible. In contrast, blends of the PIPA polymers (nylons 2-I to 10-I) were all found to be immiscible with PVME. Before we show the results of our calculations, it is again important to stress the difference between the chemical repeat of the PIPA polymers and the repeat unit used in our calculations. Below we show a schematic diagram of nylon 4-I. The translational or chemical repeat of 4-I contains one disubstituted benzene ring, two amide and four methylene groups. The *specific repeat* unit which we employ in our calculations is defined so that it contains just one functional group capable of forming specific interactions (in this case the amide group). Accordingly, the chemical repeat of Nylon 4-I contains one half of a disubstituted benzene ring, one amide and two methylene groups.

In terms of the **Miscibility Guide, Segment B** is simply defined as PIPA with the segment consisting of one N-H, one C=O and 0.5 of a disubstituted benzene ring. **Segment A** is the polyether PEO (or PVME) and **Segment C** is a methylene group. PEO and PVME have calculated solubility parameters of 9.4 and 8.5 (cal. cm^{-3})$^{0.5}$, respectively. Figure 2.55 shows the calculated solubility parameter of the nylon n-I polymers as a function of the number of methylenes in the *specific* repeat unit that we have just defined.

In keeping with our general philosophy, we can conveniently view the different nylon n-I homopolymers as equivalent "copolymers" of the PIPA segment "diluted" by methylene groups and choose **Incremental** from the alternatives under **Composition** in the **Plot** window. From the **Critical** window we choose an initial critical solubility parameter difference $(\Delta\delta)_{Crit}^{\circ}$ of 2.5 (cal. cm^{-3})$^{0.5}$, which reflects moderate to strong hydrogen bonding - a selection that is consistent with nylon - polyether blends.

The corresponding plot of χ versus the number of methylenes per chemical repeat of the PIPA "copolymers" is illustrated in figure 2.56.

PEO and Poly(PIPA-co-Methylene) Blend

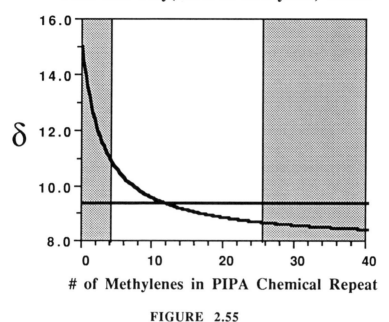

# of Methylenes in PIPA Chemical Repeat

FIGURE 2.55

PEO and Poly(PIPA-co-Methylene) Blend

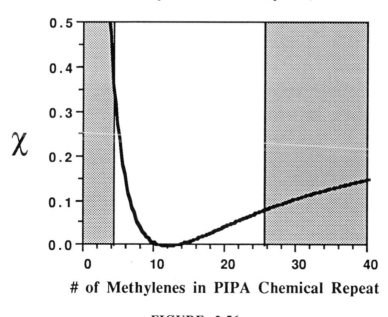

# of Methylenes in PIPA Chemical Repeat

FIGURE 2.56

A range of miscibility from nylon 8-I to approximately 50-I is predicted. (Note that in figures 2.55 and 2.56, 4 methylenes per *specific* repeat unit corresponds to nylon 8-I.) While there is good agreement with experimental observation for the lower limit of the miscibility window, the higher limit of the miscibility has not yet been experimentally tested (it is not easy to synthesize such polymers).

In the case of the PIPA - PVME blends our predictions from the solubility parameter plot (figure 2.57) suggest that one would require greater than 12 methylenes per chemical repeat (i. e. equivalent to a nylon 24-I) before the PIPA polymers would become miscible with PVME. Whether or not this prediction will be verified by experiment (at the time of writing we know experimentally that PIPA polymers up to 10-I are immiscible with PVME[50]) will have to wait until polymers with long sequences of methylenes are synthesized.

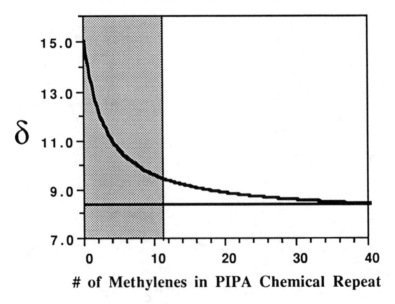

PVME and Poly(PIPA-co-Methylene) Blend

FIGURE   2.57

PVME and Poly(PIPA-co-Methylene) Blend

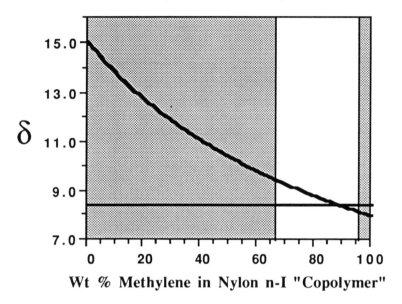

FIGURE 2.58

In addition, assuming the reader has accepted our arguments so far, he or she will not be surprised that we regard the mixing of PVME with a hypothetical random copolymer of PIPA and methylene groups to be an entirely equivalent problem. To this end, we can readily calculate the miscibility window of PVME blends with methylene-co-PIPA copolymers as a function of copolymer composition, as depicted in figure 2.58. Hence, if some long haired, unwashed, organic chemist is able to synthesize such a copolymer we predict a miscibility window for PVME blends with methylene-co-PIPA copolymers containing about 65 to 95 wt % methylene. Finally, keep in mind that the above is still a *guide* to miscibility, but at this end of the interaction scale we are able to directly calculate $\Delta G_H/RT$ (see chapters 6 and 7) and this calculation does not require a *post-facto* fitting of the data.

A similar system to that just described is amorphous polyurethane (APU) - polyether blends[51]. From thermal analysis studies we have determined that APU is miscible at 110°C with PEO and an ethylene oxide-co-propylene oxide copolymer (EPrO) containing 70% of the former monomer, but immiscible with PVME (which is isomorphous to poly(propylene oxide)), the higher homologues of the poly(vinyl n-alkyl ethers) and polytetrahydrofuran. How well then does our simple guide predict these trends?   A schematic representation of the chemical structure of APU is given below:

Amorphous Polyurethane

In a manner similar to that described for the PIPA polymers, the *specific* repeat unit of APU contains one half of a methyl, one half of a trisubstituted benzene, one urethane (an N-H and a COO) and two methylene groups with a calculated molar volume of 97.7 cm^{-3} $mole^{-1}$ and a non-hydrogen bonded solubility parameter of 11.2 (cal. $cm^{-3})^{0.5}$. Figure 2.59 shows the calculated solubility parameter of ethylene oxide-co-propylene oxide copolymers as a function of composition, together with that of APU. In calculating the miscibility window we must *activate* the **Constant Critical Value** box under the **Calculation Choices** (figure 2.18) since we are not significantly diluting ethylene oxide by copolymerization with propylene oxide but merely replacing one interacting segment with another of the same type. An initial critical solubility parameter difference of $(\Delta\delta)^{o}_{Crit} = 2.5$ (cal. $cm^{-3})^{0.5}$ is applicable for the polyurethane - polyether blends; the same as that for the polyamide - polyether blends.

APU and Poly(PEO-co-PrO) Blend

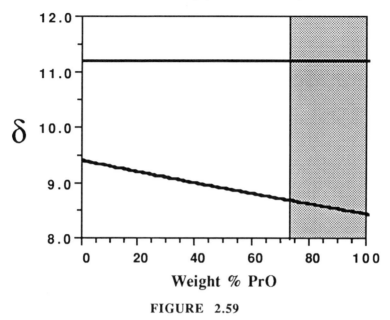

FIGURE 2.59

We arrive at the prediction that APU - PEO blends should be miscible (zero on the x-axis of figure 2.59) while APU - PPrO (or PVME) should be immiscible. Furthermore, we predict that EPrO copolymers should be miscible with APU if they contain less than about 30% propylene oxide.

Figure 2.60 shows similar calculations for APU - poly(n-alkyl ether) blends. In this case we calculate the miscibility window for APU blends with a hypothetical copolymer ether-co-methylene which we view as simply diluting the ether group with methylene groups. Accordingly, in the **Miscibility Guide** the **Incremental** button is selected and the **Constant Critical Value** box is *deactivated*. Here we predict that poly(methylene oxide) and PEO should be miscible with APU, but that the higher homologues should be immiscible.

Finally, if we perform calculations on APU blends with PVME and the higher homologues of poly(vinyl n-alkyl ether) we predict that all these blends should be immiscible. All in all, the predictions are in good accord with experimental observation.

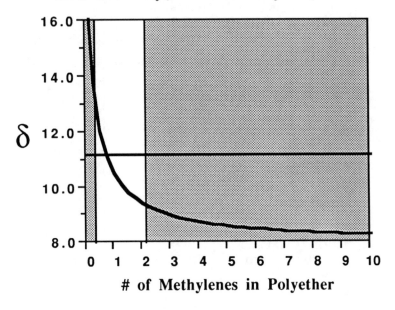

APU and Poly(Ether-co-Methylene) Blend

FIGURE 2.60

Strong Specific Interactions $(\Delta\delta)^{\circ}_{Crit} = 3.0$

We now consider the poly(vinyl phenol) (PVPh) blends, as these systems have been studied extensively and fit in the category of strong intermolecular forces. From an infrared spectroscopic point of view, studies of PVPh blends with polyacrylates, polyacetates, polyesters and polymethacrylates[52-54] have been most rewarding, because it is possible to directly measure the fraction of hydrogen bonded carbonyl groups in the blends as a function of composition and temperature. PVPh is miscible with linear polyacrylates, ranging from poly(methyl acrylate) to poly(n-butyl acrylate), but only partially miscible with poly(pentyl acrylate). The higher polyacrylates become increasingly more immiscible[54,55]. Of the linear poly-methacrylates the first three of the series, polymethyl, ethyl and n-propyl, are essentially miscible with PVPh at ambient temperature. Poly(n-butyl methacrylate), on the other hand, is on the edge of miscibility[54,55], while the higher homologues are definitely immiscible[55]. Furthermore, PVPh is miscible with poly(vinyl acetate) (PVAc) and EVA[70], "partially miscible" with EVA[45] and grossly immiscible with EVA[25][49,52,56].

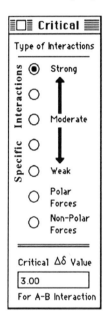

We have discussed previously in this chapter how we obtain an *initial* estimate of the *non-hydrogen bonding* solubility parameter of PVPh by eliminating the proton in the calculation, as illustrated for **Segment B** in figure 2.61. From our extensive experimental studies of the phase behavior of different PVPh blends, where we can determine the contribution from the $\Delta G_H/RT$ term of equation 1 (see chapters 6 and 7 for additional details), we have determined that this *initial* estimated value for PVPh at 11.0 (cal. cm^{-3})$^{0.5}$ is a little on the high side and a more appropriate value is 10.6 (cal. cm^{-3})$^{0.5}$. As we have seen, this is within the range of estimated error of these calculations. Thus we input a value of 10.6 into the pertinent box (figure 2.61) and strike **enter**. (Incidentally, we may also take this opportunity to correct the values of the molar volume and molecular weight before saving the segment information - this does not effect us here, but will be necessary in the phase behavior calculations discussed later in chapter 7.)

Introducing the segment information of MA (and subsequently MMA) and methylene into **Segments A** and **C**, respectively, yields figures 2.62 and 2.63, which show calculated solubility parameters for linear polyacrylates and polymethacrylates as a function of the number of methylene groups in the ester side group (e.g. zero methylenes in the side groups correspond to PMA and PMMA, respectively). In both cases the value of $\Delta\delta$ increases with the the number of methylene groups in the repeat unit, as the difference between the solubility parameters of PVPh and the polyacrylates or polymethacrylates becomes larger.

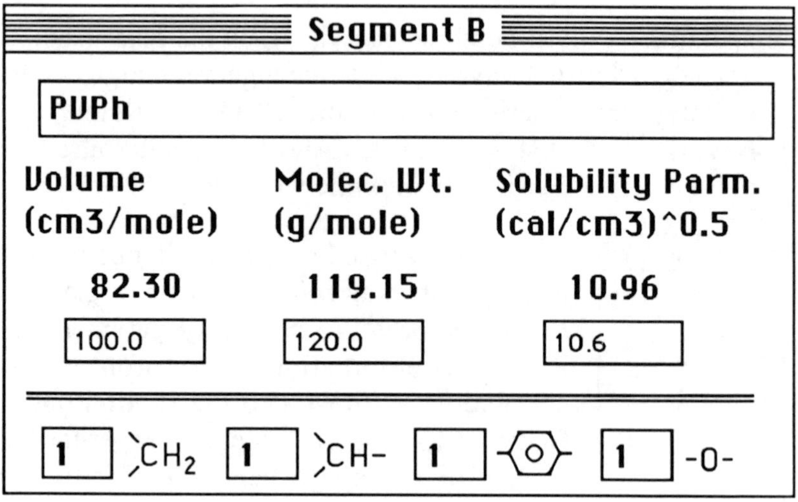

Segment B
PVPh
Volume
(cm3/mole)
Molec. Wt.
(g/mole)
Solubility Parm.
(cal/cm3)^0.5
82.30
119.15
10.96
100.0
120.0
10.6
1
CH2
1
CH-
1
1
-O-

FIGURE 2.61

PVPh and Poly(PMA-co-Methylene) Blend
δ
12.0
11.0
10.0
9.0
8.0
0
1
2
3
4
5
6
7
8
9
10
PMA-->
PBA-->
# of Methylenes in Polyacrylate Side Group

FIGURE 2.62

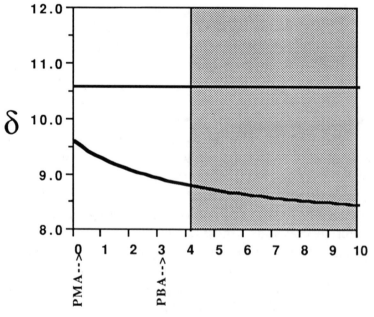

PVPh and Poly(MMA-co-Methylene) Blend

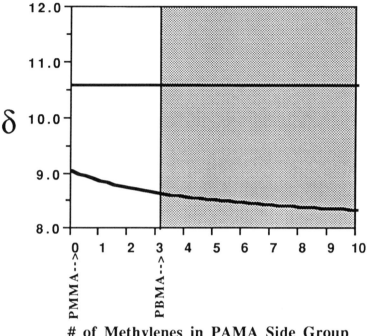

# of Methylenes in PAMA Side Group

FIGURE 2.63

Here, an initial critical solubility parameter difference of $(\Delta\delta)^{\circ}_{Crit} = 3.0$ (cal. cm^{-3})$^{0.5}$ is applicable and this corresponds to the button labelled **Strong** in the **Critical** window. This leads to a prediction that PVPh blends should be miscible with a homologous series of poly(n-alkyl acrylates) from poly(methyl acrylate) to poly(n-pentyl acrylate) and poly(n-alkyl methacrylates) from poly(methyl methacrylate) to poly(n-butyl methacrylate). This is in excellent agreement with experiment.

Not unexpectedly, this value of $(\Delta\delta)^{\circ}_{Crit}$ also applies to the PVPh - EVA copolymer blends, as it should. Figure 2.64 shows a curve representing the calculated solubility parameters of EVA copolymers as a function of the weight % of ethylene.

PVPh and Poly(VAc-co-Ethylene) Blend

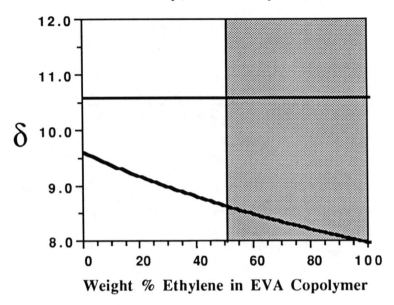

Weight % Ethylene in EVA Copolymer

FIGURE 2.64

A $(\Delta\delta)^{\circ}_{Crit} = 3.0$ (cal. cm^{-3})$^{0.5}$ leads to a prediction that a vinyl acetate content exceeding 50% is required for miscible PVPh - EVA blends; in gratifying agreement with experiment[56,57]. Note again that although this value of $(\Delta\delta)^{\circ}_{Crit}$ appears very large, the value of $\Delta G_H / RT$, calculated from experimental measurements, makes this a well established figure.

Before we leave the class of PVPh blends with polymers containing ester type carbonyl groups, consider the following: poly(methyl methacrylate) (PMMA) is *immiscible* with polystyrene (PS) (corresponding, in our terms, to a situation where there are no obvious favorable intermolecular interactions to drive miscibility), but *miscible*, as discussed above, with poly(vinyl phenol) (PVPh). The question arises, "how many vinyl phenol (VPh) units would we need to incorporate into PS to render it miscible with PMMA ?"

Chen and Morawetz recently reported that only approximately 1% VPh was necessary in the case of PMMA and poly(ethyl methacrylate) (PEMA)[58]. Our theoretical calculations of the phase diagrams of PVPh and styrene-co-vinyl phenol copolymers (STVPh) blends with poly(alkyl methacrylates) successfully predict a wide "window of miscibility"[54] and we will be considering this more rigorous approach in much more detail in chapter 7. Here, however, we wish to see if the simple guide presented in this chapter reveals such a trend. Assuming that styrene is an inert diluent, we can readily estimate the solubility parameters of the STVPh copolymers, which change in an essentially linear fashion with copolymer composition, as shown in figure 2.65. In fact, the difference in the solubility parameters of STVPh, $\overline{\delta}_B$, and PMMA, δ_A, becomes increasingly smaller with increasing styrene concentration in the copolymer. Employing the value of $(\Delta\delta)_{Crit} = 3.0$ (cal. cm^{-3})$^{0.5}$ (see preceding two paragraphs) leads to a prediction that PMMA should be completely miscible with STVPh copolymers over the entire composition range from a few % to 100% VPh units. This appears to be a remarkably good prediction given the simplistic nature of our qualitative guide.

It is important to restate that if our simple scheme is employed as we intend as a *guide* to miscibility the major *trends* will be predicted, but the results will not necessarily have a high degree of quantitative accuracy. This can be illustrated by employing the **Miscibility Guide** to calculate miscibility maps for STVPh blends with a homologous series of poly(n-alkyl methacrylates) (PAMA) or ethylene-co-methyl acrylate (EMA) copolymers. Selecting the **Copolymer - Copolymer** option from the **Options** menu the segment information for methyl methacrylate (MMA) is introduced into **Segment A** , VPh into **Segment B**, methylene into **Segment C** and styrene into **Segment D**. **Run Guide** is now selected from the **Guide** menu. In the **Copolymer A** section we need to select MMA to be incrementally "diluted" with methylene and in the **Copolymer B** section VPh to be "diluted" in units of weight % St.

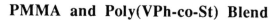

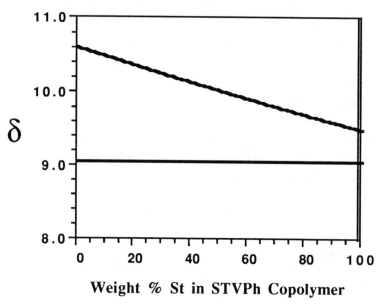

FIGURE 2.65

Since we are diluting with inert diluents in both cases, the **Constant Critical Value** boxes are *not* selected. The result is displayed in figure 2.66. On the x axis is the copolymer composition expressed in weight % of St in STVPh and on the y axis the number of methylenes in the side group of poly(n-alkyl methacrylates) (PAMA); PMMA = 0; PEMA = 1 etc. The small black circles denote blends calculated to be immiscible, where the difference in the solubility parameters of a particular STVPh copolymer composition with a particular PAMA exceeds the calculated critical value, $\Delta\delta$. As discussed above, the value of $(\Delta\delta)^{\circ}_{Crit} = 3.0$ (cal. cm^{-3})$^{0.5}$ is appropriate for the VPh - methacrylate interaction and selected from the **Critical** window.

The predicted miscible region for STVPh blends with PAMA blends is shown as the unfilled area of the miscibility map shown in figure 2.66. This figure also shows the results of experimental studies performed in our laboratories, with the larger black and unfilled filled circles corresponding to immiscible and miscible blends, respectively. The agreement is not earth shattering, but nevertheless pretty good, and it is clear that the majority of the miscible area in the map has been successfully forecasted.

Incidentally, the reader might wish to compare the miscibility map shown in figure 2.66 with that in Chapter 7 (figure 7.87). A superb prediction of the experimental observations[59] is found when the actual magnitude of the $\Delta G_H/RT$ term as a function of blend composition is determined from experimentally derived equilibrium constants and not just approximated as in the case of this **Miscibility Guide**.

We could hardly get a more impressive agreement between experimental observations and our predictions calculated using the simple **Miscibility Guide**, however, than in the case of STVPh blends with ethylene-co-methyl acrylate (EMA) copolymers, as shown in figure 2.67.

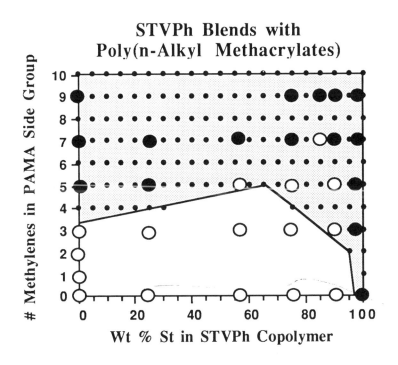

FIGURE 2.66

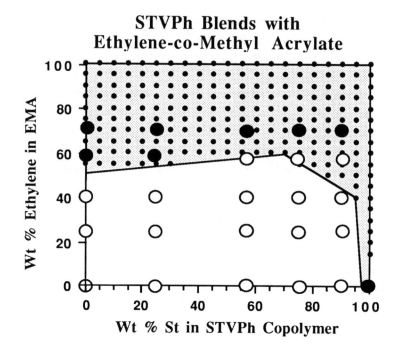

**STVPh Blends with
Ethylene-co-Methyl Acrylate**

FIGURE   2.67

The segment information for methyl acrylate (MA) is introduced into **Segment A** , VPh into **Segment B**, ethylene into **Segment C** and styrene into **Segment D** followed by **Run Guide** selected from the **Guide** menu. In both the **Copolymer A** and **Copolymer B** sections we choose units of weight % and select MA to be "diluted" with ethylene and VPh to be "diluted" with St (additionally, as discussed above, the **Constant Critical Value** boxes are *not* activated). On the x and y axes of figure 2.67 are the copolymer compositions expressed in weight % of St in STVPh and ethylene in EMA. The small black circles once more denote blends calculated to be immiscible, where in this case the difference in the solubility parameters of a particular STVPh - EMA blend exceeds the calculated critical value, $\Delta\delta$. The appropriate value of $(\Delta\delta)^{\circ}_{Crit}$ for the VPh - acrylate interaction remains 3.0 (cal. cm^{-3})$^{0.5}$ and is selected from the **Critical** window.

The predicted miscible region for STVPh blends with EMA blends - the unfilled area of the miscibility map shown in figure 2.67 - is compared to our experimental studies (the larger black (shaded) and unfilled filled circles corresponding to immiscible and miscible blends, respectively). While our innate modesty prevents us from bragging, we are forced to state that the agreement between prediction and experiment is rather good! (Again, even better results are obtained with more rigorous calculations, see figure 7.82).

Returning to homopolymer - copolymer blends involving the VPh segment, let us now consider PVPh blends with polyethers. The strength of the interaction between the phenolic hydroxyl and ether oxygen groups is considerably stronger than that occurring between PVPh and the ester type carbonyls considered above. This is readily confirmed from the shift of the hydrogen bonded phenolic hydroxyl stretching mode relative to the "free" (non-hydrogen bonded) frequency in the infrared spectrum. Frequency shifts of between 200-325 cm^{-1} are observed in the former, while typically only 100 cm^{-1} shifts are seen in the latter[49,52]. There is only a limited amount of experimental data pertaining to the phase behavior of PVPh - polyether blends and for the purposes of this work we will restrict ourselves to the trends observed in the PVPh blends with poly(vinyl alkyl ethers)[60]. In essence, PVPh is miscible at ambient temperature with poly(vinyl methyl ether) (PVME; δ = 8.5 (cal. cm^{-3})$^{0.5}$) and poly(vinyl ethyl ether) (PVEE; δ = 8.4) but immiscible with poly(vinyl butyl ether) (PVBE δ = 8.2) and poly(vinyl iso-butyl ether) (PViBE; δ = 8.0). Figure 2.68 shows the calculated solubility parameters for linear poly(vinyl alkyl ethers) as a function of methylene groups in the side group of the polymer (i.e PVME has zero methylenes in the side chain etc.). Based upon the greater relative strength of the VPh - ether interaction we know that the value of $(\Delta\delta)^{\circ}_{Crit}$ is probably somewhat greater than 3.0 (cal. cm^{-3})$^{0.5}$, but we will use this value (the **Strong** button in the **Critical** window) to reveal the major *trends,* recognizing that in the future additional experimental information may permit us to adjust this initial critical value.

PVPh and Poly(VME-co-Methylene) Blend

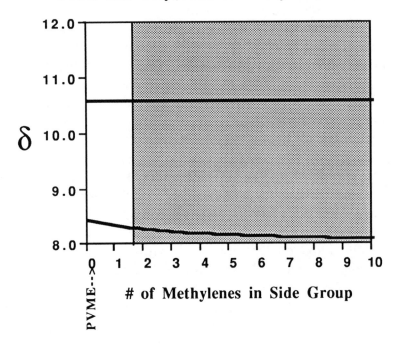

FIGURE 2.68

The calculation predicts that PVPh blends with PVME and PVEE are miscible, but that higher homologues are immiscible - a result in fine accord with the limited experimental data presently available.

Our last example of PVPh blends concerns those prepared with poly(2-vinyl pyridine) (PVPy). Here we have the unusual case of a polymer blend consisting of two homopolymers in which there are very strong favorable intermolecular interactions, but where the calculated non-hydrogen bonded solubility parameters are almost identical (δ = 10.6 and 10.9 (cal. cm^{-3})$^{0.5}$, respectively). For the purposes of a predictive rule of thumb for the molecular mixing of polymers, this blend represents the epitome of factors favoring miscibility, since we simultaneously have a large favorable $\Delta G_H/RT$ hydrogen bonding term (equation 1), and a negligible contribution from the unfavorable $\chi \Phi_A \Phi_B$ term. The interaction between the phenolic hydroxyl and pyridine nitrogen groups is much stronger than that occurring between PVPh and the ether oxygens, and is reflected in the very large shift (~ 500 cm^{-1}) of the hydrogen bonded phenolic hydroxyl stretching mode relative to the "free" (non-hydrogen bonded) frequency. Solutions of PVPh and PVPy dissolved in a common solvent (e.g.THF) immediately precipitate when mixed together, forming what is best described as a 1:1 polymer complex[61].

Meftahi and Frechet have also reported that PVPy blends are miscible with STVPh copolymers containing less than about 70-80% styrene[62]. Using the **Miscibility Guide** what do we predict? Figure 2.69 shows the effect of diluting PVPh with styrene by copolymerization. As the concentration of styrene is increased in the copolymer the solubility parameter difference increases while simultaneously the number of sites available for intermolecular interactions between the PVPh hydroxyl group and the PVPy nitrogen decreases - both factors are unfavorable to mixing.

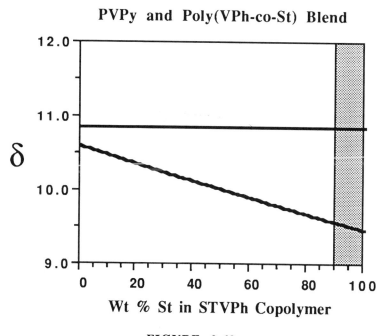

PVPy and Poly(VPh-co-St) Blend

δ

Wt % St in STVPh Copolymer

FIGURE 2.69

Nevertheless, this is offset by the strong interaction (at least a value corresponding to a $(\Delta\delta)^{o}_{Crit} = 3.0$ (cal. $cm^{-3})^{0.5}$) and blends of PVPh and STVPy are predicted to be miscible throughout a range of 0 to 90% styrene. This range appears to be somewhat greater than experimentally observed, but at the risk of being redundant, we can state that the *trend* is correct.

Finally, before we summarize this section, we would be remiss if we did not at least consider blends that involve polymers containing the carboxylic acid group, although we lack a significant body of relevant experimental data with which to compare our predictions. Systematic studies involving blends of such polymers with, for example, an homologous series of polyesters or a series of ethylene-co-vinyl acetate polymers of varying composition, where compositional ranges of miscibility have been determined, are scant. As we have seen, results of these studies were available and very helpful for the PVPh blends discussed previously. Additionally, polymers such as poly(acrylic acid) (PAA), styrene-co-acrylic acid copolymers (STAA), poly(methacrylic acid) (PMAA) and ethylene-co-methacrylic acid copolymers (EMAA) are very strongly self-associated through the formation of hydrogen bonded cyclic dimers of the type shown below:

$$2 \; \text{CH–C} \begin{smallmatrix} O-H \\ \\ O \end{smallmatrix} \rightleftharpoons \text{CH–C} \begin{smallmatrix} O-H \cdots O \\ \\ O \cdots H-O \end{smallmatrix} C-$$

Carboxylic Acid Dimer

In fact, equilibrium constants describing the self-association of polyacids are several orders of magnitude greater than those for comparable polymers containing amide, urethane or phenolic hydroxyls[63] and the intermolecular interactions between carboxylic acids and groups such as ether oxygens and pyridine nitrogens etc. are also much stronger than their phenolic counterparts. (Insoluble hydrogen bonded polymer complexes are formed between, for example, PAA - PEO and PMAA - poly(2-vinyl pyridine)).

Furthermore, the magnitudes of the equilibrium constants describing self-association and inter-association have widely disparate values and the simple function that approximates the variation of the critical value of $\Delta\delta$ with concentration of diluent (figure 2.13) breaks down. If this were not enough, we must candidly admit that we are not yet confident that we can obtain a reasonable estimate of the *non-hydrogen bonded* solubility parameter of carboxylic acid containing polymers like PAA or PMAA by simply removing the proton in the manner described previously (see pages 64 - 69) and using an ester in place of the carboxylic group. If we were of a timid nature, it might be propitious to "punt" at this point and recognize that we have reached the respectable limit of this simple guide. But, "fools rush in - - - " and since we have used this approach with good success so far we will proceed to employ the **Miscibility Guide** to calculate *trends* for selected blends containing carboxylic acid segments. Time will tell whether or not these predictions are reasonably valid.

PEO and Poly(MAA-co-Ethylene) Blend

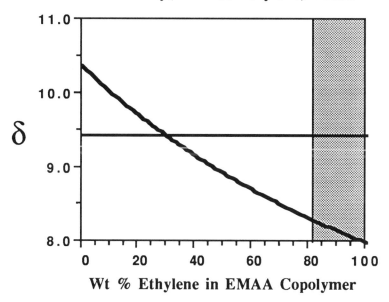

Wt % Ethylene in EMAA Copolymer

FIGURE 2.70

Let us first consider blends of ethylene-co-methacrylic acid (EMAA). We have recently published a series of papers pertaining to polymer blends involving EMAA copolymers with polyethers, poly(vinyl pyridines) and polyoxazolines[63]. These blends are more readily studied than the corresponding poly(methacrylic acid) (PMAA) blends because EMAA copolymers containing between about 30 to 60 weight % methacrylic acid are amorphous, more flexible (i.e. lower T_g's), less hydrophilic and soluble in solvents like tetrahydrofuran.

Figure 2.70 shows calculated solubility parameters for EMAA copolymers as a function of the weight % ethylene, together with that of PEO ($\delta = 9.4$ (cal. cm^{-3})$^{0.5}$). If we select the button in the **Critical** window reflecting strong hydrogen bonds, the initial value of $(\Delta\delta)^o_{Crit} = 3.0$ (cal. cm^{-3})$^{0.5}$), which may well be on the low side, predicts a miscible range for the PEO - EMAA blends that extends from pure PMAA to EMAA copolymers containing up to about 80 weight % ethylene. In the amorphous state, at temperatures above the T_m of PEO, we have previously shown experimentally that the PMAA and EMAA copolymers containing 55, 44 32 and 26% are indeed miscible with PEO. Further studies are necessary to find the lower limit of miscibility.

The final example pertains to blends of PEO with styrene-co-acrylic acid copolymers (STAA). Jo and Lee[64] have recently reported experimental results that establish that PEO is miscible with STAA copolymers containing greater that about 7 % acrylic acid. What does our simple guide predict ? The result obtained using the **Miscibility Guide** for STAA-PEO blends is shown in figure 2.71 and it essentially predicts that all compositions of STAA from 0 to 99% styrene should be miscible with PEO in the amorphous state. The predicted miscible range appears to somewhat overestimated, but note an important *trend* - one that is worthwhile seeking out when scouting for potentially miscible systems - as the concentration of styrene in the copolymer increases, the number of carboxylic acid sites decreases, but this is more than offset by the decreasing magnitude of $\Delta\delta$. (The reader might also wish to compare the distinctions between the plots given in figures 2.69 and 2.71).

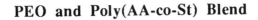

PEO and Poly(AA-co-St) Blend

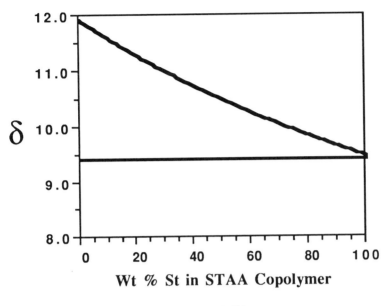

FIGURE 2.71

Summary

For the case of polymer blend systems in which there are *relatively strong favorable, specific intermolecular interactions* :

1. The presence of these interactions (usually, hydrogen bonds) between the blend components (i.e. those with dissociation energies of between about 3 and 7 kcal. mole^{-1}) further increases one's chances of finding miscible blends of polymers, compared to those that interact via dispersion forces only or are restricted to dipole - dipole or relatively weak hydrogen bonds.

2. Specific favorable intermolecular interactions again serve to counteract the unfavorable contribution to the free energy of mixing, that we express in the χ parameter. A larger value of χ_{Crit} , equivalent to upper limit of the non-hydrogen bonded solubility parameter difference of about $(\Delta\delta)^{\circ}_{Crit} = 3$ (cal. cm^{-3})$^{0.5}$, can be tolerated.

Table 2.8
Summary of the Upper Limit of the Critical Values of the
Solubility Parameter Difference, $(\Delta\delta)^{\circ}_{Crit}$

Specific Interactions Involved	Polymer Blend Examples	$(\Delta\delta)^{\circ}_{Crit}$ (cal. cm^{-3})$^{0.5}$
Dispersive Forces Only	PBD - PE	≤ 0.1
Dipole - Dipole	PMMA - PEO	0.5
Weak	PVC - BAN	1.0
Weak to Moderate	PC - Polyesters	1.5
Moderate	SAN - PMMA	2.0
Moderate to Strong	Nylon - PEO	2.5
Strong	PVPh - PVAc	3.0
Very Strong	PMAA - PEO	≥ 3.0

3. The main principle for predicting polymer miscibility still remains the matching of the non-hydrogen bonded solubility parameters. However, now the errors involved the calculation of non-hydrogen polymer solubility parameters are less important as there is a much wider range of differences to work with.

G.   FINAL WORDS

We set out in this chapter to provide the reader with a simple set of guidelines to qualitatively or semi-quantitatively predict trends in polymer - polymer miscibility. This we believe we have accomplished using ideas which are embodied in equation 2.1. It is axiomatic in our scheme that the closer the match of the two *non-hydrogen bonded* solubility parameters and the greater the relative strength of any specific intermolecular interactions present between the polymeric components of the blend, the greater the probability of miscibility. This is summarized in table 2.8 in terms of critical values of χ and the upper limits of the non-hydrogen bonded solubility parameter difference, $(\Delta\delta)^{\circ}_{Crit}$.

We cannot and do not expect such a rudimentary approach to predict all known miscible polymer blends. Again, we reiterate that in this chapter our primary objective has been to present a *practical guide* to miscibility and we are confident that the application of these guidelines can significantly reduce the time used in scouting for potential miscible polymer systems.

H. REFERENCES

1. Hildebrand, J. and Scott, R. *The Solubility of Non-Electrolytes*, 3rd Ed., Reinhold, N. Y., 1950.
2. Burrell, H. *Solubility Parameter Values*, in *Polymer Handbook*, 2nd. Ed., Brandrup, J. and Immergut, E. H., Editors, J. Wiley - Interscience, N.Y., 1975.
3. van Krevelen, P.W., *Properties of Polymers*; Elsevier: Amsterdam, 1972.
4. Hoy, K. L., *J. Paint Technol.*, 1970, **42**, 76.
5. Small, P. A., *J. Appl. Chem.*, 1953, **3**, 71.
6. Martuscelli, E., Pracella, M. and Yue, W. P., *Polymer,* 1984, **25**, 1097.
7. Coleman, M. M., Serman, C. J., Bahgwager, D. E. and Painter, P. C., *Polymer,* 1990, **31**, 1187.
8. Daubert, T. E. and Danner, R. P., *Data Compilation: Tables of Properties of Pure Compounds*, American Institute of Chemical Engineering, New York, Extant. 1989.
9. Serman, C. J., Ph.D. thesis, The Pennsylvania State University, 1989.
10. Moler, C., Herskovitz, S., Little, J. and Bangert, S., *MATLAB User's Guide*, The MathWorks, Inc., Sherborn, MA 01770, 1987.
11. Coleman, M. M., Hu, J., Park, Y. and Painter, P. C., *Polymer,* 1988, **29**, 1659.
12. Boyer, R. F. and Spencer, R. F., *J. Appl. Phys.*, 1944, **15**, 398.

13. Simha, R. and Boyer, R. F., *J. Chem. Phys.*, 1962, **37**, 1008.
14. Scott, R. L., *J. Chem Phys*, 1949, **17**, 268.
15. Scott, R. L., *J. Polym Sci,* 1952, **9**, 423.
16. Lacombe, R. H. and Sanchez, I. C., *J. Phys Chem.*, 1976, **80**, 2568.
17. Martuscelli, E., Silvestre, C. and Gismondi, G., *Makromol. Chem.,* 1983, **186**, 2161.
18. (a) Bank, M., Leffingwell, J. and Thies, C., *Macromolecules,* 1971, **4**, 44.; ibid, *J. Polym. Sci.*, 1972, **10**, 1097. (b) Nishi, T.and Kwei, T.K. *Polymer*, 1975,**16**, 285.
19. (a) Stoelting, J., Karasz, F. E. and MacKnight, W. J., *Polym. Eng. Sci.*, 1970, **10**, 133. (b) Schultz, A. R. and Gendron, B. M., *J. Appl. Polym. Sci.*, 1972, **16**, 461.
20. Kimura, M., Porter, R. S. and Salee, G., *J. Polym. Sci. - Phys. Ed.*, 1983, **21**, 367
21. (a) Cowie, J. M. G., Reid, V. M. C. and McEwen, I. J., *Polymer*, 1990, **31**, 486. (b) *ibid., Polymer,* 1990, **31**, 905.
22. Kim, J. H., Barlow, J. W. and Paul, D. R., *J. Polym. Sci. - Phys. Ed.,* 1989, **27**, 2211.
23. Varnel, D. F., M.S. thesis, The Pennsylvania State University, 1980.
24. Zakrzewski, G. A., *Polymer,* 1973, 14, 347.
25. Hall. W. J., Kruse, R. L., Meendelson, R. A. and Trementozzi, Q. A., *Am. Chem. Soc. Symp. Ser.*, 1983, **229**, 49.
26. Kim, J. H., Barlow, J. W. and Paul, D. R., *J. Polym. Sci. - Phys. Ed.,* 1989, **27**, 223.
27. Aoki, Y., *Macromolecules,* 1988, **21**, 1277.
28. Phibbs, M. K., *J. Phys Chem.*, 1955, **59**, 346.
29. Cruz, C. A., Paul, D. R. and Barlow, J. W., *J. Appl. Polym. Sci.,* 1979, **23**, 589.
30. Fernandes, A. , Barlow, J. W. and Paul, D. R., *Polymer,* 1986, **27**, 1799.
31. Coleman, M. M., Varnell, D. F. and Runt, J. P., in *Contempory Topics in Polymer Science*, Vol. 4, Bailey, W. J. and Tsurata, T., Eds., Plenum, New York, 1983.
32. Varnell, D. F., Runt, J. P. and Coleman, M. M., *Macromolecules,* 1981, **14**, 1350.
33. Coleman, M. M. and Painter, P. C., *Appl. Spectrosc. Revs.*, 1984, **20(3&4)**, 255 .
34. Vorenkamp, E. J. and Challa, G. *Polymer,* 1988, **29**, 86.

35. Woo, E. M., Barlow, J. W. and Paul, D. R. *Polymer*, 1985, **26**, 763.

36. Walsh, D. J. and McKeown, J. G., *Polymer*, 1980, **21**, 1330.

37. Walsh, D. J. and Cheng, G. L., *Polymer*, 1984, **25**, 495.

38. Rellick, G. S. and Runt, J., *J. Polym. Sci. - Phys. Ed.*, 1986, **24**, 279.

39. Cruz-Ramos, C. A. and Paul, D. R., *Macromolecules*,1989, **22**, 1289.

40. Shiomi, T., Karasz, F. E. and MacKnight, W. J., *Macromolecules*, 1986,**19**, 2274.

41. Bhagwagar, D. E., Serman, C. J., Painter, P. C. and Coleman, M. M., *Macromolecules*, 1989, **22**, 4654.

42. Suess, M., Kressler, J. and Kammer, H. W., *Polymer*, 1987, **28**, 957.

43. Fowler, M. E., Barlow, J. W. and Paul, D. R., *Polymer*, 1987, **28**, 1177.

44. Cowie, J. M. G. and Lath, D., *Makromol. Chem. Macromol. Symp.* 1988, **32**, 2055.

45. Nishimoto, M., Keskkula, H. and Paul, D. R., *Macromolecules*, 1990, **23**, 3633.

46. Chong, Y. F. and Goh, S. H. Abstract # HP2-5 *IUPAC International Symposium on Speciality Polymers*, Singapore, November 1990.

47. Neo, M. K., Lee, S. Y. and Goh, S. H., *ibid.* Abstract #HP2-6.

48. Harris, J. E., Goh, S. H., Paul, D. R. and Barlow, J. W., *J. Appl. Polym. Sci.*, 1982, **27**, 839.

49. Coleman, M. M. and Moskala, E. J., *Polymer*, 1983, **24**, 251.

50. Hu, J., Painter, P. C., Coleman, M. M. and Krizan, T. D., *J. Polym Sci, Phys Ed.*, 1990, **28**, 149.

51. (a) Coleman, M. M., Skrovanek, D. J., Hu, J. and Painter, P. C., *Macromolecules*, 1988,**21**, 59.
    (b) Painter, P. C., Park, Y. and Coleman, M. M., *ibid.*, 1988, **21**, 66.

52. Moskala, E. J., Varnell, D. F. and Coleman, M. M., *Polymer*, 1985, **26**, 228.

53. Coleman, M. M., Lichkus, A. M. and Painter, P. C. *Macromolecules*, 1989, **22**, 586.

54. Serman, C. J., Xu, Y., Painter, P. C. and Coleman, M. M., *Macromolecules*, 1989, **22**, 2015.

55. Serman, C. J., Painter, P. C. and Coleman, M. M., *Polymer*, in press.

56. Moskala, E. J., Howe, S. E., Painter, P. C. and Coleman, M. M., *Macromolecules*, 1984, **17**, 1671.

57. Moskala, E. J., Runt, J. P. and Coleman, M. M., Chapter 5 in *Multi-*

component Polymer Materials, Paul, D. R. and Sperling, L. H., Eds., Advances in Chemistry Series, #211, ACS, 1986.

58. Chen, C-T. and Morawetz, H. *Macromolecules*, 1989, **22**, 159.
59. Xu, Y., Painter, P. C. and Coleman, M. M., *Polymer*, in press.
60. Serman, C. J., Painter, P. C. and Coleman, M. M., *Polymer*, in press.
61. Lee, J. Y., Moskala, E. J., Painter, P. C. and Coleman, M. M., *Appl. Spectrosc.*, 1986, **40**, 991.
62. de Meftahi, M. V. and Frechet, J. M. J., *Polymer*, 1988, **29**, 477.
63. (a) Lee, J. Y., Painter, P. C. and Coleman, M. M., *Macromolecules*, 1988, **21**, 346. (b) *ibid.*, 1988, **21**, 954. (c) Lichkus, A. M., Painter, P. C. and Coleman, M. M., *ibid.*, 1988, **21**, 2636 (d) Coleman, M. M., Lee, J. Y., Serman, C. J., Wang, Z. and Painter, P. C., *Polymer*, 1989, **30**, 1298.
64. Jo, W. H. and Lee, S. C., *Macromolecules*, 1990, **23**, 2261.

## CHAPTER 3

The Nature of the Hydrogen Bond

A. INTRODUCTION

In the first two chapters of this book we have discussed simple theories of polymer mixing and presented a guide to predicting miscibility that is essentially based on making informed guesses concerning the strength of any specific interactions that may be present. The guesses are "informed" in the sense that there is a wide body of literature concerning interactions between various functional groups commonly found in polymers, particularly those that hydrogen bond (e.g., enthalpies of hydrogen bond formation). We believe these interactions are of particular significance in forming miscible polymer mixtures and also lend themselves to more rigorous quantitative treatments, because in many systems the *number* of such interactions can be measured as a function of composition and temperature using infrared spectroscopy. Consequently, a discussion of hydrogen bonding in amorphous polymer mixtures will take up much of the rest of this book. We commence in this chapter with a brief review of the nature of the hydrogen bond.

Most chemists and even some physicists have at least a passing notion of the nature of hydrogen bonding and its effect on the structure and physical properties of various materials. Classic examples of systems where the elucidation of the role of hydrogen bonding led to an understanding of structure and function are found in molecular biology, where Pauling et al.[1,2]

157

identified the α- helix and β-sheet structures of polypeptides and, of course, Watson and Crick[3], proposed that the double stranded DNA molecule is determined by the complementary nature of the hydrogen bonds formed between pyrimidine and purine residues. Another intriguing example is water. It is now recognized that the anomalous properties of this ubiquitous material (its increase in volume upon freezing, high viscosity and thermal conductivity, etc.) are a result of the arrangement of the molecules in a dynamic three dimensional network, made possible by the ability of each molecule to hydrogen bond to up to four different neighbors. In general, if a material hydrogen bonds there is often a profound effect on various physical properties, including melting points, boiling points, the location of a glass transition, dielectric constant, choice of crystal structure and, the central concern of this book, its solubility or ability to form miscible mixtures with another material.

There is a wealth of literature concerning hydrogen bonds in all their manifestations, summarized in various major reviews and books[4-13]. This work principally concerns small molecules, however, and *synthetic* polymers are largely ignored. Biopolymers can be considered a separate class and there are numerous studies of the role of hydrogen bonding in the structure and function of biological materials. The neglect of polymer materials by most chemists and physicists that study hydrogen bonding is unsurprising, much of the work cited above is concerned with obtaining a fundamental understanding of the nature of the hydrogen bond rather than its role in determining the properties of specific materials, so that it makes sense to study simpler, better defined systems. What is startling is the magnificent disregard for this literature *occasionally* displayed by polymer scientists that have studied hydrogen bonding in long chain molecules. In our view there are three problems. The first is the use and abuse of infrared spectroscopy; more on this later. The second is the disregard for the established body of work in the low molecular weight literature concerning the nature and geometry of the hydrogen bond and the properties of specific types of hydrogen bonds (in various studies extraordinary enthalpies of hydrogen bond formation or extremely distorted, non-linear hydrogen bonds have been proposed). Finally, there is a persistent temptation to see hydrogen bonds as something akin to an on/off switch, so that when these interactions are present they have been viewed as static things that act like permanent cross links. Thus, in some papers it has been proposed that hydrogen bonds are responsible for a "memory effect," maintaining certain structures for a significant period of time, or, from the opposite view but in the same vein, one can also come across statements that suggest most, if not all, hydrogen bonds become suddenly broken in the melt. (For some reason nylons have been particularly victimized by such unprovoked attacks). The dynamic nature of this interaction and the systematic variation in the number of hydrogen bonds present as a function of temperature and composition is often ignored.

Here, our principal aim is not a detailed discussion of the nature of hydrogen bonding. In our opinion one cannot make a better start than a reading of the book by Pimentel and McClellan[4], still a classic after thirty years. Any efforts of ours would pale in comparison. Our purpose is to address the narrower, albeit important, subject of the role of hydrogen bonding in the mixing of polymers. Nevertheless, it is useful to review some fundamental aspects of hydrogen bonding, if for no other purpose but to introduce various terms and definitions. That will take up the rest of this chapter. We will then proceed to discussion of the stoichiometry of hydrogen bonding and the role of infrared spectroscopy in measuring the number of hydrogen bonded groups present in a system. This leads us to the heart of this book, a consideration of the role of hydrogen bonding in the thermodynamics of mixing.

B. WHAT IS A HYDROGEN BOND?

There is no simple, universally accepted definition of a hydrogen bond, but the description given by Pauling[13] comes close to capturing its essence;

> ....under certain circumstances an atom of hydrogen is
> attracted by rather strong forces to two atoms instead of
> only one, so that it may be considered to be acting as a
> bond between them. This is called a hydrogen bond.

Most other authors define a hydrogen bond in one of two ways; by its effect on the properties of a material or by its molecular characteristics. The former is a relatively straightforward approach, qualitatively, and best accomplished by comparing hydrogen bonds to chemical bonds and dispersion forces. The former link the atoms of a molecular together into a distinct three dimensional arrangement. To break these bonds normally requires large energies. Dispersion interactions, in contrast, are an order of magnitude weaker and are responsible for the condensation of non-polar molecules from the vapor into liquid and then solid state as the temperature (and hence kinetic energy of the molecules) is lowered. Hydrogen bonds lie somewhere between these extremes. Covalent bonds have strengths of the order of 50 kcal. mole^{-1}; van der Waals attractions may be of the order of 0.2 kcal. mole^{-1}, while hydrogen bonds most often lie in the range 1-10 kcal. mole^{-1}. This range of energies is such that in "liquids" (i.e. materials in the non-crystalline state at temperatures above their T_g) at room temperature there is a dynamic situation, with hydrogen bonds constantly breaking and reforming at the urgings of thermal motion. This situation is perhaps best described as in the preceding chapters in terms of the magnitude of energies involved. To reiterate, for interaction strengths of about 3 kcal. mole^{-1}, approximately 1% of the interacting units would have sufficient energy to dissociate at any particular instant of time; this is the realm of hydrogen bonding interactions typical of phenolic OH groups, carboxylic acids, urethane and amide groups found in a number of important polymers and will be of most concern to us in this work. The crucial point to keep in mind is the dynamic character of the interaction, however. In water, for example, it has been estimated that the mean lifetime of a hydrogen bond is of the order of 10^{-11} seconds[14]. Covalent polymer chains are thus not "cross-linked" by the presence of such bonds. A nylon in the melt still has a large fraction of its amide groups hydrogen bonded to one another at any instant (see Chapter 5), but flows in the usual viscoelastic manner. This is not to say that fundamental properties, such as the rate of diffusion of individual chains, are not altered by the association of hydrogen bonded groups into "aggregates", but such processes are not prevented, as they would be if the hydrogen bonds were not dynamic and behaved like covalent cross links.

The alternative description of a hydrogen bond in terms of the atoms involved in the system seems paradoxically both simple and complicated. In a qualitative sense a hydrogen bond can be described in a very straightforward manner, with Pimentel's[4] description being as good a starting point as any:

A hydrogen bond exists between a functional group A-H and an atom or a group of atoms B in the same or a different molecule when:

(a) there is evidence of bond formation (association or chelation),

(b) there is evidence that this new bond linking A-H and B specifically involves the hydrogen atom already bonded to A.

The proton (or deuteron) usually lies on a line joining the A,B atoms, i.e., the hydrogen bond is linear; A-H - - - B, and the distance between the nuclei of the A and B atoms is considerably less than the sum of the van der Waals radii of A and B and the diameter of the proton, i.e. the formation of the hydrogen bond leads to a contraction of the A-H - - - B system. The atoms A and B are usually only the most electronegative, i.e. F, O and N. Chlorine is as electronegative as nitrogen, but because of its larger size only forms weak hydrogen bonds. Hydrogen bonds involving S, some C-H groups and the Π electrons on aromatic rings have also been invoked, but these are much weaker, they are less easily identified, and the evidence for their presence is not always unambiguous. Finally, the strength of a hydrogen bond also depends upon the atoms to which A and B are attached, i.e. the nature of the molecule or molecules involved. For example, highly charged species can give rise to very short and hence very strong hydrogen bonds.

Descriptions of the hydrogen bond become more complicated when quantitative descriptions of forces are attempted. Very early work considered a hydrogen bond to be a result of the formation of two covalent bonds, but this model was abandoned once it was recognized that the hydrogen atom, with only one stable orbital (1s), can form only one covalent bond[13]. This approach was superseded by the postulation that the hydrogen bond is largely ionic in character and initial efforts focused on attempts to reproduce the physical properties of ice and water through a simple electrostatic model of point charges, arranged so as to give overall neutrality and reproduce the observed dipole moment. Thanks to an apparent cancellation of two large but neglected terms, this treatment was partially successful in that certain properties were reproduced. Others were not, however, and it became recognized that a quantum mechanical approach was required. Various models and procedures have been applied to the problem[15,16] but for our purposes the qualitative theoretical view given by Coulson[16] thirty years ago remains enlightening. Four contributory effects are recognized as being central to the formation of hydrogen bonds

      a) Electrostatic interactions
      b) Delocalization effects
      c) Repulsive forces
      d) Dispersion forces

Coulson pointed out that this separation is artificial in that all these forces have their origin, in one way or another, in electrostatic interactions, but this approach at least provides some insight and each of the factors listed above can be associated with a simple physical model, allowing calculations of the relative contributions of each to the overall energy.

C. EXPERIMENTAL CHARACTERIZATION OF HYDROGEN BONDS

Hydrogen bonds are not easily characterized. In highly crystalline solids the positions of the hydrogen atom or, more precisely, pattern of electron density, can theoretically be determined by x-ray diffraction, but the size of the proton and the amplitude of thermal vibrations makes this an extraordinarily demanding task. In certain materials neutron diffraction has been used to locate the position of the proton. Even if these methods were easily applied to the whole range of crystalline solids that have hydrogen bonds, however, this would still leave a yawning gap. Many materials of interest are either liquids or amorphous solids at ambient temperature. Many synthetic polymers have a bit of everything and, more crucially, true single crystals have not been prepared, so that there is insufficient crystallographic data for the required detailed analysis of these materials. In principle, this leaves two types of experimental studies that could be applied to the characterization of polymers:

a) Thermodynamic
b) Spectroscopic

Thermodynamic measurements depend upon changes in a system as a whole and can be related to molecular properties through the methods of statistical mechanics, but the results are often model dependent and sensitive to the various assumptions that have to be made. Furthermore, direct measurements of quantities such as the heat of mixing cannot be performed on polymers directly, and studies are often confined to low molecular weight analogues. Accordingly, this leaves the art of spectroscopy as the (presently) most powerful probe of the nature of hydrogen bonding. The most widely used methods are:

1. Infrared (ir) and Raman techniques, which provide information about the stretching and deformation vibrations of A−H bonds and acceptor groups;

2. electronic absorption and fluorescence spectroscopy in the ultraviolet and visible regions, which show the effect of hydrogen bond formation on the electronic levels of the participating molecules, and

3. proton magnetic resonance spectroscopy, which can be used to study the effect of hydrogen bond formation on the chemical shift of the A−H proton.

Of these, by far the most sensitive is infrared spectroscopy. In addition, it is at the present moment far more readily applied to solid state samples, such as polymers, than proton magnetic resonance. Consequently, we will confine our discussion of experimental measurements to this technique. As we will show, vibrational spectroscopy not only allows a measure of the strength or enthalpy of hydrogen bond interactions, but more crucially allows a determination of the number of free and bonded groups. As we will discuss in Chapter 6, it is these latter quantities that allow a calculation of the entire range of thermodynamic parameters. The determination of the enthalpy of hydrogen bond formation then allows a calculation of the variation of these parameters with temperature.

D. HYDROGEN BONDING IN POLYMERS

Hydrogen bonding plays an important role in the crystal structures of polymers such as the nylons, but in this book we are not concerned with polymer morphology. Our interest is the effect of hydrogen bond formation on the free energy of mixing and for that purpose it is sufficient for us to focus our attention here on the *types* of hydrogen bonds that can form These can be classified according to the functional groups involved, whether they are between like or unlike functional groups, and by the geometry of the complexes that are formed.

We will first consider polymers that *self-associate*. These contain functional groups that hydrogen bond to one another in the form of chains or cyclic complexes. Polymers containing amide, urethane and alkyl or aromatic hydroxyl groups associate in the form of linear chains, as illustrated in figure 3.1. It has also been proposed that alcohols form various cyclic structures, such as the dimer illustrated in figure 3.2. Molecular orbital calculations[18] indicate that these types of cyclic and bifurcated structures are not preferred in these systems, however, and that normal hydrogen bonds are linear in their minimum energy configurations. Larger cyclic complexes (trimers etc.) where the hydrogen bonds are linear are undoubtedly present in low molecular weight alcohols, but these occur principally in dilute

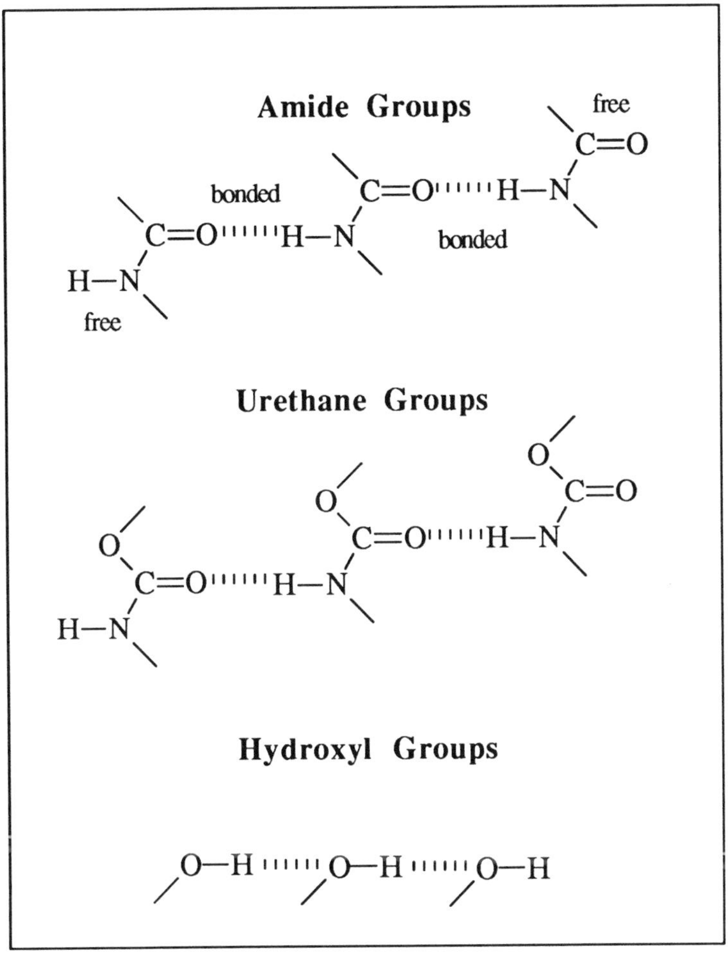

FIGURE 3.1

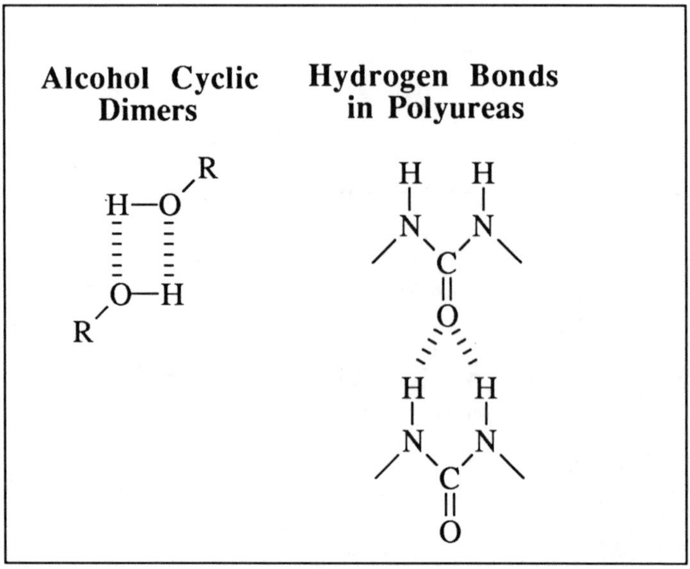

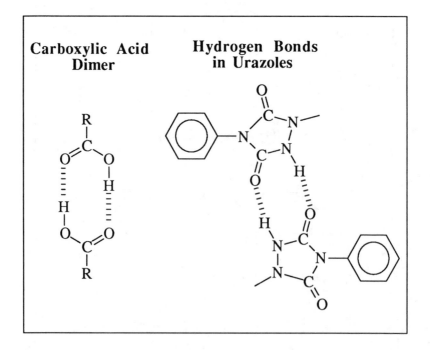

Alcohol Cyclic Dimers

Hydrogen Bonds in Polyureas

FIGURE 3.2

Carboxylic Acid Dimer

Hydrogen Bonds in Urazoles

FIGURE 3.3

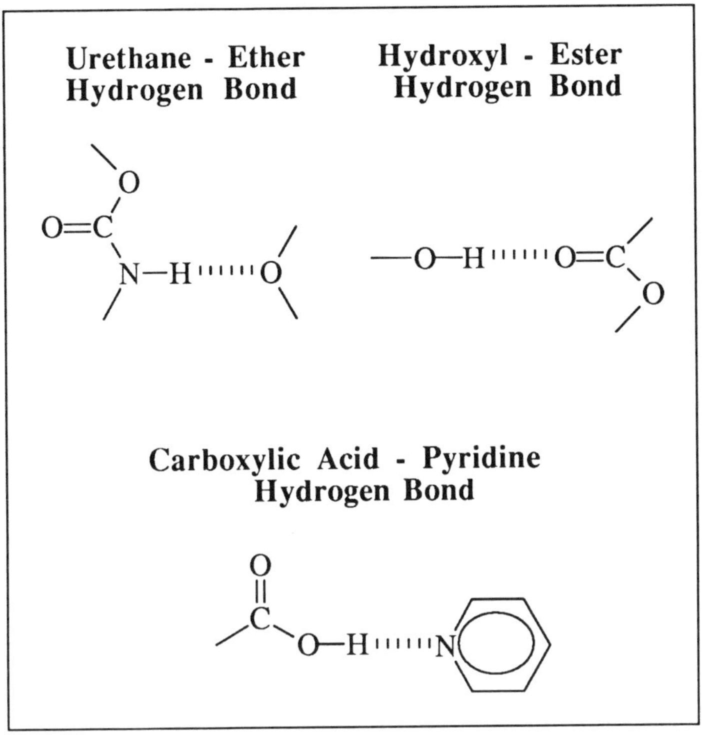

Urethane - Ether Hydrogen Bond

Hydroxyl - Ester Hydrogen Bond

Carboxylic Acid - Pyridine Hydrogen Bond

FIGURE 3.4

solutions and open chain structures predominate at higher concentrations (see discussion in reference 17, page 178). We will assume that chain entanglements, etc., would reinforce this trend in polymers and mainly concern ourselves with open chain association in materials containing OH functional groups. There is a large literature on this subject, but unfortunately most arguments rely on comparing how well various models fit experimental data of a very general kind, such as heat of mixing data. Various models provide practically equally acceptable fits to this type of data, so we believe a choice should be made on other grounds, such as the calculations quoted above. Bifurcated and cyclic structures certainly occur in molecules containing other types of functional groups, however, such as the polyureas also illustrated in figure 3.2. For these structures, it appears that more energy can be gained by forming two (weaker) non-linear hydrogen bonds that one (stronger) non-linear interaction.

Cyclic hydrogen bonded structures are favored in molecules containing carboxylic acid and urazole functional groups, as illustrated in figure 3.3. Carboxylic acids have been studied extensively and it is well known that the six membered cyclic ring formed by pairs of these functional groups form particularly strong, linear hydrogen bonds (some linear open chain structures may occur at very high concentrations of acid groups). Urazoles have been less widely studied, but there is an important body of work by Stadler and co-workers concerning the effect of hydrogen bonding on the properties of polymers containing such groups and it is considered that cyclic hydrogen bonded structures are energetically favored in these materials[18].

Hydrogen bonds can also form between "unlike" groups. Of particular importance in our studies of polymer blends are mixtures where one component self-associates (i.e., has functional groups of the type described above), while the second does not, but has a functional group capable of forming a hydrogen bond with the A-H group of the self-associating polymer. Examples of this second type of functional group are ethers, esters and nitrogen containing heterocyclic rings (e.g., pyridine), as illustrated in figure 3.4.

This type of association has been labeled "adduct formation" or "mutual association," but we prefer "inter-association" (for no good reason, we just like the sound of it). Obviously, hydrogen bonds can form between two self-associating species also (e.g., amide - urethane) and in low molecular weight molecules and certain polymer - solvent mixtures this surely occurs. We believe that this is unlikely to occur in polymer - polymer mixtures, however, for the following reason. When a self-associating polymer is mixed with a non self-associating polymer, free energy can be gained from the balance between breaking like hydrogen bonds and forming unlike ones. In mixing two self-associating polymers, however, two different types of like hydrogen bonds are broken and a distribution of unlike hydrogen bonds formed. This, in general, will be unfavorable to mixing, but in low molecular weight materials or polymer - solvent mixtures it would occur because of the favorable contribution of combinatorial entropy to the free energy. The mixing of high polymers involves a very small combinatorial entropy of mixing. Accordingly, it has been our experience that mixing polymers that hydrogen bond generally involves one component that self-associates and a second that does not, but has a functional group capable of hydrogen bonding with the A-H group of the first. We now proceed to a consideration of the stoichiometry of hydrogen bonding in such systems.

E. REFERENCES

1. Pauling, L., Corey, R. B. and Branson, H. R., *Proc. Nat. Acad. Sci.*, 1951, **37**, 205.
2. Pauling, L. and Corey, R. B., *Proc. Nat. Acad. Sci.*, 1951, **37**, 729.
3. Watson, J. D. and Crick, F. H., *Nature*, 1953, **171**, 737; 964.
4. Pimentel, G. C. and McClellan, A. L., *The Hydrogen Bond*, W. H. Freeman and Co., San Francisco and London (1960).
5. Pimentel, G. C. and McClellan, A. L., *Ann. Rev. Phys. Chem.*, 1971, **22**, 347.
6. Novak, A., *Structure and Bonding*, 1974, **18**, 177.
7. Schuster, P., Zundel, G. and Sandorfy, C., *The Hydrogen Bond. Recent Developments in Theory and Experiments.*, Vol. I-III, New York, Oxford (1976).
8. Hadzi, D. (Editor), *Hydrogen Bonding*, Pergamon Press, 1959.
9. Hadzi, D., *Pure and Applied Chem.*, 1965, **11**, 435.
10. Odinokov, S. E. Mashkovsky, A. A. Glazunov, V. P. Iogansen, A. V. and Rassadin, B. V. *Spectrochim Acta*, 1976, **32A**, 1355.
11. Speakman, J. C., *Structure and Bonding*, 1972, **12**, 141.
12. Vinogradov, S. N. and Linnell, R. H., *Hydrogen Bonding*. Van Nostrand Reinhold Co., New York (1971).
13. Pauling, L. *The Nature of the Chemical Bond*. Third Edition. Cornell University Press, Ithaca, new York, 1960.
14. Conde, O. and Teixerira, J., *J. Phys (Paris)*,1983, **44**, 525.
15. Schuster, P., *Energy surfaces for Hydrogen Bonded Systems* in *The Hydrogen Bond I Theory*. Schuster, P., Zundel, G., and Sandorfy, C., North Holland Publishing Co., 1976, Chapter 2.

16. Coulson, C. A., in *Hydrogen Bonding* Hadzi, D., editor. Pergamon Press, 1959, p. 339.
17. Marcus, Y., *Introduction to Liquid State Chemistry*, John Wiley and Sons, London, 1977.
18. Kerns, R. C. and Allen, L. C., *J.A.C.S.*, 1978, **100**, 6587.
19. Stadler, R. and Freitas, L., *Polym. Bull.*, 1986, **15**, 173.

## CHAPTER 4

Equilibrium Constants and the Stoichiometry of Hydrogen Bonding

A. INTRODUCTION

In order to account for the effect of hydrogen bonding upon the miscibility of polymer blends we will need to establish a relationship between the number and types of hydrogen bonded groups present and the thermodynamics of mixing. One approach is through the arbitrary definition of additional interaction parameters that separately account for the enthalpic and entropic components specific interactions, but in the final analysis, we believe this is unsatisfactory. One can always obtain a fit to observed properties if a sufficient number of terms of this type are defined, but such quantities do not always have direct physical meaning and they are usually system dependent, by which we mean they would not, for example, be applicable to amide groups in a wide range of nylons, but would only be empirically derived constants characteristic of a specific blend.

Our aim is more ambitious; to establish a relationship between *direct* experimental measurements of the hydrogen bonded species present and the contribution of such interactions to the free energy of mixing. This can be accomplished through the definition of equilibrium constants that are capable of describing the stoichiometry of hydrogen bonding and this will form the subject matter of this chapter. As we will see, these equilibrium constants provide a direct link between stoichiometry, experimental infrared spectroscopic measurements (next chapter) and the free energy of mixing (the chapter after that).

171

In a hypothetical system where simple 1:1 hydrogen bonded complexes are formed an equilibrium can be simply represented by:

$$A\text{-}H + B \rightleftharpoons A\text{-}H\text{-}\text{-}B \qquad (4.1)$$

and an association equilibrium constant defined as

$$K = \frac{[A\text{-}H\text{-}\text{-}B]}{[A\text{-}H]\,[B]} \qquad (4.2)$$

Then, from the known initial concentrations of the species, c_{AH}^o and c_B^o and, for example, a measurement by infrared spectroscopy of the fraction of A-H groups that are "free" (i.e., not hydrogen bonded), the equilibrium constant can be directly determined in the usual manner. This equilibrium is, of course, dynamic, but it is perfectly valid to proceed to derive a statistical mechanical model on the basis that the equilibrium distribution of A-H, B, and A-H - - B species are distinguishable (i.e. the system consists of three components). This will be considered in more detail later, in Chapter 6. Here, we must first take into account the fact that real systems are much more complicated, in the sense that more often than not we have the situation where at least one component *self-associates* to form chains of hydrogen bonds, as illustrated for a simple molecule containing an amide group at the top of figure 4.1 (following the arguments given in the preceding chapter we will neglect the formation of cyclic species, with certain exceptions that we will discuss later). Clearly, we need to be able to define an equilibrium constant or constants that defines this situation. Then, if we consider a system where such molecules are mixed with a second component that does not self-associate, but has a functional group capable of hydrogen bonding with the first (e.g., ether, ester, basic nitrogen as in pyridine), as illustrated at the bottom of figure 4.1, we would have to define an additional equilibrium constant describing this interaction. Finally, we could have mixtures of two different types of molecules that each self associate (e.g., one contains an amide group and the other an OH) and, in principle, could then form "mixed" chains, each consisting of a distribution of various hydrogen bonded groups

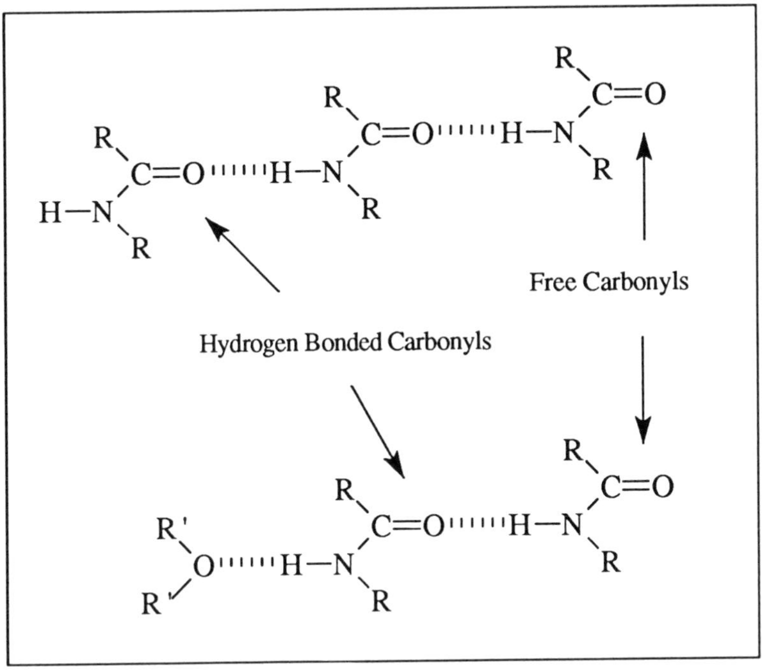

FIGURE 4.1

analogous to the various distributions found in (covalent) statistical copolymers.

At first sight, these problems might just seem to be those associated with a moderate increase in algebraic complexity, together with perhaps the more imposing experimental difficulties of measuring (by infrared spectroscopy) the proportions of the various hydrogen bonded species that are present. There is a much more fundamental difficulty, however, concerning the choice of the most appropriate concentration scale and reference states for the definition of equilibrium constants. To illustrate, we start with the thermodynamic relationship:

$$\mu_i = \mu_i^\circ + RT \ln a_1 \qquad (4.3)$$

where μ_i is the chemical potential of component i, a_i is its activity, and μ_1° is the standard chemical potential (value of μ_i when $a_i = 1$). This equation can be written in terms of two common concentration scales, mole fractions (x) and molarity (c):

$$\mu = \mu_x^o + RT \ln \gamma_x \, x \tag{4.4}$$

$$\mu = \mu_c^o + RT \ln \gamma_c \, c \tag{4.5}$$

where γ_x, γ_c are the activity coefficients. For the hydrogen bonding equilibrium shown in equation 4.1 we can write for each species:

$$\mu_{AH} = \mu_{x_{AH}}^o + RT \ln \gamma_{AH} \, x_{AH} \tag{4.6}$$

$$\mu_B = \mu_{x_B}^o + RT \ln \gamma_B \, x_B \tag{4.7}$$

$$\mu_{AHB} = \mu_{x_{AHB}}^o + RT \ln \gamma_{AHB} \, x_{AHB} \tag{4.8}$$

We can define for the general case:

$$\Delta\mu = \mu_{AHB} - \mu_B - \mu_{AH} \tag{4.9}$$

so that in terms of mole fraction:

$$\Delta\mu_x^o = \mu_{x_{AHB}}^o - \mu_{x_B} - \mu_{x_{AH}} \tag{4.10}$$

and we obtain:

$$\Delta\mu = \Delta\mu_x^o + RT \ln \left[\frac{x_{AHB}}{x_{AH} \, x_B} \right] \tag{4.11}$$

Similarly, in terms of molarity:

$$\Delta\mu = \Delta\mu_c^o + RT \ln \left[\frac{c_{AHB}}{c_{AH} \, c_B} \right] \tag{4.12}$$

where it is assumed that the activity coefficients are unity over the range of compositions chosen, or reference states are defined so to as near as possible to fulfill this condition. At equilibrium $\Delta\mu = 0$ and the quantities $\Delta\mu_x^o$ and $\Delta\mu_c^o$ are the standard free energy changes for converting a mole of reactants into products. They are related to equilibrium constants through the fundamental equations:

$$\Delta\mu_x^o = - RT \ln K_x \tag{4.13}$$

$$\Delta\mu_c^o = - RT \ln K_c \tag{4.14}$$

where it follows that:

$$K_x = \frac{x_{AHB}}{x_{AH}\, x_B} \qquad (4.15)$$

$$K_c = \frac{c_{AHB}}{c_{AH}\, c_B} \qquad (4.16)$$

Both mole fraction and molar concentration scales have been used in studies of hydrogen bonding, but which of these is the most appropriate?

The choice depends upon how we describe the thermodynamics of mixing. If we assume that mixing is ideal or use regular solution theory (i.e., assume the species A–H, B and A–H - - B are all approximately the same size), then we obtain the mole fraction result, and it is useful to briefly demonstrate this here. Ignoring (for now) interactions (London dispersion forces, etc.) between the components of a mixture of AH, B and AHB "molecules", the free energy of mixing is written:

$$\frac{\Delta G_{mix}}{RT} = n_{AH} \ln x_{AH} + n_B \ln x_B + n_{AHB} \ln x_{AHB} \qquad (4.17)$$

where n_i is the number of moles of species i. The partial molar free energies and hence chemical potentials are given by:

$$\frac{\partial (G_{mix}/RT)}{\partial n_{AH}} = \ln x_{AH} \qquad (4.18)$$

$$\frac{\partial (G_{mix}/RT)}{\partial n_B} = \ln x_B \qquad (4.19)$$

$$\frac{\partial (G_{mix}/RT)}{\partial n_{AHB}} = \ln x_{AHB} \qquad (4.20)$$

Hence,

$$(\mu_{AHB} - \mu_{AHB}^\circ) - (\mu_{AH} - \mu_{AH}^\circ) - (\mu_B - \mu_B^\circ)$$
$$= RT\,(\ln x_{AHB} - \ln x_{AH} - \ln x_B) \qquad (4.21)$$

and, at equilibrium:

$$- (\mu^{\circ}_{AHB} - \mu^{\circ}_{AH} - \mu^{\circ}_{B}) = - \Delta\mu^{\circ}_{x} = RT \ln \left[\frac{x_{AHB}}{x_{AH} \, x_{B}} \right] \tag{4.22}$$

or:

$$\Delta\mu^{\circ}_{x} = - RT \ln \left[\frac{x_{AHB}}{x_{AH} \, x_{B}} \right] = - RT \ln K_{x} \tag{4.23}$$

If we confine our attention to simple systems where association occurs between A–H and B molecules only, then the use of mole fractions might be a reasonable approximation, as the difference in size of the species is not too great. In many materials and mixtures of interest, however, at least one component self-associates to form chains of hydrogen bonded molecules (those containing carboxylic acid groups and urazoles perhaps being an exception), so that volume fractions and hence molar concentrations are more appropriate variables. We will demonstrate this in the following section, where we will simply present Flory's result in the same manner as we assumed a knowledge of the regular solution result for the entropy of mixing given above. Following this discussion, however, we will consider Flory's methodology in some detail, with the aim of extending the treatment so as to demonstrate that, in principle, we can obtain a "transferable" equilibrium constant, by which we mean one that is characteristic of a particular functional group in a series of structurally similar molecules (e.g., the n-alcohols). This will also prove useful when we consider the free energy of mixing in Chapter 6. Those of our readers who like to work from first principles, or who consider heavy dollops of algebra good for the soul, may wish to jump ahead and consider this section first.

B. ASSOCIATION MODELS AND EQUILIBRIUM CONSTANT DEFINITIONS

There are a number of important papers that describe the application of so-called association models to the problem of hydrogen bonding[1-7], and a summary can be found in the book by Acree[8]. The seminal work is due to Flory, however, who used a lattice model to derive an expression for the entropy of mixing polymers with a distribution of molecular weights[9]. In an appendix to this paper he demonstrated how this could be applied to reversible association in polymers species and in a later note[10], commenting on the work of Tobolsky and Blatz[11], specifically addressed hydrogen bonding. Flory[9] assumes that self-association in the form of chains can be described by a series of equilibria of the type:

$$B_h + B_1 \rightleftharpoons B_{h+1} \qquad (4.24)$$

where the standard free energy change $\Delta G°$ is independent of h. This is not an essential assumption, but one that leads to particularly simple solutions. The situation where, for example, formation of a dimer is described by a different equilibrium constant to subsequent h-mers is particularly important and will be considered later. Here our aim is to establish a fundamental basis for the definition of equilibrium constants and hence the choice of an association model, so we will for now maintain the Flory assumption.

There are two widely used association models that each claim to rest on the assumption that $\Delta G°$ is independent of h, the Mecke Kempter model[5,7] and the Kretschmer-Wiebe model[1]. The difference between them can be shown to depend upon the choice of reference states for mixing heterogeneous polymers. In Flory's original paper on this subject[9] the "proper" standard reference state is identified as one where the individual molecules (i.e., the various h-mers) are separate and oriented. The entropy change upon forming a mixture from solvent and n_{B_1}, n_{B_2}, n_{B_3}, etc., moles of monomer, dimer, trimer, etc., each initially separated according to chain length into ordered "crystalline" states, is given by:

$$-\frac{\Delta S_m}{R} = n_A \ln \Phi_A + \sum_{h=1}^{\infty} n_{B_h} \ln (\Phi_{B_h}/h)$$
$$- \sum_{h=1}^{\infty} [(h-1) n_{B_h}][\ln (z-1) - 1] + \sum_{h=1}^{\infty} n_{B_h} \ln \sigma \qquad (4.25)$$

where n_A is the number of moles of inert (i.e., non-hydrogen bonding solvent), n_{B_h} is the number of hydrogen bonded chains of B units that each contain h "monomers", z and σ are the lattice coordination number and a symmetry number, respectively, and Φ_i is the volume fraction of component i. As mentioned above, we will review the derivation of this equation in a modified form later, here we simply state Flory's result and note the implicit assumption that the molar volumes of A and non-hydrogen bonded B units are assumed equal. This expression is the sum of three contributions: the entropy of disorientation of the individual (h-mer) pure components; the entropy of mixing these with one another; and the entropy of dilution of the mixture. An enthalpy term describing physical interactions between the h-mers and the solvent can be included, if desired, but cancels from the final expression for the equilibrium constant (see page 216). If, for simplicity, athermal mixing of the components is therefore assumed, then the partial molar free energy of an h-mer is given by:

$$\frac{\Delta\overline{G}_h}{RT} = \ln\frac{\Phi_{B_h}}{h} + h\,\Phi_B\left(1-\frac{1}{h}\right) - (h-1)\ln(z-1) + \ln\sigma$$

(4.26)

where Φ_B is the volume fraction of all polymer (h-mer) molecules $(=\Sigma\Phi_{B_h})$. For the equilibrium described in equation 4.24 Flory[9] demonstrates that:

$$\Delta\overline{G}_{h+1} - \Delta\overline{G}_h - \Delta\overline{G}_1 + \Delta G^\circ = 0 \qquad (4.27)$$

where in Flory's nomenclature ΔG° is the standard free energy for formation of a linkage (hydrogen bond) and is independent of the length of the chains being joined. Substituting from equation 4.26 yields:

$$\frac{\Delta G^\circ}{RT} = -\ln\left[\frac{\Phi_{h+1}}{\Phi_h\,\Phi_1}\left(\frac{h}{h+1}\right)\right] + \ln\sigma(z-1) \quad (4.28)$$

Because z is a function of the liquid structure only and σ is a constant, an equilibrium constant K_B can be defined as:

$$\ln K_B = -\frac{\Delta G^\circ}{RT} + \ln [\, \sigma\,(z-1)\,] \qquad (4.29)$$

where:

$$K_B = \frac{\Phi_{h+1}}{\Phi_h \Phi_1} \cdot \frac{h}{h+1} \qquad (4.30)$$

This is a dimensionless quantity equal to the definition of a equilibrium constant in terms of molar concentrations (c) divided by the molar volume per structural unit, V_B, i.e.

$$K_B = \frac{c_{h+1}}{c_h c_1} \cdot \frac{1}{V_B} = \frac{K'}{V_B} \qquad (4.31)$$

As Flory points out, because V_B is a constant characteristic of the molecule concerned, either K or K' is an acceptable equilibrium constant.

The derivation of the Mecke-Kempter equilibrium constant follows precisely the same procedure but takes the separate but disoriented species as the reference state, rather than the condition where the molecules are separate and oriented. As a result, the following expressions for the partial molal free energy is obtained[7]:

$$\frac{\Delta \overline{G}_h}{RT} = \ln \Phi_{B_h} - (h-1) + \Phi_B h \left(1 - \frac{1}{h}\right) \qquad (4.32)$$

and an equilibrium constant can be derived in the same manner as above to give:

$$K^{MK} = \frac{\Phi_{h+1}}{\Phi_h \Phi_1} \qquad (4.33)$$

Both the Kretschmer-Weibe and Mecke-Kempter model have been successfully applied to the description of the excess properties of alcohols in hydrocarbon solvents[8]. It is not possible to choose between the models on the basis of this work, however, because equilibrium constants have almost invariably been determined from binary data reduction, so that their values are strongly influenced by the criteria adopted to separate the "physical" from the "chemical" (i.e., hydrogen bonding) contributions. Nevertheless, we believe that the Kretschmer-Wiebe model is the proper choice. Flory[9] argues that for the treatment of chemical equilibria between polymer species the most appropriate and least ambiguous reference state is that one in which the individual species are separate and oriented. We will comment on this point further in section E (page 209), after we have considered Flory's lattice model in more detail.

C. THE STOICHIOMETRY OF HYDROGEN BONDING AND THE RELATIONSHIP TO INFRARED SPECTROSCOPIC MEASUREMENTS

In the first parts of this chapter we have discussed in some detail Flory's arguments that the use of mole fractions to describe association is "irreconcilable with the statistical mechanics of these mixtures"[9] and concluded that a molar concentration or volume fraction concentration scale is more appropriate. We have, in addition, decided to follow Flory's arguments concerning the choice of reference state, and will therefore use the equilibrium constant for self association defined in equation 4.30. Given this model, we can now consider the relationships between the concentrations of the components of a mixture, infrared spectroscopic measurements and the distributions of species present. We will start with simple systems, by which we mean "small" molecules each of which contains only one functional group capable of hydrogen bonding, then proceed step-by-step to consider systems (and models) of greater complexity, including those molecules that each contain a large number of functional groups capable of hydrogen bonding, i.e., polymers such as the nylons or polyurethanes.

Simple Self-Association

The first system we will discuss consists of only one component that is capable of hydrogen bonding, e.g., each molecule contains a single OH or amide group. Such groups have both "donor" and "acceptor" parts and can thus take part in up to two hydrogen bonds, forming chains of molecules, as illustrated earlier in figure 4.1. The second component will be an "inert" diluent, by which we mean one that does not contain a functional group capable of forming a hydrogen bond. We let the volume fraction of the self-associating component be Φ_B, while that of the inert diluent is Φ_A. There is a distribution of hydrogen bonded chains present at a particular concentration and temperature and we can describe this condition as:

$$\Phi_B = \sum_{h=1}^{\infty} \Phi_{B_h} \qquad (4.34)$$

where Φ_{B_h} is the volume fraction of the chains of length h present at any instant in time. The equilibrium between species can be written:

$$B_1 + B_1 \overset{K_2}{\rightleftharpoons} B_2$$

$$B_2 + B_1 \overset{K_3}{\rightleftharpoons} B_3$$

$$| \quad | \qquad \qquad |$$
$$| \quad | \qquad \qquad |$$
$$| \quad | \qquad \qquad |$$
$$| \quad | \quad K_h \quad |$$
$$B_{h-1} + B_1 \rightleftharpoons B_h$$

$$(4.35)$$

where the equilibrium constants for each successive step are defined according to equation 4.33, i.e.

$$K_2 = \frac{\Phi_{B_2}}{\Phi_{B_1}^2} \left(\frac{1}{2}\right)$$

$$K_3 = \frac{\Phi_{B_3}}{\Phi_{B_2} \Phi_{B_1}} \left(\frac{2}{3}\right)$$

$$| \qquad \qquad | \qquad \qquad |$$

$$\begin{array}{ccc} | & | & | \\ | & | & | \end{array}$$

$$K_h = \frac{\Phi_{B_h}}{\Phi_{B_{h-1}} \Phi_{B_1}} \left(\frac{h-1}{h} \right) \tag{4.36}$$

It then follows that:

$$\begin{aligned} \Phi_{B_h} &= K_h \left(\Phi_{B_{h-1}} \Phi_{B_1} \right) \frac{h}{h-1} \\ &= K_h K_{h-1} \left(\Phi_{B_{h-2}} \Phi_{B_1}^2 \right) \frac{h}{h-2} \\ &= (K_h K_{h-1} - - - - - K_2) (\Phi_{B_1})^h h \end{aligned} \tag{4.37}$$

The simplest model corresponds to the assumption that $K_2 = K_3 = - - - - - K_h = K_B$, so that we can write:

$$\Phi_{B_h} = h K_B^{h-1} \Phi_{B_1}^h \tag{4.38}$$

The quantity Φ_{B_1}, the volume fraction of "monomers", those molecules that have no hydrogen bonded partners at all, will be the crucial concentration variable in all our derivations, providing the link between the stoichiometry of hydrogen bonding, infrared spectroscopic measurements and the thermodynamics of mixing. The volume fraction of B molecules can now be written:

$$\Phi_B = \sum_{h=1}^{\infty} h \Phi_{B_1}^h K_B^{h-1} = \Phi_{B_1} \sum_{h=1}^{\infty} h (K_B \Phi_{B_1})^{h-1} \tag{4.39}$$

If $K_B \Phi_{B_1} < 1$, then the summation converges and is given by:

$$\sum_{h=1}^{\infty} h (K_B \Phi_{B_1})^{h-1} = \frac{1}{(1 - K_B \Phi_{B_1})^2} \tag{4.40}$$

so that:

$$\Phi_B = \frac{\Phi_{B_1}}{(1 - K_B \, \Phi_{B_1})^2} \tag{4.41}$$

In pure B we have $\Phi_B = 1$ and we let the volume fraction of monomers be $\Phi^{\circ}_{B_1}$, so that

$$\Phi^{\circ}_{B_1} = (1 - K_B \, \Phi^{\circ}_{B_1})^2 \tag{4.42}$$

In general, in pure B, or any mixture of known composition, we have two unknowns, K_B and Φ_{B_1}, and one stoichiometric equation linking them. The second equation required for a solution follows from the characteristic infrared spectra of these systems[12]. In such spectra a bond due to free or non-hydrogen bonded groups can be observed (e.g., "free" OH or "free" carbonyls). Each of these "free" groups is at *one* end of a chain (see figure 4.1) and thus represents a measure of the number of chains present (in mixtures where the second component (A) can hydrogen bond to functional groups of the first this is not always the case, but we will consider this later). If we determine from our spectroscopic measurements a "fraction free" (e.g., the fraction of C = O, or OH groups that are not hydrogen bonded) then this quantity is simply equal to the total number of chains divided by the total number of units or B molecules present in the system:

$$f_F = \frac{\displaystyle\sum_{h=1}^{\infty} n_{B_h}}{\displaystyle\sum_{h=1}^{\infty} n_{B_h} h} \tag{4.43}$$

where f_F is the fraction free and n_{B_h} is the number of chains that have h hydrogen bonded B units. We can express this in terms of volume fractions using

$$\Phi_{B_h} = \frac{n_{B_h} V_{B_h}}{V} = \frac{n_{B_h} h V_B}{V} \tag{4.44}$$

where V_{B_h} is the molar volume of an h-mer, V_B is the molar volume of an individual B molecules and V is the total volume of the system. In applying equation 4.44 we assume that there is no change in volume upon forming a hydrogen bond. It follows that:

$$\sum_{h=1}^{\infty} n_{B_h} = \frac{V}{V_B} \sum_{h=1}^{\infty} \frac{\Phi_{B_h}}{h} = \frac{V}{V_B} \Phi_{B_1} \sum (K_B \Phi_{B_1})^{h-1} \qquad (4.45)$$

and:

$$\sum_{h=1}^{\infty} n_{B_h} h = \frac{V}{V_B} \sum_{h=1}^{\infty} \Phi_{B_h} = \frac{V}{V_B} \Phi_{B_1} \sum_{h=1}^{\infty} h (K_B \Phi_{B_1})^{h-1} \qquad (4.46)$$

Again, if $K_B \Phi_{B_1} < 1$ we can use:

$$\sum_{h=1}^{\infty} h (K_B \Phi_{B_1})^{h-1} = \frac{1}{(1 - K_B \Phi_{B_1})^2} \qquad (4.47)$$

and, in addition:

$$\sum_{h=1}^{\infty} (K_B \Phi_{B_1})^{h-1} = \frac{1}{(1 - K_B \Phi_{B_1})} \qquad (4.48)$$

to obtain:

$$f_F = (1 - K_B \Phi_{B_1}) \qquad (4.49)$$

In pure B we let the fraction free be f_F°, so that:

$$f_F^\circ = (1 - K_B \Phi_{B_1}^\circ) \qquad (4.50)$$

[Kehiaian[13] has given a proof that $K_B \Phi_{B_1}$ is indeed less than 1].

We therefore have, from the equation describing the stoichiometry of the system and infrared measurements, the ability to determine K_B and the value of Φ_{B_1} at any composition. As we will see, this will allow us to obtain a measure of the free energy change associated with the change in hydrogen bonding as a function of composition.

Those familiar with the statistics of linear polycondensation reactions have probably already noticed a direct correspondence with this simple model of association. This is a natural consequence of the assumption that all steps in the association (equation 4.35) are governed by the same equilibrium constant, equivalent to the assumption in a polycondensation that the reactivity of end groups are independent of the size of the chain to which they are attached. The quantity we have defined as the fraction free is equal to the reciprocal of the number average "degree of polymerization", or, in this case degree of association, which can be written:

$$\frac{1}{f_F} = \bar{h} = \frac{1}{(1 - K_B \Phi_{B_1})} = \frac{1}{(1 - p_{BB})} \qquad (4.51)$$

where p_{BB} is the probability that a group has "reacted" and in our problem is equal to the fraction of groups that have formed hydrogen bonds, $K_B \Phi_{B_1}$.

Competing Equilibria

The analogy with polycondensation reactions brings us naturally to the next degree of complexity. In such reactions a "chain stopper" can be added to control the molecular weight of the chains. These have but one functional group and thus once reacted terminate the polymerization at that end of the chain. Similarly, molecules with ether, ester, ketone, basic nitrogen (e.g., pyridine) etc. functional groups can form hydrogen bonds with an OH group or , for example, the N - H part of an amide or urethane group, but are incapable of forming any subsequent hydrogen bonds, as illustrated above in figure 4.1 for an amide mixed with an ether. We will again assume that there is only one "active" (hydrogen bonding) functional group per molecule, but now at any particular concentration we have "competing" equilibria, between self association of the B units and the formation of hydrogen bonds between the B and A molecules. This can be written:

$$B_h + B_1 \underset{}{\overset{K_B}{\rightleftharpoons}} B_{h+1}$$

$$B_h + A_1 \underset{}{\overset{K_A}{\rightleftharpoons}} B_hA \qquad (4.52)$$

Now, of course, the distribution of h-mers is not modulated by dilution alone, but by the ability of B units to hydrogen bond to A and the strength of this interaction, measured by the equilibrium constant K_A, relative to self association, measured by K_B. These equilibrium constants are defined following Flory[9] as:

$$K_B = \frac{\Phi_{B_{h+1}}}{\Phi_{B_h} \Phi_{B_1}} \frac{h}{(h+1)} \qquad (4.53)$$

$$K_A = \frac{\Phi_{B_hA}}{\Phi_{B_h} \Phi_{A_1}} \frac{hr}{(h+r)} \qquad (4.54)$$

The equilibrium constant K_B is, of course, identical to that defined earlier and was obtained from a Flory lattice model. The constant K_A can be obtained in the same way (by considering mixing of the various B_h and B_hA species), but because in general the size of the A molecule differs from that of B (i.e., occupies a different number of lattice sites) we have to include the factor r equal to the ratio of the molar volumes of the A and B "monomer" (non-hydrogen bonded) molecules, V_A/V_B. The stoichiometry of the mixture is now given by:

$$\Phi_B = \sum_{h=1}^{\infty} \Phi_{B_h} h + \sum_{h=1}^{\infty} \Phi_{B_hA} \left(\frac{h}{h+r} \right) \qquad (4.55)$$

$$\Phi_A = \Phi_{A_1} + \sum_{h=1}^{\infty} \Phi_{B_hA} \left(\frac{r}{h+r} \right) \qquad (4.56)$$

where Φ_{A_1} is the volume fraction of A molecules that are *not* hydrogen bonded at any instant and like its counterpart, Φ_{B_1}, will form the basis for a description of the thermodynamics of mixing. Accordingly, we must relate these quantities to composition (Φ_A and Φ_B) and infrared spectroscopic measurements, so that they can be experimentally determined.

If we again assume that all steps in the self association of B are governed by the same equilibrium constant, K_B, we obtain for equation 4.55:

$$\Phi_B = \Phi_{B_1} \sum_{h=1}^{\infty} h \, (K_B \, \Phi_{B_1})^{h-1} + \frac{K_A \, \Phi_{A_1} \, \Phi_{B_1}}{r} \sum_{h=1}^{\infty} h \, (K_B \, \Phi_{B_1})^{h-1}$$

$$(4.57)$$

and by substituting from equations (4.47) and (4.48) we obtain:

$$\Phi_B = \frac{\Phi_{B_1}}{(1 - K_B \, \Phi_{B_1})^2} \left[1 + \frac{K_A \, \Phi_{A_1}}{r} \right] \qquad (4.58)$$

$$\Phi_A = \Phi_{A_1} \left[1 + \frac{K_A \, \Phi_{B_1}}{(1 - K_B \, \Phi_{B_1})} \right] \qquad (4.59)$$

We now need to determine four quantities, K_A, K_B, Φ_{A_1} and Φ_{B_1}, so we need at least two experimental infrared measurements to be used in conjunction with equations (4.58) and (4.59). The first measurement should be made in pure B. As we will see later, the fraction of free carbonyl groups is relatively easily measured and thus provides a measure of the number of end groups relative to the total number of molecules present. Hydroxyl groups present imposing problems as we will discuss in the following chapter. If we can make a suitable measurement using pure B, however, we obtain the result given above in equation 4.50. This, together with the equation for the stoichiometry of hydrogen bonding in pure B (4.42), can be used to obtain:

$$K_B = \frac{(1 - f_F^\circ)}{(f_F^\circ)^2} \qquad (4.60)$$

For our second measurement we must now distinguish between three situations.[12,14] First, we can measure the concentration of that part of a functional group that does not hydrogen bond to the competing A species. This is only possible in molecules like amides or urethanes, where we have an N-H and C=O component of the functional group. When mixed with, for example, an ether, hydrogen bonds between an N-H and ether oxygen can form at the expense of N-H to amide carbonyl hydrogen bonds (see figure 4.1), thus increasing the concentration of free carbonyls. The fraction of carbonyl groups that are free is given by:

$$f_F^I = \frac{\sum n_{B_h} + \sum n_{B_h A}}{\sum n_{B_h} h + \sum n_{B_h A} h} \tag{4.61}$$

which is the total number of B chains divided by the total number of B units, because there is always a free carbonyl at one end of the chain. We now make the following substitutions:

$$\sum_{h=1}^{\infty} n_{B_h} = \frac{V}{V_B} \sum_{h=1}^{\infty} \frac{\Phi_{B_h}}{h} = \left(\frac{V}{V_B}\right) \Phi_{B_1} \sum_{h=1}^{\infty} (K_B \Phi_{B_1})^{h-1} \tag{4.62}$$

$$\sum_{h=1}^{\infty} n_{B_h} h = \frac{V}{V_B} \sum_{h=1}^{\infty} \Phi_{B_h} = \left(\frac{V}{V_B}\right) \Phi_{B_1} \sum_{h=1}^{\infty} h (K_B \Phi_{B_1})^{h-1} \tag{4.63}$$

$$\sum_{h=1}^{\infty} n_{B_h A} = \frac{V}{V_B} \sum_{h=1}^{\infty} \frac{\Phi_{B_h}}{(h+r)} = \left(\frac{V}{V_B}\right) \Phi_{B_1} \frac{K_A \Phi_{A_1}}{r} \sum_{h=1}^{\infty} (K_B \Phi_{B_1})^{h-1} \tag{4.64}$$

$$\sum_{h=1}^{\infty} n_{B_h A} h = \frac{V}{V_B} \sum_{h=1}^{\infty} \frac{\Phi_{B_h} h}{(h+r)}$$

$$= \left(\frac{V}{V_B}\right) \Phi_{B_1} \frac{K_A \Phi_{A_1}}{r} \sum_{h=1}^{\infty} h (K_B \Phi_{B_1})^{h-1} \tag{4.65}$$

and using the relationships given in equations (4.47) and (4.48) we obtain:

$$f_F^I = (1 - K_B \Phi_{B_1}) \tag{4.66}$$

This is identical in form to equation (4.49) describing the mixing of B with inert diluent. In this case Φ_{B_1} depends upon K_A as well as K_B, however (equations 4.58 and 4.59). The equilibrium constant K_B is assumed to be the same in the mixture as in pure B, so that equations 4.58 and 4.59, describing the stoichiometry, and equations 4.60 and 4.66, relating infrared spectroscopic measurements to the concentrations of species, are in principle sufficient to define K_A, K_B, Φ_{A_1} and Φ_{B_1} (in practice, measurements at different compositions are made in order to improve the precision of the results).

The second type of measurement we can make would apply to a system containing hydroxyl groups mixed with, say, an ether, so that the only band amenable to analysis would be the free OH. This would be an exceedingly difficult measurement to make with any accuracy in the mid infrared range, for reasons we will discuss in Chapter 5 and also because the band due to free groups becomes vanishingly small as the concentration of the second component increases. The fraction of free groups is in this case just equal to the number of chains that *do not* have an A unit attached to one end, divided by the total number of B units present;

$$f_F^{II} = \frac{\displaystyle\sum_{h=1}^{\infty} n_{B_h}}{\displaystyle\sum_{h=1}^{\infty} n_{B_h} h + \sum_{h=1}^{\infty} n_{B_h A} h} \qquad (4.67)$$

Making the same substitutions as above (equations 4.62-4.65) we obtain:

$$f_F^{II} = \frac{(1 - K_B \Phi_{B_1})}{\left(1 + \dfrac{K_A \Phi_{A_1}}{r} \right)} \qquad (4.68)$$

Finally, we have the situation where the functional group on the competing species, i.e. the A molecule, is suitable for spectroscopic analysis (e.g., is the carbonyl of an ester or ketone). Experimentally, this type of measurement is usually relatively easy and in practice very useful. Here, the fraction free is simply the number of non-hydrogen bonded A molecules divided by the total number of A molecules and, converting to volume fractions, we obtain:

$$f_F^{III} = \frac{n_{A_1}}{n_A} = \frac{\Phi_{A_1}}{\Phi_A} \tag{4.69}$$

Temperature Dependence of the Equilibrium Constants

Obviously, the equilibrium constants, and hence the stoichiometry of hydrogen bonding, vary with temperature. This is simply described through the usual dependence on the enthalpy of hydrogen bond formation:

$$\frac{\partial(\ln K_B)}{\partial(1/T)} = -\frac{h_B}{R} \tag{4.70}$$

$$\frac{\partial(\ln K_A)}{\partial(1/T)} = -\frac{h_A}{R} \tag{4.71}$$

hence:

$$K_B = K_B^{\circ} \exp\left[-\frac{h_B}{R}\left(\frac{1}{T} - \frac{1}{T^{\circ}}\right)\right] \tag{4.72}$$

$$K_A = K_A^{\circ} \exp\left[-\frac{h_A}{R}\left(\frac{1}{T} - \frac{1}{T^{\circ}}\right)\right] \tag{4.73}$$

where K_B° and K_A° are the values of the equilibrium constants determined at T° K (usually 298°K). This assumes that the enthalpies of hydrogen bond formation are independent of temperature. For polymer mixtures we are concerned with the relatively limited range between the T_g and the degradation temperature and for these materials this seems to be a good assumption, as we will show in the following chapter.

Two Equilibrium Constant Model for Self-Association

It has been argued on theoretical grounds that for molecules that self-associate the equilibrium constant describing dimer formation should be different from that describing subsequent h-mer formation[15]. Experimental infrared spectroscopic studies of, for example, phenol[16,17] support this, as it is often not possible to reproduce the data obtained as a function of composition using a single equilibrium constant. Two equilibrium constants seem to suffice, however, one describing dimer formation (K_2) and one subsequent h-mer formation (K_B). This appears to be predominantly a problem in systems that contain OH groups. We speculate that this could be due to the direct connection of the donor and receiver groups in the chain of bonded units:

$$\begin{array}{ccc} & \text{O—H}\,''''''\text{O—H}\,''''''\text{O—H} \\ R^{/} & R^{/} & R^{/} \end{array}$$

where it is easy to envisage that there is a cooperative effect, such that the distribution of charges and hence hydrogen bond strength changes successively with each OH group that is added to the chain, but that this effect is quickly attenuated and it is an adequate approximation to assume that $K_3 = K_4 = K_5 - - = K_B$, but $K_2 \neq K_B$ (see equations 4.35 and 4.36 for a representation of each of these equilibria and a definition of the constants). In molecules containing amide or urethane groups for example, where the donor and acceptor atoms are separated by additional atoms:

$$\begin{array}{ccc} R & R & R \\ \diagdown & \diagup & \diagdown \\ \text{C}={O}''''''\text{H—N} & & \text{C}={O} \\ \diagup & \diagdown & \diagup \\ \text{H—N} & \text{C}={O}''''''\text{H—N} \\ \diagdown & \diagup & \diagdown \\ R & R & R \end{array}$$

one would anticipate that such affects are smaller, and this seems to be the case, a single self-association constant usually fits the data (see Chapter 5). The equations describing the stoichiometry of binary mixtures when dimer formation in the self-associating component is described by a different equilibrium constant than the addition of each succeeding unit is straightforward and has been used by a number of authors to describe the self-association of alcohols in various solvents (see ref 8 and citations therein). We write the equilibria describing dimer and subsequent h-mer (h > 2) formation in the self-associating species as:

$$B_1 + B_1 \;\xrightleftharpoons{K_2}\; B_2 \tag{4.74}$$

$$B_h + B_1 \;\xrightleftharpoons{K_B}\; B_{h+1} \;\; (h \geq 2) \tag{4.75}$$

where K_2 and K_B are defined as above:

$$K_2 = \frac{\Phi_{B_2}}{2\,\Phi_{B_1}^2} \tag{4.76}$$

$$K_B = \frac{\Phi_{B_{h+1}}}{\Phi_{B_h}\,\Phi_{B_1}} \cdot \frac{h}{h+1} \tag{4.77}$$

For the competing equilibrium:

$$B_h + A_1 \;\xrightleftharpoons{K_A}\; B_hA \tag{4.78}$$

where the A unit makes no distinction between forming a hydrogen bond to a dimer or an h-mer, we have:

$$K_A = \frac{\Phi_{B_hA}}{\Phi_{B_h}\,\Phi_{A_1}} \cdot \frac{hr}{(h+r)} \tag{4.79}$$

where $r = V_A/V_B$.

The stoichiometric relationships are again simply obtained from materials balance considerations. The total volume fraction of all B units present in the mixture is given by:

$$\Phi_B = \Phi_{B_1} + \sum_{h=2}^{\infty} \Phi_{B_h} + \sum_{h=1}^{\infty} \Phi_{B_h A}\left(\frac{h}{h+r}\right) \quad (4.80)$$

We note that for formation of dimer, trimer, ..., h-mer, according to equation 4.76 and 4.77 we have:

$$\Phi_{B_2} = 2 K_2 \Phi_{B_1}^2$$

$$\Phi_{B_3} = \frac{3}{2} K_B \Phi_{B_2} \Phi_{B_1}$$

$$\Phi_{B_h} = \frac{h}{h-1} K_B \Phi_{B_{h-1}} \Phi_{B_1} \quad (4.81)$$

By successive substitution of $\Phi_{B_{h-1}}, \Phi_{B_{h-2}}, ..., \Phi_{B_2}$, we obtain:

$$\Phi_{B_h} = h K_B^{h-2} \Phi_{B_1}^{h-2} K_2 \Phi_{B_1}^2 \quad (4.82)$$

hence:

$$\Phi_{B_h} = \frac{K_2}{K_B^2} h (\Phi_{B_1} K_B)^h \quad (4.83)$$

and:

$$\sum_{h=2}^{\infty} \Phi_{B_h} = \frac{K_2}{K_B^2} \sum_{h=2}^{\infty} h (K_B \Phi_{B_1})^h \quad (4.84)$$

However, for $K_B \Phi_{B_1} < 1$:

$$\sum_{h=1}^{\infty} h (K_B \Phi_{B_1})^{h-1} = \frac{1}{(1 - K_B \Phi_{B_1})^2} \quad (4.85)$$

So we rewrite equation 4.84 as:

$$\sum_{h=2}^{\infty} \Phi_{B_h} = \frac{K_2}{K_B^2} \Phi_{B_1} K_B \sum_{h=1}^{\infty} h \left(K_B \Phi_{B_1}\right)^{h-1} - \frac{K_2}{K_B^2} 1 \left(K_B \Phi_{B_1}\right)^1$$

(4.86)

Hence:

$$\sum_{h=2}^{\infty} \Phi_{B_h} = - \frac{K_2}{K_B} \Phi_{B_1} + \frac{K_2}{K_B} \Phi_{B_1} \left[\frac{1}{\left(1 - K_B \Phi_{B_1}\right)^2} \right]$$

(4.87)

Similarly:

$$\sum_{h=1}^{\infty} \Phi_{B_h A} \left(\frac{h}{h+r} \right) = \sum_{h=1}^{\infty} K_A \Phi_{B_h} \Phi_{A_1} \left(\frac{h+r}{hr} \right) \left(\frac{h}{h+r} \right)$$

$$= \frac{K_A \Phi_{A_1}}{r} \left[\Phi_{B_1} + \sum_{h=2}^{\infty} \Phi_{B_h} \right]$$

(4.88)

Substituting:

$$\Phi_B = \Phi_{B_1} \left[\left(1 - \frac{K_2}{K_B} \right) + \frac{K_2}{K_B} \left(\frac{1}{\left(1 - K_B \Phi_{B_1}\right)^2} \right) \right] \left[1 + \frac{K_A \Phi_{A_1}}{r} \right]$$

(4.89)

Similarly, from:

$$\Phi_A = \Phi_{A_1} + \sum_{h=1}^{\infty} \Phi_{B_h A} \left(\frac{r}{h+r} \right)$$

(4.90)

and using:

$$\sum_{h=1}^{\infty} \left(K_B \Phi_{B_1}\right)^{h-1} = \frac{1}{\left(1 - K_B \Phi_{B_1}\right)}$$

(4.91)

we obtain:

$$\Phi_A = \Phi_{A_1} + K_A \Phi_{A_1} \Phi_{B_1} \left[\left(1 - \frac{K_2}{K_B} \right) + \frac{K_2}{K_B} \left(\frac{1}{\left(1 - K_B \Phi_{B_1}\right)} \right) \right]$$

(4.92)

We note that for $K_2 = K_B$, equation 4.89 and 4.92 reduce to equation 4.58 and 4.59 as they must.

As before, we can relate infrared spectroscopic parameters to the stoichiometry of hydrogen bonding. We first make the substitutions:

$$\Gamma_1 = \left(1 - \frac{K_2}{K_B}\right) + \frac{K_2}{K_B}\left(\frac{1}{(1 - K_B \Phi_{B_1})}\right) \tag{4.93}$$

$$\Gamma_2 = \left(1 - \frac{K_2}{K_B}\right) + \frac{K_2}{K_B}\left(\frac{1}{(1 - K_B \Phi_{B_1})^2}\right) \tag{4.94}$$

and note that for $K_2 = K_B$ we obtain:

$$\frac{\Gamma_1}{\Gamma_2} = (1 - K_B \Phi_{B_1}) \tag{4.95}$$

Deriving expressions for the quantities f_F^o, f_F^I, f_F^{II} and f_F^{III} in exactly the same manner as in the previous section we obtain:

$$f_F^o = \frac{\Gamma_1^o}{\Gamma_2^o} \tag{4.96}$$

$$f_F^I = \frac{\Gamma_1}{\Gamma_2} \tag{4.97}$$

$$f_F^{II} = \frac{\Gamma_1}{\Gamma_2}\left[\frac{1}{\left(1 + \dfrac{K_A \Phi_{A_1}}{r}\right)}\right] \tag{4.98}$$

$$f_F^{III} = \frac{\Phi_{A_1}}{\Phi_A} \tag{4.99}$$

Equation 4.99 is, of course, identical to that obtained previously, while the correspondence of equations 4.96 - 4.99 to the single equilibrium constant self-association model is apparent from equation 4.95.

Although it would seem that we could obtain values of the equilibrium constants fairly easily, given that we could obtain values of the fraction free as a function of composition, thus having many more data points than unknowns, introduction of a second equilibrium constant for self-association enormously magnifies the effect of inevitable experimental errors and makes the determination of equilibrium constants a more complicated problem, not least because K_2 and K_B can no longer be fixed by data from the pure self-associating polymer alone. This is discussed in more detail in Chapter 5.

Carboxylic Acids and Urazoles–Formation of Cyclic Species

Among the self-associating functional groups we are considering, carboxylic acids and urazoles are unique in that in most situations they hydrogen bond in closed rings to form pairs, as illustrated for carboxylic acids in figure 4.2 (open chains may also be present at high concentrations of acids). Also illustrated in this figure is a hydrogen bond formed between an acid group and an ether, which is characterized by the formation of a free carbonyl. The stoichiometry of hydrogen bonding and the relationship to infrared spectroscopic measurements follows in the same manner as above.[18] We first define equilibrium constants for the competing equilibria:

$$B_1 + B_1 \quad \overset{K_B}{\rightleftharpoons} \quad B_2 \tag{4.100}$$

$$B_1 + A \quad \overset{K_A}{\rightleftharpoons} \quad BA \tag{4.101}$$

where for consistency we let B represent the self-associating species, while in the example shown in figure 4.2, A would represent the ether containing molecules. The equilibrium constants, K_A and K_B, are defined in terms of volume fractions as:

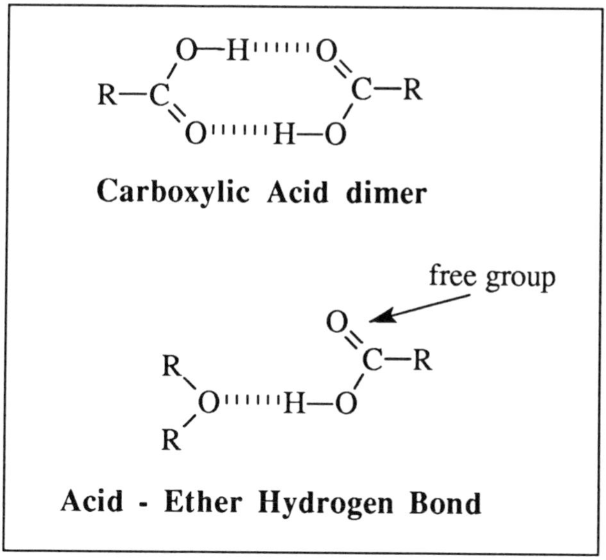

Carboxylic Acid dimer

free group

Acid - Ether Hydrogen Bond

FIGURE 4.2

$$K_B = \frac{\Phi_{B_2}}{\Phi_{B_1}^2} \frac{1}{2}$$ (4.102)

$$K_A = \frac{\Phi_{BA}}{\Phi_{B_1} \Phi_{A_1}} \left[\frac{r}{1+r} \right]$$ (4.103)

where Φ_{B_1}, Φ_{B_2}, Φ_{A_1} and Φ_{BA} are the volume fractions of the carboxylic acid monomers, carboxylic acid dimers, unassociated ethers and associated carboxylic acid-ether groups, respectively. As above the parameter r is the ratio of the molar volumes of the molecules, V_A/V_B. Using the expression describing the mass balance:

$$\Phi_{B_1} + \Phi_{B_2} + \Phi_{A_1} + \Phi_{BA} = 1$$ (4.104)

we arrive at the following equations describing the stoichiometry of the system:

$$\Phi_B = \Phi_{B_1}\left[1 + \frac{K_A \, \Phi_{A_1}}{r}\right] + 2\,K_B\,(\Phi_{B_1})^2 \qquad (4.105)$$

$$\Phi_A = \Phi_{A_1}\,(1 + K_A\,\Phi_{B_1}) \qquad (4.106)$$

where Φ_A and Φ_B are the volume fractions of non self-associating species A and self-associating species B, respectively. For the pure acid (superscript °), $\Phi_B = 1$ and equation (4.105) reduces to:

$$1 = \Phi_{B_1}^{\circ} + 2\,K_B\,(\Phi_{B_1}^{\circ})^2 \qquad (4.107)$$

Accordingly, since the negative root has no physical meaning:

$$\Phi_{B_1}^{\circ} = \frac{-1 + \sqrt{1 + 8K_B}}{4\,K_B} \qquad (4.108)$$

The equilibrium constants, K_A and K_B, are again determined directly from infra-red spectroscopic measurements of the fraction of "free" (non-hydrogen bonded) carboxylic acid carbonyl groups, f_F. The fraction of "free" carbonyl groups in the pure acid, assuming no volume change on forming an acid dimer, is simply given by:

$$f_F^{\circ} = \frac{n_{B_1}}{n_{B_1} + 2n_{B_2}} \qquad (4.109)$$

where n_{B_1} is the number of "monomers" and n_{B_2} is the number of dimers present at a particular instant. Substituting volume fractions for these quantities ($\Phi_{B_1} = n_{B_1} V_B/V$, $\Phi_{B_2} = 2n_{B_2} V_B/V$) we obtain:

$$f_F^{\circ} = \frac{\Phi_{B_1}^{\circ}}{\Phi_{B_1}^{\circ} + \Phi_{B_2}^{\circ}} \qquad (4.110)$$

Since, $\Phi_{B_1}^{\circ} + \Phi_{B_2}^{\circ} = 1$, the fraction of "free" carbonyls is given by $f_F^{\circ} = \Phi_{B_1}^{\circ}$ and using equation (4.102):

$$K_B = \frac{\left[1 - f_F^o\right]}{2\left[f_F^o\right]^2} \qquad (4.111)$$

In the blend the fraction of free carbonyls is given by:

$$f_F = \frac{n_{AB} + n_{B_1}}{2n_{B_2} + n_{AB} + n_{B_1}} \qquad (4.112)$$

which in terms of volume fractions is:

$$f_F = \frac{\Phi_{B_1} + \dfrac{\Phi_{AB}}{(1+r)}}{\Phi_{B_1} + \Phi_{B_2} + \dfrac{\Phi_{AB}}{(1+r)}} \qquad (4.113)$$

Using the equilibrium constant definitions:

$$f_F = \frac{\left(1 + \dfrac{K_A \Phi_{A_1}}{r}\right)}{\left[2 K_B \Phi_{B_1} + \left(1 + \dfrac{K_A \Phi_{A_1}}{r}\right)\right]} \qquad (4.114)$$

so that once more the equations for the stoichiometry of hydrogen bonding can be used with infrared measurements to obtain values of the crucial parameters, K_A, K_B, Φ_{B_1}, and Φ_{A_1}. [Note that there is a typographical error in equation 11 of reference 18].

D.   HYDROGEN BONDING IN POLYMERS

In the preceding sections we have used an association model to describe the stoichiometry of hydrogen bonding in molecules that each contain just one functional group. The materials we are primarily interested in, however, are polymeric, containing a multitude of such groups arranged in a regular manner along a homopolymer chain, or interspersed with other units in a copolymer. Describing the associations of entire covalent chains, with their multiplicities of interacting groups and diverse morphologies, might at first sight seem a hopeless task, but immediately becomes manageable if first we focus attention on a chemical repeat unit, or some arbitrary portion of the covalent chain defined so as to contain just one hydrogen bonding functional group; and second, we only consider amorphous polymers at temperatures above their glass transitions. The units of a particular covalent chain are then immersed in a jostling sea of other chains and their nearest neighbors are most often elements of these other chains. It would therefore seem reasonable to assume that each functional group of a polymer in this state can form hydrogen bonds according to its intrinsic proclivities. By this we mean that its associations can be described by the equilibrium constants defined in the preceding sections, now defined in terms of the concentration of a repeat unit. Indeed, later in this book we will argue on theoretical grounds that *as long as the chain is flexible*, then the equilibrium constant describing the association of its repeat units is equal to that of a "small" molecule (one containing a single functional group) of equal molar volume and is independent of the covalent polymer chain length. This, in turn, would seem to be a reasonable assumption for many polymers of interest, in so much as the ability to hydrogen bond is localized in a particular functionality, such as the amide group of a nylon illustrated in figure 4.3, and such groups are separated from each other by a number of backbone bonds (in this case linking methylene units) that taken together confer considerable flexibility, as in this example measured from amide group to amide group.

Even if this were not the case, just the fact of segmental motion in an amorphous material would mean that some sort of equilibrium will be attained. Accordingly, we can still apply an association model, but the equilibrium constants describing the distribution of species present would differ from those describing hydrogen bonding in a small molecule similar in size to the polymer repeat unit. It follows that the general description of the stoichiometry of hydrogen bonding and the relationship to experimental measurements given in the preceding section can still be applied to segments of polymer chains, each defined so as to contain one functional group. This will be justified later, both on theoretical grounds and by experimental studies of a large number of systems. First, however, we wish to return to a consideration of a description of the association of "small" molecules using a Flory lattice model. There are a number of reasons for this, but the most immediate can be ascertained from an examination of figure 4.3, where a nylon repeat unit containing 5 CH_2 units is depicted. Assuming for the moment that we can obtain an equilibrium constant describing the self-association of such units, how would this compare to the equilibrium constant describing the self-association of a nylon containing 10 CH_2 units in every repeat? The interaction strength between the amide groups should not differ significantly in these two polymers, but in the second the number of amide groups per unit volume is considerably less (i.e., is diluted). As one might therefore expect, the equilibrium constant should be modified according to a ratio of the molar volumes of the two repeat units.

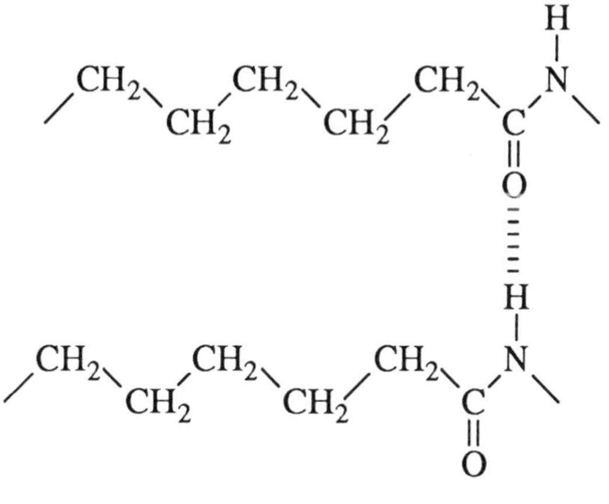

Self-Association in Nylons

FIGURE 4.3

We can use the Flory lattice model to derive this explicitly and at the same time derive an equation for the free energy of mixing associating species that will later form an element of our treatment of polymers.

E. THE FLORY LATTICE MODEL FOR MIXING HETEROGENEOUS POLYMERS

Association models have been successful in describing the behavior of alcohols in hydrocarbon solvents[1-8], but in these studies equilibrium constants have usually been obtained by a reduction of phase equilibria and enthalpy data. As a result, different values of self-association constants have been obtained, including some that varied with the nature of the diluent. Brandani[19] pointed out that this is inconsistent with the statistical mechanics of these mixtures, as the equilibrium constant should be a property of the associating component (of course, this assumes we are not comparing diluents of different character, say cyclohexane and carbon tetrachloride; this latter solvent can apparently form a specific interaction thus affecting experimentally determined equilibrium constants considerably - see Chapter 5). This author then presented a treatment, based on the Flory lattice model[9], where each molecule is assumed to behave as a set of segments, and found that this solved a number of difficulties. We will now consider in some detail a minor extension of the Flory treatment, as this will allow us to define equilibrium constants that are, in principle, transferable between similar molecules, providing that we account for differences in molar volume. Furthermore, this will give us an initial expression for the free energy of mixing, which we will later modify to a more useful form.

Lattice Model for Simple Self-Association

In the Flory treatment the entropy of mixing is obtained by a lattice filling procedure and a mean field interaction term is then added to obtain the free energy, as discussed in Chapter 1. The chemical potentials can then determined and this leads to a definition of equilibrium constants for polymers that form by reversible association, as described in section 4B.

Because interaction terms, normally expressed in terms of a χ parameter, cancel from this final expression we will initially neglect them and simply concentrate on an expression for the entropy of mixing. The situation we will consider is a set of h-mers (i.e., molecules consisting of h units hydrogen bonded to form a chain), but we will modify the Flory methodology to account for the attachment of an equal number of segments to each repeat unit, as illustrated schematically in figure 4.4.

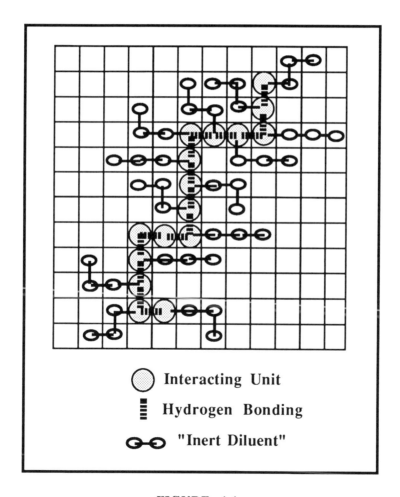

FIGURE 4.4

This allows us to define a lattice cell size in terms of the molar volume of an "interacting unit". For example, figure 4.5 shows a chain of hydrogen bonded ethyl alcohol molecules. We presume that the "interacting unit" in this case is the hydroxyl group, or perhaps the ($-CH_2-OH$) group. *The precise molecular identity of this species is unimportant,* as ultimately we will obtain parameters in terms of the molar volumes of the molecules under consideration. This definition is a useful device in the derivation, however, and allows us to determine how equilibrium constants change in, for example, the n-alcohol series, so that experimental values determined for ethanol could be applied to propanol, butanol, etc., assuming, of course, that the model is correct.

Accordingly, if the molar volume of the self-associating molecule under consideration is V_B, and the molar volume of the "interacting unit" is V_β, then each molecule has $s_B = V_B/V_\beta$ segments and in a hydrogen bonded chain of length h there are hs_B segments, as illustrated in figure 4.6. There are two things to note; first that we are confining our treatment to "simple" molecules, by which we mean that there is only one interacting functional group per molecule. Second, we will treat the hydrogen bonded chain of molecules as if it consists of covalently linked segments, ignoring the dynamic nature of the bonds characteristic of behavior at ambient temperatures.

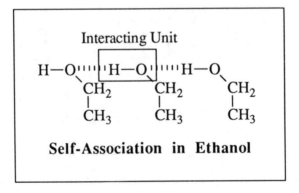

Self-Association in Ethanol

FIGURE 4.5

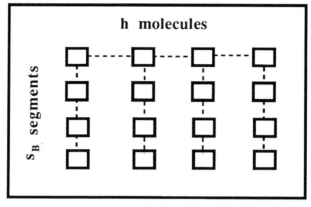

FIGURE 4.6

We believe that this is perfectly valid as long as we are dealing with the equilibrium distribution of chains or h-mers present at a particular temperature and composition and we will return to this point later. We will assume, for initial simplicity, that the h-mers are mixed with "inert" (i.e. non-hydrogen bonding) solvent molecules (A) that are equal in size to the interacting unit, i.e., each occupies one lattice site. The situation where the components differ in size and the species A have functional groups that can also hydrogen bond will be treated later.

The Flory methodology[9] is based on a determination of the number of configurations available to the jth chain, after $j-1$ chains have previously been placed randomly on the lattice. This procedure involves various approximations and assumptions, discussed in the book by Guggenheim[19]. It is beyond the scope and purpose of this book to consider these in any detail. We will mention some key assumptions in passing and for the rest assume that the Flory result is a reasonable approximation.

After $j-1$ chains have been added to the lattice the number of sites available for the placement of the first unit of the jth chain is:

$$n_o - \sum_{i=1}^{j-1} h_i s_B$$

where n_o is the total number of lattice sites and h_i is the number of molecules in the ith hydrogen bonded chain (recall that there are s_B segments per molecule). We first (arbitrarily) consider the placement of the remaining (s_B -1) segments of the first molecule before proceeding to the second molecule. If there are z adjacent lattice sites (z = lattice coordination number), then the number of ways of placing the second segment is equal to z times the probability that a site adjacent to the first is vacant. In the Flory approximation the probability that a site is vacant is simply assumed to be equal to the fraction of empty sites, which after the first segment is placed is equal to:

$$\left[\frac{n_o - \sum_{i=1}^{j-1} h_i s_B - 1}{n_o} \right]$$

The number of ways of placing the third segment is:

$$(z - 1) \left[\frac{n_o - \sum_{i=1}^{j-1} h_i s_B - 2}{n_o} \right]$$

and so on to the final segment of the first molecule:

$$(z - 1) \left[\frac{n_o - \sum_{i=1}^{j-1} h_i s_B - (s_B - 1)}{n_o} \right]$$

The first segment of the second molecule can be placed in

$$(z - 1) \left[\frac{n_o - \sum_{i=1}^{j-1} h_i s_B - s_B}{n_o} \right]$$

ways, but the first segment of the <u>third</u> molecule in only

$$(z - 2) \left[\frac{n_o - \sum_{i=1}^{j-1} h_i s_B - 2s_B}{n_o} \right]$$

ways. For simplicity of presentation we will assume $(z-1) \sim (z-2)$, although this is not necessary in so much as terms in z do not appear in the final result. Accordingly, the probability of finding a vacant site after $(\xi-1)$ segments have been added to the lattice is:

$$\frac{\left(n_o - \sum_{i=1}^{j-1} h_i s_B - (\xi - 1) \right)}{n_o}$$

so that we can write an expression for the number of configurations available to the jth chain in terms of this last factor, with ξ running from 2 to $h_j s_B$:

$$\upsilon_j = \frac{z}{(z-1)} (z-1)^{h_j s_B - 1} \times \frac{1}{\sigma}$$

$$\times \left(n_o - \sum_{i=1}^{j-1} h_i s_B \right)^{h_j s_B} \prod_{\xi=2}^{} \left[\frac{n_o - \sum_{i=1}^{j-1} h_i s_B - (\xi-1)}{n_o} \right] \tag{4.115}$$

where a factor σ equal to the symmetry number of the chain has been added.

This expression can be written in a more useful form by including the $(n_o - \Sigma h_i s_B)$ term in the product term, so that its numerator now reads:

$$\left(n_o - \sum_{i=1}^{j-1} h_i s_B \right) \left(n_o - \sum_{i=1}^{j-1} h_i s_B - 1 \right) - - - - \left(n_o - \sum_{i=1}^{j-1} h_i s_B - (h_j s_B) \right)$$

which is equal to:

$$\frac{\left(n_o - \sum_{i=1}^{j-1} h_i s_B\right)!}{\left(n_o - \sum_{i=1}^{j-1} h_i s_B - h_j s_B\right)!} = \frac{\left(n_o - \sum_{i=1}^{j-1} h_i s_B\right)!}{\left(n_o - \sum_{i=1}^{j} h_i s_B\right)!}$$

Hence, assuming $z \sim (z-1)$ we obtain:

$$\upsilon_j = \frac{(z-1)^{h_j s_B - 1}}{\sigma} \left\{ \frac{1}{n_o^{(h_j s_B - 1)}} \frac{\left(n_o - \sum_{i=1}^{j-1} h_i s_B\right)!}{\left(n_o - \sum_{i=1}^{j} h_i s_B\right)!} \right\} \qquad (4.116)$$

The total number of configurations available to the $\sum_{h=1}^{\infty} n_{B_h}$ molecules (i.e. all the h-mers present in the system) is then given by:

$$\Omega = \frac{1}{\prod_{h=1}^{\infty} n_{B_h}!} \left[\prod_{j=1}^{\Sigma n_{B_h}} \upsilon_j \right] \qquad (4.117)$$

We substitute equation 4.116 into equation 4.117 and note that the factorial terms can be simplified once their product over all h-mer molecules is considered:

$$\frac{\left(n_o - \sum_{i=1}^{0} h_i s_B\right)! \left(n_o - \sum_{i=1}^{1} h_i s_B\right)! - - - - - -}{\left(n_o - \sum_{i=1}^{1} h_i s_B\right)! - - - - - - \left(n_o - \prod_{i=1}^{\Sigma n_{B_h}} h_i s_B\right)!}$$

so that:

$$\Omega = \frac{1}{\prod_{h=1}^{\infty} n_{B_h}!} \left\{ \frac{(z-1)^{\Sigma b}}{\sigma^{\Sigma a}} \times \frac{1}{n_o^{\Sigma b}} \times \frac{n_o!}{(n_o - \sum_{i=1}^{\Sigma a} h_i s_B)!} \right\} \qquad (4.118)$$

where

$$\Sigma a = \sum_{h=1}^{\infty} n_{B_h} \text{ and } \Sigma b = \sum_{j=1}^{\Sigma n_{B_h}} (h_j s_B - 1)$$

It will prove useful to change the summation terms by making the following substitutions:

$$\sum_{j=1}^{\Sigma n_{B_h}} h_j s_B = \sum_{i=1}^{\Sigma n_{B_h}} h_i s_B = \sum_{h=1}^{\infty} n_{B_h} h s_B \qquad (4.119)$$

and noting that:

$$n_o = n_A + \Sigma n_{B_h} h s_B \qquad (4.120)$$

where the limits of the summation terms are now omitted, as they invariably run from 1 to ∞. The entropy of mixing is given by:

$$\frac{\Delta S_m}{R} = \ln \Omega \qquad (4.121)$$

so that by using Stirling's approximation ($\ln N! = N \ln N - N$) we can obtain:

$$-\frac{\Delta S_m}{R} = \Sigma n_{B_h} \ln \frac{n_{B_h}}{n_o} + n_A \ln \frac{n_A}{n_o}$$
$$+ \Sigma n_{B_h}(h s_B - 1) - \left[\Sigma n_{B_h}(h s_B - 1)\ln(z-1) - \Sigma n_{B_h} \ln \sigma \right] \qquad (4122)$$

In terms of volume fractions:

$$-\frac{\Delta S_m}{R} = \sum n_{B_h} \ln \left(\frac{\Phi_{B_h}}{h \, s_B} \right) + n_A \ln \Phi_A$$

$$-\sum \left\{ n_{B_h} (h \, s_B - 1) \left[\ln (z-1) - 1 \right] \right\} + \sum n_{B_h} \ln \sigma \tag{4.123}$$

If, for now, we neglect "physical" interactions (e.g., van der Waals forces) between the h-mer chains, this expression is equal to $\Delta G_m/RT$. The enthalpy of hydrogen bond formation does not appear as a separate term in this equation because of the way it is derived (we will consider enthalpy charges in more detail later, in Chapter 6). The above equation is with respect to a reference state where the various h-mers (monomers, dimers etc) are *separate* (according to length) and *oriented* [9] and can thus be considered to consist of three parts:

1. an entropy of disorientation of the individual pure h-mers,
2. an entropy of mixing the various h-mers with one another, and
3. an entropy of mixing the h-mers with solvent.

The free energy of mixing the distribution of h-mers found in the mixture with solvent with respect to a reference state where the various polymers molecules are mixed with one another and disoriented can be obtained from equation 4.122 by letting $n_A = 0$ and by assuming that this *does not change* the distribution of h-mers (not a good assumption):

$$\frac{\Delta S_{a,b}}{R} = \sum n_{B_h} \ln (n_{B_h} / \sum n_{B_h} h s_B)$$

$$- \sum (n_{B_h} (h s_B - 1)) (\ln(z-1)-1) + \sum n_{B_h} \ln \sigma \tag{4.124}$$

Upon subtracting, we then obtain the familiar Flory expression for the combinatorial entropy of mixing:

$$-\frac{\Delta S_c}{R} = \sum n_{B_h} \ln \Phi_{B_h} + n_A \ln \Phi_A \tag{4.125}$$

where the volume fractions, Φ_B and Φ_A are simply:

$$\Phi_B = \frac{\sum_h n_{B_h} h \, s_B}{n_o}$$

$$\Phi_A = \frac{n_A}{n_o} \tag{4.126}$$

As we noted in section 4A, association models have been based on *both* reference states, the so-called Kretschmer-Wiebe[1] model on the reference state where the h-mers are separate and oriented (equation 4.123), while the Mecke-Kempter model[5,7] uses the disoriented and mixed molecules as a reference state (equation 4.125). On the basis of Flory's arguments[9], illustrated above, we prefer the Kretschmer-Wiebe model as this accounts for the variation of the distribution of species with concentration in a more consistent manner by including the variation in the entropy of disorientation as the degree of association changes with composition.

The chemical potentials can now be obtained in the usual way by differentiating with respect to composition. We require the chemical potential of the h-mer species:

$$\frac{\partial(\Delta G_m)}{\partial n_{B_h}} = \Delta\overline{G}_h = \mu_{B_h} - \mu_{B_h}^{ref} \tag{4.127}$$

Choosing the separate and oriented species as our reference state we can obtain by simple differentiation (keeping in mind that all terms in the summation have to be considered, because they depend on n_{B_h} through $n_o = n_A + \Sigma n_{B_h} h s_B$):

$$\frac{\mu_{B_h} - \mu_{B_h}^*}{RT} = \ln(\Phi_{B_h}/h\,s_B) + 1 + (h\,s_B - 1) - \frac{h\,s_B\,n_A}{n_o}$$

$$- \frac{h\,s_B\,\Sigma n_{B_h}}{n_o} - (h\,s_B - 1)\ln(z-1) + \ln\sigma \tag{4.128}$$

If we let the number average degree of polymerization of the hydrogen bonded chain be:

$$\bar{h} = \frac{\Sigma n_{B_h} h}{\Sigma n_{B_h}} \qquad (4.129)$$

then we obtain:

$$\frac{\mu_{B_h} - \mu_{B_h}^*}{RT} = \ln (\Phi_{B_h} / h \, s_B) + \Phi_B \, h \, s_B \left(1 - \frac{1}{s_B \, \bar{h}} \right)$$

$$- (h \, s_B - 1) \ln (z - 1) + \ln \sigma \qquad (4.130)$$

With respect to the reference state where the h-mers are mixed with one another and disoriented (i.e., from equation 4.125) we obtain the result most often used to describe *covalent* chains or used in the Mecke-Kempter model for reversible association:

$$\frac{\mu_{B_h} - \mu_{B_h}^\circ}{RT} = \ln \Phi_{B_h} + (1 - h \, s_B) + \Phi_B \, h \, s_B \left(1 - \frac{1}{s_B \, \bar{h}} \right)$$

$$(4.131)$$

Lattice Model for Competing Equilibria

Many of the systems that are of interest consist of mixtures where one component self-associates while the second does not, but has a functional group capable of hydrogen bonding with those of the first component. Accordingly, we will now briefly consider a lattice model for this situation. This also provides a basis for the definition of the equilibrium constant K_A, defined above, in terms of a segment size defined by the molar volume of the "interacting unit" of the self-associating molecule.

A representation of a chain of B molecules which is hydrogen bonded at one end to an A molecule is shown in figure 4.7. In general, this latter molecule will not be the same size as a B molecule and we let its number of segments be equal to s_A. The lattice cell size is defined by the volume of the "interacting unit", whatever we define that to be. Flory's lattice filling procedure can now be applied by considering the mixture to

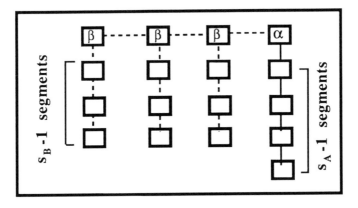

FIGURE 4.7

consist of three classes of distinguishable species; chains of self-associating B units, chains of B units with an A unit at one end, and molecules of A units that are not hydrogen bonded.

By filling the lattice successively with the molecules of these species we obtain:

$$
-\frac{\Delta S_m}{R} = \Sigma n_{B_h} \ln \frac{n_{B_h}}{n_o} + \Sigma n_{B_h A} \ln \frac{n_{B_h A}}{n_o} + n_{A_1} \ln \frac{n_{A_1}}{n_o}
$$

$$
+ \Sigma n_{B_h} (hs_B - 1) + \Sigma n_{B_h A} (hs_B + s_A - 1)
$$

$$
+ \{terms\ in\ z, \sigma\} \tag{4.132}
$$

where the number of lattice sites, n_o is equal to $\Sigma n_{B_h} hs_B + \Sigma n_{B_h A} (hs_B + s_A) + n_{A_1} s_A$, n_{B_h} is the number of chains of B units of length h, and $n_{B_h A}$ is the number of chains that have h B units and one A unit attached at an end.

In terms of volume fractions:

$$
-\frac{\Delta S_m}{R} = \Sigma n_{B_h} \ln (\Phi_{B_h}/hs_B) + \Sigma n_{B_h A} \ln (\Phi_{B_h A}/(hs_B + s_A))
$$

$$
+ n_{A_1} \ln (\Phi_{A_1}/s_A) + \Sigma n_{B_h} (hs_B - 1) + \Sigma n_{B_h A} (hs_B + s_A - 1)
$$

$$
+ \{terms\ in\ z, \sigma\} \tag{4.133}
$$

The chemical potentials of the various species (with respect to Flory's reference state) is given by:

$$\frac{\mu_{B_h} - \mu_{B_h}^*}{RT} = \ln(\Phi_{B_h}/hs_B) - hs_B \left\{ \frac{\Sigma n_{B_h} + \Sigma n_{B_h A} + n_{A_1}}{n_o} \right\}$$

$$+ hs_B + \{\text{terms in } z,\sigma\} \tag{4.134}$$

$$\frac{\mu_{B_h A} - \mu_{B_h A}^*}{RT} = \ln(\Phi_{B_h A}/(hs_B + s_A)) - (hs_B + s_A) \left\{ \frac{\Sigma n_{B_h} + \Sigma n_{B_h A} + n_{A_1}}{n_o} \right\}$$

$$+ (hs_B + s_A) + \{\text{terms in } z,\sigma\} \tag{4.135}$$

and:

$$\frac{\mu_{A_1} - \mu_{A_1}^*}{RT} = \ln(\Phi_{A_1}/s_A) - s_A \left\{ \frac{\Sigma n_{B_h} + \Sigma n_{B_h A} + n_{A_1}}{n_0} \right\}$$

$$+ s_A + \{\text{terms in } z,\sigma\} \tag{4.136}$$

F. "TRANSFERABLE" EQUILIBRIUM CONSTANTS

In the preceding section we have considered at some length the derivation of expressions for the mixing of hydrogen bonded chains. We divided the associated molecules into segments according to the molar volume of an "interacting unit", which in an amide, for example, might consist of the -CONH- group. The precise identity of this unit is ultimately immaterial, as we are simply using some arbitrary unit to define the lattice cell size. As we will now show, this is an extremely useful device for demonstrating that we can define an equilibrium constant characteristic of a particular hydrogen bonding functional group, which in a series of *similar* molecules will then simply vary with molar volume. This will be an essential tool in our treatment of copolymers.

We first assume that we can define an equilibrium constant according to Flory[9], but in terms of the "interacting units" only, depicted as β and α in figure 4.7. In other words, the attached segments are "cut free" and act as "inert" diluent.

We then have:

$$\beta_h + \beta_1 \;\underset{\longleftarrow}{\overset{K_\beta}{\longrightarrow}}\; \beta_{h+1}$$

$$K_\beta = \frac{\Phi_{\beta_{h+1}}}{\Phi_{\beta_h} \Phi_{\beta_1}} \cdot \frac{h}{h+1} \qquad (4.137)$$

and:

$$\beta_h + \alpha \;\underset{\longleftarrow}{\overset{K_\alpha}{\longrightarrow}}\; \beta_h\alpha$$

$$K_a = \frac{\Phi_{\beta_h\alpha}}{\Phi_{\beta_h} \Phi_\alpha} \cdot \frac{hr_i}{h+r_i} \qquad (4.138)$$

where $r_i = V_\alpha/V_\beta$.

We now convert the volume fraction terms from "interacting units" to molecules. Starting with the following definitions of volume fractions:

$$\Phi_{\beta_h} = \frac{n_{\beta_h} V_{\beta_h}}{V} \qquad (4.139)$$

$$\Phi_{B_h} = \frac{n_{B_h} V_{B_h}}{V} \qquad (4.140)$$

where Φ_{β_h} is the volume fraction of self-associating interacting units present as h-mers, Φ_{B_h} is the corresponding volume fraction of molecules present as these h-mers, n_{β_h} is the number of hydrogen-bonded chains of interacting units of length h, n_{B_h} is the corresponding number of chains of molecules associated in chains of length h, V is the total volume of the sample, finally, V_{β_h} and V_{B_h} are the molar volumes of a chain of h interacting units or chemical repeat units, respectively. Assuming no change in volume upon association:

$$V_{\beta_h} = h \, V_\beta \tag{4.141}$$

$$V_{B_h} = h \, V_B \tag{4.142}$$

Because we are confining our discussion to the most common situation of one interacting unit per molecule or chemical repeat:

$$n_{\beta_h} = n_{B_h} \tag{4.143}$$

Hence:

$$\Phi_{\beta_h} = \Phi_{B_h} \frac{V_{\beta_h}}{V_{B_h}} = \Phi_{B_h} \frac{V_\beta}{V_B} = \frac{\Phi_{B_h}}{s_B} \tag{4.144}$$

Substituting equivalent definitions for $\Phi_{\beta_{h+1}}$ and Φ_{β_1} into equation 4.137, we obtain:

$$\frac{K_\beta}{s_B} = \frac{\Phi_{B_{h+1}}}{\Phi_{B_h} \Phi_{B_1}} \cdot \frac{h}{h+1} \tag{4.145}$$

Our previous definition of the equilibrium constant in terms of chemical repeat units, K_B, is thus equal to the constant K_β/s_B defined here. A similar result is obtained for the equilibrium constant K_α describing interactions with the competing species:

$$\Phi_{\beta_h A} = \frac{n_{\beta_h A} (V_{\beta_h A} + V_\alpha)}{V} \tag{4.146}$$

$$\Phi_{B_h A} = \frac{n_{B_h A} (V_{B_h A} + V_A)}{V} \tag{4.147}$$

Using $n_{\beta_h \alpha} = n_{B_h A}$:

$$\Phi_{\beta_h \alpha} = \Phi_{B_h A} \left[\frac{h + r_i}{h \, s_B + V_A/V_B} \right] \tag{4.148}$$

Writing similar expressions for Φ_{β_h} and Φ_{α_1}, in terms of Φ_{B_h} and Φ_{A_1}, it follows that:

$$K_\alpha = \frac{\Phi_{B_h A}}{\Phi_{B_h} \Phi_{A_1}} \cdot \frac{s_B \, s_A \, h \, r_i}{h \, s_B + V_A/V_\beta} = \frac{\Phi_{B_h A}}{\Phi_{B_h} \Phi_{A_1}} \cdot \frac{s_A \, h \, r_i}{h + V_A/V_B}$$

$$(4.149)$$

Noting that:

$$\frac{s_A r_i}{s_B} = \frac{V_A}{V_\alpha} \cdot \frac{V_\alpha}{V_\beta} \cdot \frac{V_\beta}{V_B} = \frac{V_A}{V_B} \qquad (4.150)$$

Then:

$$\frac{K_\alpha}{s_B} = \frac{\Phi_{B_h A}}{\Phi_{B_h} \Phi_{A_1}} \cdot \frac{h \, (V_A/V_B)}{h + (V_A/V_B)} \qquad (4.151)$$

Substituting $r = V_A/V_B$ and writing the equations for the equilibrium constants side-by-side for comparative purposes, we obtain:

$$K_\alpha = \frac{\Phi_{\beta_h \alpha}}{\Phi_{\beta_h} \Phi_{\alpha_1}} \left[\frac{r_i \, h}{h + r_i} \right] \qquad K_A = \frac{K_\alpha}{s_B} = \frac{\Phi_{B_h A}}{\Phi_{B_h} \Phi_{A_1}} \left[\frac{hr}{h + r} \right]$$

$$(4.152)$$

$$K_\beta = \frac{\Phi_{\beta_{h+1}}}{\Phi_{\beta_h} \Phi_{\beta_1}} \left[\frac{h}{h+1} \right] \qquad K_B = \frac{K_\beta}{s_B} = \frac{\Phi_{B_{h+1}}}{\Phi_{B_h} \Phi_{B_1}} \left[\frac{h}{h+1} \right]$$

$$(4.153)$$

Clearly, the equilibrium constants determined in section 4A are the quantities defined above as K_B and K_A and are equal to the constants K_α/s_B and K_β/s_B. Note that both these latter equilibrium constants are adjusted by the same quantity, s_B, the number of segments in the self-associating chemical repeat. This is because we (implicitly) defined a lattice cell size in terms of this interacting unit. Differences in size of the self-associating and inter-associating units are then accounted for by the factors r_i and r in the definition of the equilibrium constants.

This result will prove useful, as we will presently see, but its derivation in this manner is unsatisfactory because of the assumption that the attached non hydrogen bonding segments of the molecules can be treated as an inert diluent in which the hydrogen bonded chains are randomly dispersed. Starting from the chemical potentials derived in the preceding section, however, we can obtain the same result directly. We presented the more roundabout derivation first because it clearly illustrates what we are about; defining an equilibrium constant in terms of some arbitrarily defined "interacting unit". In any event, as in section 4A we can write:

$$\Delta\mu^\circ = \ln K_B = \mu_{B_{h+1}} - \mu_{B_h} - \mu_{B_1} \qquad (4.154)$$

and substituting from equation 4.130 we obtain:

$$K_\beta = \frac{\Phi_{B_{h+1}}}{\Phi_{B_h} \Phi_{B_1}} \cdot \frac{(h\, s_B)\, (s_B)}{(h+1)\, s_B} \qquad (4.155)$$

or:

$$\frac{K_\beta}{s_B} = K_B = \frac{\Phi_{B_{h+1}}}{\Phi_{B_h} \Phi_{B_1}} \left[\frac{h}{h+1} \right] \qquad (4.156)$$

We can use equation 4.135 to obtain the corresponding relationship between K_α and K_A.

At various points in this chapter we have made the statement that terms in χ, describing "physical" interactions between the covalent chains, are eliminated from the final result. For those of our readers who are skeptical, it is a trivial exercise to show that the contribution of this term to the chemical potential of an h-mer is $h\, \Phi_A^2 \chi$, so that in the expression for the standard free energy change associated with formation of an $(h + 1)$ mer from an h-mer and a monomer we would have:

$$(h+1)\, \Phi_A^2 \chi - h\, \Phi_A^2 \chi - \Phi_A^2 \chi$$

[Of course, this assumes that the dispersion force interactions involving B units in a hydrogen bonded chain are the same as those that involve units that are not hydrogen bonded].

If we now consider the mixing of a particular self-associating molecule, say a urethane, with a range of "competing" components, say ethers, of different size (molar volume V_A), we would use the *same* equilibrium constant K_A to describe the stoichiometry (and thermodynamics) of all these mixtures, because the equilibrium constant K_α characteristic of the interacting unit is adjusted by the *same* parameter s_B to give the values defined in terms of the molar volumes of the whole molecules:

$$K_B = \frac{K_\beta}{s_B} \qquad K_A = \frac{K_\alpha}{s_B} \qquad (4.157)$$

The differences in size of the ether molecules are accounted for in the stoichiometric and free energy equations by the known factor $r = V_A/V_B$, the ratio of the molar volumes of the ether and urethane molecules.

Conversely, the equilibrium constants for a range of self-associating molecules, in this example urethanes, will vary, but can also be simply determined given that a value for one is known, as follows:

For urethane (a):

$$K_B^a = \frac{K_\beta}{s_B^a} = \frac{K_\beta}{V_B^a} V_\beta \qquad (4.158)$$

For urethane (b)

$$K_B^b = \frac{K_\beta}{s_B^b} = \frac{K_\beta}{V_B^b} V_\beta \qquad (4.159)$$

So:

$$K_B^b = K_B^a \frac{V_B^a}{V_B^b} \qquad (4.160)$$

In fact, this simple equation should allow a test of the validity of our assumption that a "true interacting unit" can be defined for a particular functional group in a range of different but structurally similar molecules or polymers. Values of K_B can be calculated from infrared measurements and plotted against $1/V_B$. An experimental test of this kind will be considered in the following chapter. The result should be a straight line of slope $K_\beta V_\beta$. Obviously, this treatment should only be applied to a series of molecules that vary in a simple manner, such as by differences in the number of CH_2 groups that separate the interacting units. If substituents that give rise to steric hinderance or that change the electronic structure of the interacting unit are introduced then we would expect the equilibrium constants to change.

G. REFERENCES

1. Kretschmer, C.B. and Wiebe, R. *J. Chem. Phys.* 1954, **22**, 1697.
2. Renon, H. and Prausnitz, J.M. *Chem. Eng. Sci.* 1967, **22**, 299; *Erratum* 1967, **22**, 1891.
3. Nagata, I. Z. *Phys. Chem. (Leipzig)* 1973, **252**, 305.
4. Hwa, S.C.P. and Ziegler, W.T. *J. Phys. Chem.* 1966, **70**, 2572.
5. Wiehe, L.A. and Bagley, E.B. *Ind. Eng. Chem. Fundam.* 1967, **6**, 209.
6. Kehiaian, H. and Treszczanowicz, A. *Bull. Soc. Chim. Fr.* 1969, **19**, 1561.
7. Kempter, H. and Mecke, R. *Z. Phys. Chem.* 1940, **B46**, 229.
8. Acree, W.E., *Thermodynamic Properties of Nonelectrolyte Solutions*; Academic: New York, 1984.
9. Flory, P. J., *J. Chem. Phys.* 1944, **12**, 425.
10. Flory, P.J., *J. Chem. Phys.* 1946, **14**, 49.
11. Tobolsky, A.V. and Blatz, P.J. *J. Chem. Phys.* 1945, **13**, 379.
12. Painter, P.C., Park, Y. and Coleman, M.M. *Macromolecules*, 1988, **21**, 66.
13. Kehiaian, H., *Bull. Acad. Polon. Sci. Ser. Sic. Chim.*, 1964, **12**, 497.
14. Coleman, M.M., Lichkus, A.M. and Painter, P.C., *Macromolecules*, 1989, **22**, 586.
15. Sarolea-Mathot, L. *Trans. Farad Soc.*, 1953, **49**, 8.
16. Coggeshall, N.D. and Saier, L.E., *J. Am. Chem. Soc.* 1951, **73**, 5414.
17. Whetsel, K.B. and Lady, J.H. in *Spectrometry of Fuels*, Friedel, H., Editor, Plenum, London, 1970, p. 259.
18. Coleman, M.M., Lee, J.Y., Serman, C.J., Wang, Z. and Painter, P.C., *Polymer*, 1989, **30**, 1298.
19. Guggenheim, E.A., *Mixtures*. Claredon Press, Oxford, 1952.

## CHAPTER 5

Vibrational Spectroscopy and the Hydrogen Bond

The Carbonyl Band: God's Gift to the Incompetent Spectroscopist[#]

A. INTRODUCTION

Infrared spectroscopy is an old and familiar friend to the polymer chemist. It is a technique that for many years was the workhorse of most analytical laboratories and found routine use in the analysis and identification of polymer materials. In this task it can no longer compete with the detail provided by nuclear magnetic resonance (nmr), but it nevertheless remains a powerful probe of molecular structure. As we will see in this chapter, infrared spectroscopy is perhaps uniquely sensitive to hydrogen bonding and, in fact, has often been considered a diagnostic test of the presence of such bonds.

Infrared spectroscopy detects transitions between the vibrational energy levels of a molecule. As such, it is not unique. Raman spectroscopy, based on the inelastic scattering of light, also involves the so-called normal modes of vibration of a molecule. The information provided by the Raman spectrum of a polymer is not redundant if one already has the infrared spectrum, however, as the two techniques are sensitive to different types of bonds (i.e. different electronic characteristics). Furthermore, if a molecule has elements of symmetry in its structure quantum mechanical selection rules can result in certain vibrations being "disallowed" in the Raman, or infrared, or both. By determining the symmetry properties of a molecule it

---

[#] A disparaging summary of our work made by our friend and mentor, Professor J. L. Koenig.

is possible to determine the number of active normal modes using the methods of group theory[1,2]. This can be a powerful method of analysis and until normal coordinate calculations of large molecules became feasible it provided the most fundamental knowledge of the normal modes that could be achieved[3]. Our concern here is limited to the subject of hydrogen bonding, where infrared spectroscopy is in general far more sensitive than the Raman effect, and in addition is experimentally less demanding. We will therefore not discuss Raman spectroscopy in any detail. We will commence with some background material that we hope will be sufficient to obtain an understanding of the vibrational spectrum of the hydrogen bond. The reader who wishes to pursue this topic in more detail should refer to the extensive literature in this field (we kind of like reference 2).

B. ORIGIN OF THE VIBRATIONAL SPECTRUM

Theoretical Background

A change in the total energy of a molecule occurs upon interaction with electromagnetic radiation. This change is reflected in the observed spectrum of the material. In order to describe this interaction and formulate a useful mathematical model certain assumptions are usually made. The total energy of a molecule consists of contributions from the rotational, vibrational, electronic and electromagnetic spin energies. This total energy is usually approximated to a sum of the individual components and any interaction treated as a perturbation. The separation of the electronic and nuclear motions, known as the Born-Oppenheimer approximation, depends upon the large difference in mass between the electrons and nuclei. Since the former are much lighter they have relatively greater velocities and their motion can be treated by assuming fixed positions of the nuclei. Conversely, small nuclear oscillations (compared to inter-atomic distance) occur in an essentially averaged electron distribution. The change in energy of these nuclear vibrations upon interaction with radiation of suitable frequency is the origin of the vibrational spectrum.

Experimentally, in order to observe an infrared spectrum a beam of radiation in the appropriate frequency range is passed through a sample and those frequencies corresponding to *allowed* infrared transitions of so-called normal modes of vibration are absorbed. In general, an infrared absorption band or Raman line nearly always corresponds to discrete vibrational transitions in the ground electronic state. Absorption of higher energy visible or UV light is required to produce changes in electronic energy.

Normal Modes of Vibration and Normal Coordinate Analysis

Any vibration of an atom can be resolved into components parallel to the x, y, and z axis of a Cartesian system. The atom is described as having three degrees of freedom. A system of N nuclei therefore has 3N degrees of freedom. For a non-linear molecule, however, six of the total degrees of freedom correspond to translations and rotations of the molecules as a whole and have zero frequency, leaving 3N-6 vibrational degrees of freedom, each of which is associated with a normal mode of vibration. For strictly linear molecules, such as carbon dioxide, rotation about the molecular axis does not change the position of the atoms and only two degrees of freedom are required to describe any rotation. Consequently, linear molecules have 3N-5 normal vibrations. For an infinite, ordered polymer chain, only the three translations and one rotation have zero frequency and there are 3N-4 degrees of vibrational freedom. Each normal mode consists of vibrations (although not necessarily significant displacements) of all the atoms in the system.

The normal modes of vibration can be calculated by assuming that the nuclei behave like point masses and the forces acting between them are springs that obey Hooke's Law. The motion of each atom is also assumed to be simple harmonic. Even with these assumptions, it is intuitively obvious that a system of N atoms is capable of innumerable different complex vibrations, each involving a range of displacements of (and phase relationships between) the various nuclei, rather than a number equal to the vibrational degrees of freedom. In the harmonic approximation, however, any motion of the system can be resolved into a sum of the so-called fundamental normal modes of vibration, just as displacements can be represented by components parallel to a set of Cartesian coordinates. In a normal vibration, each particle carries out a simple harmonic motion of the same frequency and in general these oscillations are in phase; the amplitude may be different from particle to particle, however. It is these normal modes of vibration that are excited upon infrared absorption or Raman scattering. Naturally, different types of vibrations will have different energies and so absorb or inelastically scatter radiation at different frequencies.

A discussion of normal mode calculations is beyond the scope of this book, but it is possible to briefly demonstrate the major factors that influence vibrational frequencies using a simple model system consisting of the harmonic oscillations of a one dimensional molecule, illustrated in figure 5.1, consisting of two masses m_1 and m_2 connected by a spring. The displacements of the two atoms relative to one another is given by :

$$z = (z_1' - z_2') - (z_1 - z_2) \qquad (5.1)$$

where $z_1 - z_2$ is the equilibrium separation between m_1 and m_2 and $z_1' - z_2'$ is the distance after a finite extension or compression. Assuming that the spring obeys Hooke's law, the exerted force is equal to $-fz$, where f is the force constant or ("stiffness") of the spring. It is also convenient at this stage to define m_r, the reduced mass, equal to $(m_1 m_2)/(m_1+m_2)$.

Since force is defined as mass times acceleration, for a conservative field (i.e., no frictional forces) the following equation holds:

$$-fz = m_r \ddot{z} \qquad (5.2)$$

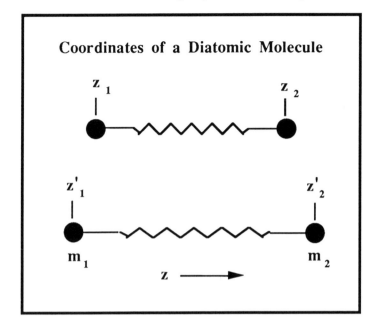

Coordinates of a Diatomic Molecule

FIGURE 5.1

where $\ddot{z} = d^2z/dt^2$. Hence:

$$m_r \ddot{z} + f z = 0 \qquad (5.3)$$

This is a second-order differential equation that has a periodic solution of the form:

$$\ddot{z} = A \cos (2\pi \upsilon t + \varepsilon) \qquad (5.4)$$

where A is an amplitude, ε is a phase angle, and υ is the vibrating frequency. Differentiating twice with respect to time leads to:

$$\ddot{z} = -4 \pi^2 \upsilon^2 z \qquad (5.5)$$

Substituting back into equation 5.2 we obtain:

$$(4 \pi^2 \upsilon^2 m_r + f) z = 0 \qquad (5.6)$$

Assuming that $z \neq 0$, we obtain the familiar equation:

$$\upsilon = \frac{1}{2\pi} \sqrt{\frac{f}{m_r}} \qquad (5.7)$$

This is the basic vibrational equation for the harmonic oscillator. Although the model considered is extremely simple, it does demonstrate that the vibrational frequency depends inversely on mass and directly on the force constant. Consider, as an example, the isolated stretching vibrations of the O-H bond. This occurs near 3600 cm^{-1}. If we now substitute a deuteron for the proton, the O-D stretching mode is now found near 2500 cm^{-1}. Similarly, if the O-H bond force constant is reduced, as it is when a hydrogen bond is formed, the stretching mode again shifts to lower frequencies.

The absorption of infrared radiation or inelastic scattering of light (Raman effect) changes the vibrational energy of a system. The total energy E of a classical harmonic oscillator is constant and equal to the sum of the potential and kinetic energies at any point in the oscillation. The potential energy (PE) is given by:

$$PE = \int_0^z f\,z\,dz = \frac{1}{2} f\,x^2 \qquad (5.8)$$

A plot of PE as a function of displacement z is a parabola, and a typical curve is shown in figure 5.2. At the maximum displacement (z_m) in the vibration there is no movement for an instant in time, and the kinetic energy is zero. Hence, the total energy is given by:

$$E = \frac{1}{2} f\,z_m^2 \qquad (5.9)$$

In the classical solution with z being continuous, E is continuously variable from 0 to $\frac{1}{2} f\,z_m^2$.

The quantum mechanical solution requires discrete quantized energy levels, rather than a continuum. To obtain this solution in a simple manner the Schrodinger equation is expressed in terms of normal coordinates rather than the usual Cartesian set.

Harmonic Diatomic Oscillator

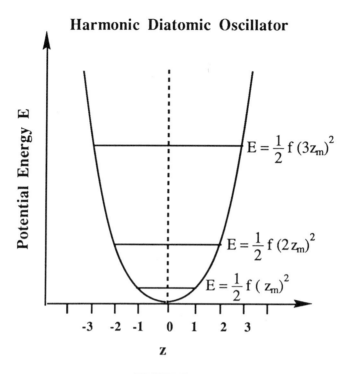

FIGURE 5.2

As a result of using normal coordinates the total energy of a system can be expressed as the sum of the energies of the individual normal modes. In general:

$$E = \sum_i E_i = \sum_i (n_i \pm \frac{1}{2}) h \, \upsilon_i \qquad (5.10)$$

where υ_i is the classical vibrational frequency for the ith normal mode and n_i is the vibrational quantum number. Additionally, quantum mechanical selection rules for the harmonic oscillator only allow $n_i = \pm 1$. For our simple example:

$$E_i = (n_i + \frac{1}{2}) h \, \upsilon_i \qquad (5.11)$$

$$\Delta E = (n + 1 + \frac{1}{2}) h \, \upsilon_i - (n + \frac{1}{2}) h \, \upsilon_i = h \, \upsilon_i \qquad (5.12)$$

where ΔE, is the energy change for absorption or emission of radiation. This is an extremely important result since it allows the calculation of vibrational frequencies by the standard methods of classical mechanics, provided that the harmonic approximation holds.

Figure 5.3 illustrates the absorption of various energy quanta according to equation 5.12. A molecule in the ground state E_0 can absorb an energy quantum $h\upsilon$ to reach the first excited state E_1. The subsequent absorption of another quantum $h\upsilon$ would allow the second excited state E_2 to be reached. Similarly, the absorption of $2h\upsilon$ would allow the second excited state to be reached directly. For a complex system, quanta of $h\upsilon_i + h\upsilon_j$ could be absorbed, where υ_i and υ_j are the frequencies of different normal modes. These processes are the basis for the observation of overtone and combination bands, respectively. For a harmonic oscillator, however, a fundamental condition of quantum mechanics is that only transitions between adjacent levels are allowed. Nevertheless, these selection rules break down and combination and overtone bands are observed, but only with relatively weak intensities. These observations reflect the limitations of the theoretical model.

Energy Levels of a Diatomic Harmonic Oscillator

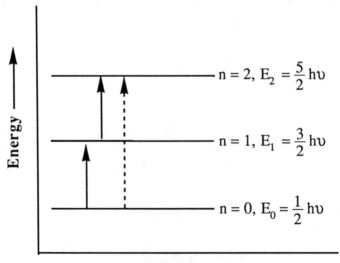

FIGURE 5.3

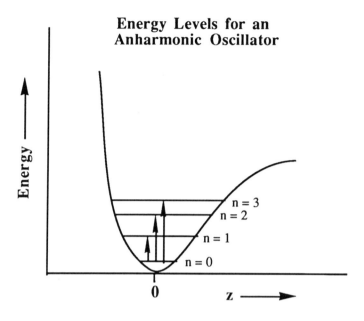

FIGURE 5.4

Molecular vibrations are not strictly harmonic and the potential energy curve is closer to that shown in figure 5.4 than to the parabola of figure 5.2. Low energy transitions approximate the harmonic model closely, while the higher energy levels, including the second, are more accurately given by:

$$E = \left[(n + \frac{1}{2}) - x (n + \frac{1}{2})^2\right] h \upsilon \qquad (5.13)$$

where x is known as the anharmonicity constant and is usually a small fraction of unity. In most cases the anharmonicity is small, but sufficient for overtone and combination bands to appear. Their intensity is weak because of the small number of transitions of this type that occur. As we will see, anharmonic effects appear to play a key role in the infrared spectra of hydrogen bonded systems.

Conditions for Infrared Absorption

We have briefly discussed one basic and inherent condition for infrared absorption, namely that the frequency of the absorbed radiation must correspond to the frequency of a normal mode of vibration and hence a transition between vibrational energy levels. This condition is not by itself sufficient. There are additional so-called *selection rules* that determine activity. Naturally, an understanding of the selection rules can only be attained through the methods or theories capable of successfully describing the interaction of radiation with matter, i.e. quantum mechanics. It is easier to obtain a mental picture of these interactions by first considering the classical interpretation, however, and this will be given briefly here. The quantum mechanical description will then be simply stated and the results compared.

Infrared absorption is simply described by classical electromagnetic theory; an oscillating dipole is an emitter or absorber of radiation. Consequently, the periodic variation of the dipole moment of a vibrating molecule results in the absorption or emission of radiation of the same frequency as that of the oscillation of the dipole moment. The requirement of a change in dipole moment with molecular vibration is fundamental, and it is simple to illustrate that certain normal modes do not result in such a change.

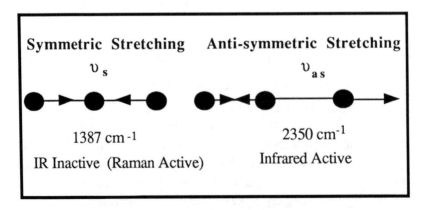

Symmetric Stretching Anti-symmetric Stretching

υ_s υ_{as}

1387 cm^{-1} 2350 cm^{-1}

IR Inactive (Raman Active) Infrared Active

FIGURE 5.5

Consider the two in-plane stretching vibrations of CO_2 shown in figure 5.5. The net dipole moment of the symmetrical unperturbed molecule is zero. In the totally symmetric vibration, υ_s, the two oxygen atoms move in phase successively away and towards the carbon atom. The symmetry of the molecule is maintained in this vibration and there is no net change in dipole moment, hence no interaction with infrared radiation. Conversely, in anti-symmetric vibration, υ_{as}, the symmetry of the molecule is perturbed and there is a change in dipole moment with consequent infrared absorption.

Infrared selection rules are expressed in terms of quantum mechanics by the following condition:

$$\int \Psi_i \, \mu \, \Psi_f \, d\tau = 0 \qquad (5.14)$$

This integral (which is over the whole configurational space of the molecule) is called the transition moment and its square is a measure of the probability of the transition occurring; μ is the dipole moment vector and Ψ_i and Ψ_f are wavefunctions describing the initial and final states of a molecule. The dipole moment, μ, can be expressed as the sum of three components μ_x, μ_y and μ_z in a Cartesian system, so that the condition for absorption is that any one of the three corresponding transition moment integrals does not vanish. Note that the size of the integral need not be calculated in order to determine activity, we only need to know whether or not its value is zero.

The selection rules for the symmetric and antisymmetric stretching modes of CO_2 were determined above by inspection. For this simple molecule it is easy to see that the symmetric stretch does not result in a change in the dipole moment. It is intuitively clear that this behavior is due to the symmetry of the CO_2 molecule and it turns out that the number and activity of the normal modes of any molecule can be predicted from symmetry considerations alone. All molecules can be classified into groups according to the symmetry operations (mirrors, rotation axes, etc.) that can be used to characterize their structure. Each quantity in the transition moment integral also has a clearly defined behavior with respect to these symmetry operations.

Consequently, the vanishing or non-vanishing of the integrals is the same for all transitions between states of two particular symmetry classes. The activity of each class of vibrations of a given symmetry group has already been determined, so that the infrared activity of the normal modes of a molecule can be determined solely from a knowledge of its symmetry[1,2]. Similarly derived selection rules apply to the Raman spectrum.

Apart from unusually strong and symmetric hydrogen bonds, symmetry conditions do not play a significant role in the vibrational spectrum of hydrogen bonding functional groups. In general, OH, NH and C=O stretching vibrations lead to a change in dipole moment and hence, usually result in relatively intense infrared absorption bands.

C. THE PRACTICE OF INFRARED SPECTROSCOPY

Introduction

In the preceding sections of this chapter we have discussed some of the fundamentals of infrared spectroscopy and with simple models attempted to relate molecular structure to the appearance of the spectrum. Spectroscopy also involves a measure of art and acquired craft, however, and certain skills and experience are still necessary in order to obtain high quality spectra and also to differentiate between fact and artefact when applying some of the data manipulating programs that are packaged with contemporary instruments. Fortunately, in the study of hydrogen bonding in polymers we most often want to know one simple thing; the fraction of groups that are bonded (equal to $1-f_F$, where f_F is the fraction of groups that are "free", or not hydrogen bonded). Even though this seems straightforward, there are complications. Absorption coefficients, those quantities relating the intensity of a mode to the concentration of functional groups associated with it, differ for free and bonded bands, which more often than not seriously overlap. Here we will briefly outline some aspects of practical spectroscopy that apply to the problems that are of concern in the study of hydrogen bonding. This subject is covered in more depth in the books by Griffiths[4,5] and a review article by Koenig[6].

Instrumentation

Dispersive infrared instruments, once commonplace in chemical laboratories, are gradually becoming extinct. Fourier transform machines, utilizing an interferometer rather than a system of gratings and slits, have done them in.

As the name implies, dispersive instruments use either gratings of prisms to separate the radiation into its component wavelengths. Resolution is obtained by a set of slits. The narrow openings of the entrance and exit slits, required for high resolution, severely limit the amount of energy reaching the detector. Consequently, if good signal-to-noise ratio's are to be achieved, slow scan rates (high measurement times per frequency interval) are required.

The use of interferometers produces two principle advantages, high energy throughput and multiplexing. There are now various types of interferometers in use, but the general principles of the technique are clearly illustrated from the simplest (and still most widely used) design, shown in figure 5.6. This consists of two mirrors, one fixed and one moveable, at right angles to one another. A beam-splitter bisects the angles between these two mirrors. There are no slits to limit the incoming energy and, in general, the energy throughput is 80 to 200 times greater than that of a dispersive instrument (depending upon the resolution). The incoming light is divided at the beam-splitter and is reflected to the two mirrors. Upon recombination at the beam-splitter constructive or destructive interference will occur, depending upon the pathlength difference of the light and its wavelength. Upon driving the moveable mirror an interference pattern is obtained. Each frequency will have an interference pattern independent of all other frequencies. If this mirror starts at a point equal in distance from the beam-splitter and the fixed mirror, then all of the frequencies are characterized by initial constructive interference.

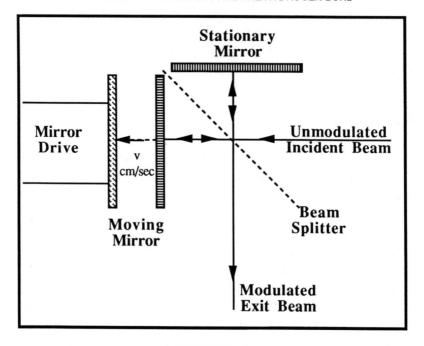

FIGURE 5.6

A typical interferogram with this characteristic strong "centerburst" is shown in figure 5.7. The detector views all of the frequencies all of the time during the mirror displacement, giving rise to what is termed the multiplex advantage (relative to dispersive instruments). The multiplex advantage is proportional to the number of frequency elements scanned and is given by $(N)^{1/2}$, where N is the number of frequency elements[4,5].

The principle disadvantage of FTIR is that a generally unrecognizable interferogram rather than a spectrum is produced. The intensity of radiation, I(x) falling on the detector is given by:

$$I(x) = \int_{-\infty}^{\infty} B(\upsilon) \cos(2\pi\upsilon_x)\, d\upsilon \qquad (5.15)$$

where $B(\upsilon)$ is the spectrum of the source and x is the mirror displacement. The spectrum, $B(\upsilon)$ is obtained by calculating the Fourier transform pair of the above equation:

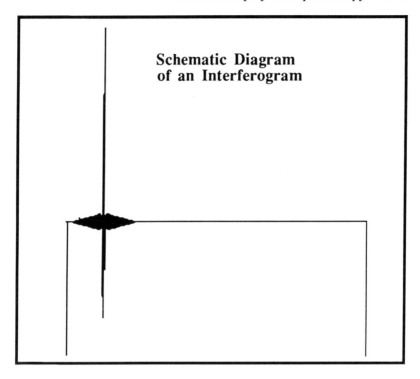

FIGURE 5.7

$$B\,(\upsilon) = \int_{-\infty}^{\infty} I\,(x)\cos\,(2\,\pi\,\upsilon_x)\,dx \qquad (5.16)$$

Although the optical advantages of interferometric systems were recognized by Michaelson at the turn of the century, routine laboratory application depended upon three developments. The introduction of "fast" detectors (DTMS) allowed interferometry to be extended from the far infrared into the mid infrared. The Cooley-Tukey algorithm, published in 1965, reduced the computation time for Fourier transforms several orders of magnitude. Finally, at approximately the same time, minicomputers were becoming available. These developments allowed the introduction of "off the shelf" commercial instruments, which typically, consist of a number of basic components; an interferometer, minicomputer (often including a processor for specific calculations, in addition to the central processing unit), a magnetic disk storage device and various input-output interfaces and peripheral devices (terminals, plotters, etc.). Software is included, but there are various programs developed by and still indigenous to specific laboratories. For most of the new "quantitative analysis" packages we have nothing but contempt and scorn, believing that it is too easy to generate artefacts suitable for over-interpretation. (Perhaps we are getting old and set in our ways.) We will review the methods we believe to be the most appropriate for our problems below.

Sample Preparation

When FTIR instruments were first introduced the widely touted optical advantages of the instrument sometimes gave the impression that one could place a brick in the sample chamber and still obtain a spectrum. Initially, the quality of the spectra that could be obtained were actually not that superior to top-of-the-line dispersive spectrometers. The major advantages were in the realm of data manipulation. Accordingly, old and familiar polymer sample preparation techniques, most commonly the casting of thin films from solvent or compression molding, were still the methods of choice and remain so today. If anything, the ability to subtract one spectrum from another and perform other manipulations imposes new limitations and requires greater care in sample preparation. For example, it is now a "rule of thumb" that if the spectrum of one polymer film is to be subtracted from another, then the two should have comparable film thickness. In addition, the films should be thin enough to ensure that the strongest band in the region of interest should have an absorbance of less than 1.0 units. (In fact, it is common practice in our laboratory to aim to produce films exhibiting maximum absorbances of 0.6 absorbance units). These prerequisites are still important (but often ignored), even though in contemporary FTIR instruments advances in infrared sources, detectors, etc., have led to a considerable enhancement in sensitivity compared to first generation spectrometers.

The details of various sample preparation methods can be found in numerous standard texts. Koenig[6] has described various sampling devices. Studies of hydrogen bonding are almost invariably performed on solutions of low molecular weight liquids and thin films of polymers, so that many of the newest techniques do not offer advantages that need to be considered here.

Quantitative Analysis and the Use of Curve-Resolving

The quantitative analysis of polymers in transmission studies is based on the Beer-Lambert law. This is usually written:

$$A = abc \qquad (5.17)$$

where A is the intensity (in absorption, not transmission) of the band of interest; integrated areas or peak heights can be used, but the units will change accordingly; a is the absorptivity or extinction coefficient, b is the sample thickness and c is the concentration of the component of interest. For multicomponent systems, the absorption of any particular band can be written:

$$A = \sum_i a_i b_i c_i \qquad (5.18)$$

if it is assumed that there is no interaction between the i components.

Polymer spectra generally consist of broad overlapping bands and this remains true even when we simply concentrate our attention on those modes associated with hydrogen bonding functional groups. For quantitative analysis of the proportion of free and bonded species, key parameters in our formulation of a thermodynamic model, it is therefore necessary to separate the components of a spectral profile. This can be accomplished by curve-resolving, but the method is not without problems and the results obtained are not always unambiguous. It is usually a simple task to obtain a good fit between a synthesized spectrum consisting of a sum of a sufficient number of bands and an observed spectral profile. The validity of the results are open to question, however, unless a number of key problems are addressed.

The first problem is the choice of mathematical function most suitable for characterizing the observed band shapes. Gauss-Lorentz sum and product functions have often been used, for historical reasons, as with dispersive instruments a shift from Lorentzian to Gaussian bandshapes was found with increasing slit width. Although FTIR instruments do not have a set of slits, bandshapes can be affected by instrument factors, such as the apodization function used in the Fourier transformation, and by factors associated with the material under consideration; for example, interactions between the constituents of a complex multicomponent system. Consequently, an empirically determined bandshape would seem to offer the best choice. We use a function that is a sum of Gaussian and Lorentzian bandshapes in the proportion g to (1-g);

$$
A = g A_0 \exp\left[-\ln 2 \left(\frac{2(x - x_0)}{\Delta x_{1/2}} \right)^2 \right] + \frac{(1 - g) A_0}{1 + \left[2(x - x_0) / \Delta x_{1/2} \right]^2}
$$

$$(5.19)$$

where A is the peak height, x is the wavenumber coordinate of the peak and $\Delta x_{1/2}$ is the bandwidth at half height. Initial estimates of these parameters form the input to a program that fits the parameters of a set of bands to the experimental spectral profile by a least squares optimization procedure.

Other problems with curve resolving are associated with the initial choice of parameters. It is often crucial to have a prior knowledge of the number of bands in the region of the spectrum that is to be resolved, together with good initial estimates of the peak position of each band and its width at half height. In some hydrogen bonded systems this is relatively straightforward. For example, in blends of ethylene-co-methacrylic acid copolymers containing 44% methacrylic acid (EMAA[44]) with poly(vinyl methyl ether) (PVME) two carbonyl bands, one assigned to hydrogen bonded carboxylic acid dimers at 1700 cm^{-1} and one assigned to "free" or non-hydrogen bonded species, can be clearly seen, because their overlap is only moderate, as shown in figure 5.8. Actually, the 1728 cm^{-1} band is due to the free C=O of COOH groups where the OH group is hydrogen bonded to ether oxygens. The carbonyl band of -COOH groups where *both* functional groups are free appears near 1750 cm^{-1}.

This and other subtleties will be discussed later, but fitting this profile to two bands is clearly warranted (the 1750 cm^{-1} band only has detectable intensity at elevated temperatures)[7].

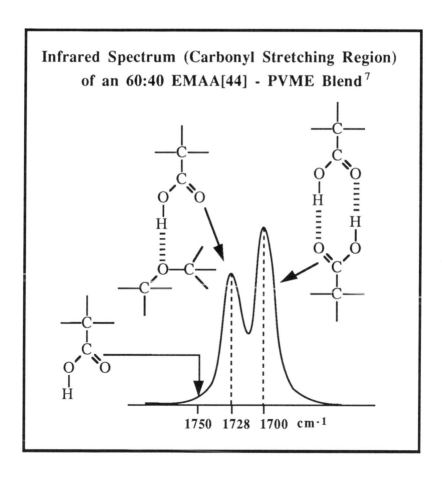

Infrared Spectrum (Carbonyl Stretching Region) of an 60:40 EMAA[44] - PVME Blend[7]

1750 1728 1700 cm^{-1}

FIGURE 5.8

Mixtures of poly(vinyl phenol) (PVPh) with ethylene-co-vinyl acetate (EVA)[8] provide another example of a hydrogen bonded system where clear-cut decisions pertaining to the number, position and breadth of the pertinent infrared bands is readily established. Figure 5.9 shows infrared spectra obtained from films of miscible PVPh - EVA[70] blends of varying compositions acquired at 120°C.

Two distinct and well separated bands attributed to free and hydrogen bonded carbonyl groups are observed at 1737 and 1708 cm^{-1}, respectively. Initial curve fitting parameters for the free band are obtained from the spectrum of pure EVA[70]. A Gaussian shaped band with a width of half height of 19 cm^{-1} matches the experimental data very well. On the other hand, the curve fitting parameters for the hydrogen bonded band are established from a PVPh rich blend where the contribution from the 1708 cm^{-1} band dominates (e.g. the top spectrum of figure 5.9). Not unexpectedly, here we find that an appreciably broader Gaussian band ($w_{1/2} = 30$ cm^{-1}) is necessary to adequately match the experimental data. These initial estimates of the curve fitting parameters for the free and hydrogen bonded bands are very important, especially when a spectrum such as that obtained from the PVPh - EVA[70] blend containing 19:81 volume % (second from the bottom in figure 5.9) is subjected to a least-squares curve fitting procedure. The hydrogen bonded band (1708 cm^{-1}) appears as a shoulder and is poorly resolved. In this case, we hold the shape and breadth of this band *constant* and do not permit the computer to vary these parameters in the least-squares curve fitting procedure. More on this below.

An example of a system where the overlap of the bands due to various hydrogen bonded species starts to become a serious problem is nylon 11[9]. The carbonyl, or amide I, region of the spectrum of this polymer recorded as a function of temperature, is shown in figure 5.10. At 30°C this region of the spectrum is characterized by a relatively sharp band skewed to the high-frequency side and centered at about 1638 cm^{-1}. Upon increasing the temperature in a stepwise fashion to 170°C, the amide I mode shifts systematically to higher frequency and appears to decrease in absolute intensity.

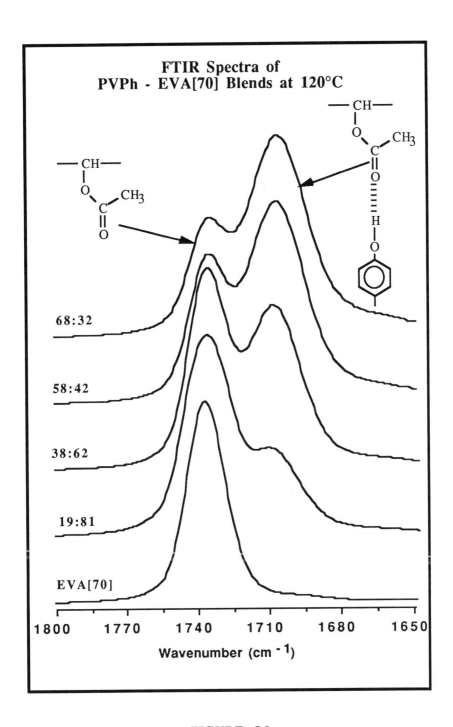

FIGURE 5.9

Concurrently, there is increasing evidence of spectral contributions at approximately 1650 and 1680 cm[-1]. At 190°C, before this semicrystalline nylon has completely melted, one can discern three contributions to the amide I mode. Increasing the temperature further to 210°C, which is above the crystalline melting point of nylon 11, yields a spectrum in the amide I region that is characteristic of a completely amorphous nylon.

Only two obvious amide I modes are seen, near 1656 cm[-1] and 1680 cm[-1]. These judgements, based on a simple subjective examination of the bandshape, can be supported by obtaining the derivatives of the spectrum, using programs routinely available on most instruments.

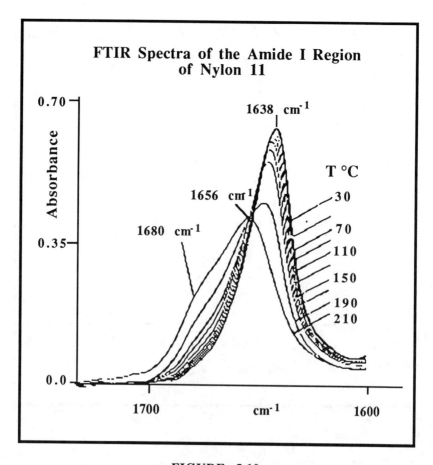

FIGURE 5.10

Minima in the second derivative indicate the presence of bands, as illustrated in figure 5.11 for the spectrum obtained at 190°C. There are limits to such methods, however, discussed in an excellent review article by Maddams[10]. In general, very broad bands and bands separated by less than half their half-width are not detected by this procedure. This is a real problem in systems where there may be weak hydrogen bonds (e.g., PVC / polyesters), but fortunately in blends where there are moderate to strong hydrogen bonds it is usually a relatively straightforward matter to identify the species present. In this respect, common sense, experience and a degree of chemical intuition are also important.

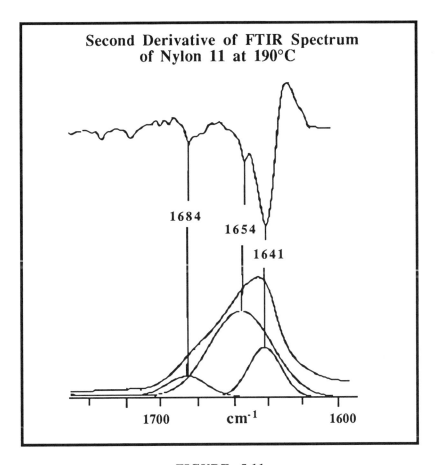

Second Derivative of FTIR Spectrum of Nylon 11 at 190°C

1684

1654

1641

1700   cm⁻¹   1600

FIGURE 5.11

For example, in the spectrum of nylon 11 we would anticipate distinct spectral features in the amide I region attributable to hydrogen bonded carbonyl groups in ordered ("crystalline") domains, in addition to those hydrogen bonded carbonyl groups associated with disordered (amorphous) conformations and non-hydrogen bonded (free) carbonyl groups. Combining these expectations with the results of the second derivative analysis, one would be reasonably confident in fitting this region of the spectrum obtained below the melting point to three bands, while the spectra of amorphous samples obtained above the melting point are characterized by two carbonyl bands.

Even with a knowledge of the number of bands, good prior estimates of band positions, shapes and widths at half-height are necessary prerequisites for obtaining valid solutions. In fact, our first attempts at least-squares curve fitting the amide I region of nylon 11 to three bands yielded an excellent fit, but the results were physically meaningless.[9] Because of this experience, we have taken a somewhat different but, in our view, more logical approach to the problem of curve fitting. Rather than permitting the computer to obtain a best fit by changing a multitude of variables, we severely restrict the options by making a number of justifiable assumptions; for the specific case of nylon 11:

1. The amide I region of the semicrystalline nylon was assumed to be composed of a maximum of three bands, attributable to hydrogen bonded carbonyl groups in ordered domains, hydrogen bonded carbonyl groups in disordered conformations, and "free" carbonyl groups, based on the criteria described above.

2. We first determined the frequency and breadth of the bands in the spectra of the samples obtained at temperatures above the melting point, and used the values so calculated as first estimates for the corresponding bands in the spectra of the semi-crystalline samples. For this particular study we were fortunate to have data available from a preceding study of a completely amorphous nylon[11], so that the values of the parameters obtained in each case could be compared. It is essential to take the time to make such comparisons, if data is

available, as errors will usually show up very quickly (in, for example, anomalous values of the band width at half-height).

3. The frequency of the band attributed to the ordered hydrogen bonded amide I band was established from a second-derivative plot of the spectra obtained below 190°C. From experience, we know that this band will be considerably narrower than the corresponding disordered amide I band. An initial estimation of $w_{1/2}$ equal to approximately half that of the disordered band was employed.

4. The band shape was assumed to be Gaussian. Previous studies of an amorphous nylon indicated that the shape of the two bands in the amide I region were best approximated by Gaussian band shapes[11]. Furthermore, in initial attempts at curve fitting the amide I region by allowing the computer to vary band shape, an essentially pure Gaussian band shape was consistently determined by a least-squares fit.

5. A linear base line was assumed from 2000 to 850 cm^{-1}, where there are not significant underlying absorbances, and then fixed, or held constant. Although the identification of the "correct" base line is rather subjective, the perils of a wrongly chosen base line cannot be overstated.

6. Curve fitting was limited to the spectral data available from between 1626 and 1700 cm^{-1} *which is acceptable if the "true" baseline has been fixed using a wider spectroscopic range*. This circumvents the problems of overlapping contributions in the wings of the amide I region.

In summary, each spectrum recorded at the different temperatures was resolved into three components (two above the crystalline melting point). The base line and the band shape (Gaussian) were fixed. Initial frequencies, band widths, and intensities were estimated according to the assumptions detailed above. A least-squares iterative procedure was then employed to obtain a best fit. Representative results are shown in figure 5.12. At 220°C, above the crystalline melting point, only two bands are necessary to obtain a satisfactory fitting of the experimental data. This is entirely consistent with the results obtained previously for the amorphous nylon[11].

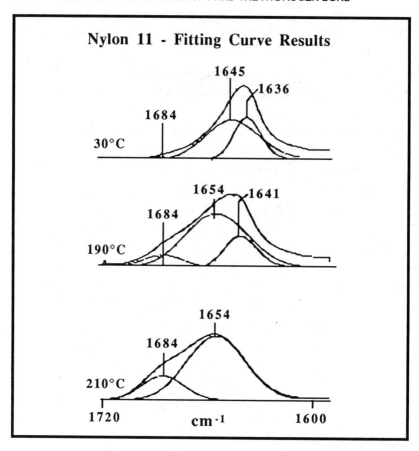

FIGURE 5.12

The disordered hydrogen bonded amide I band of nylon 11, which occurs at 1654 cm⁻¹ at 210°C, has a width at half-height almost identical with that of the amorphous nylon studied previously. The "free" band at 1684 cm⁻¹ is significantly narrower than this, but again consistent with the results obtained for the amorphous nylon. At 190°C, in the transition region, three bands are necessary to satisfactorily fit the data. In addition to the two bands at approximately 1654 and 1684 cm⁻¹, a third at about 1640 cm⁻¹ is required. This is pleasingly consistent with the second-derivative data (see figure 5.11). At 30°C, three bands are again necessary. However, the relative intensities have changed and there have been some subtle but important frequency shifts.

Table 5.1
Curve Fitting Results of the Amide I Region of Nylon 11

| Temp °C | Hydrogen Bonded Carbonyl Bands | | | | | |
| | Ordered | | | Disordered | | |
	υ cm^{-1}	$w_{1/2}$ cm^{-1}	A_o	υ cm^{-1}	$w_{1/2}$ cm^{-1}	A_d
30	1636	17	30.0	1645	34	55.0
50	1637	18	30.2	1646	35	56.5
70	1637	18	29.8	1646	35	57.5
90	1638	18	29.9	1647	36	58.6
110	1638	18	28.4	1647	38	60.6
130	1639	18	26.7	1648	40	63.2
150	1640	18	25.6	1649	41	64.8
170	1641	18	23.2	1650	42	66.5
180	1641	18	21.9	1651	42	66.4
190	1641	21	19.8	1654	41	64.8
200	-	-	-	1653	39	73.8
210	-	-	-	1653	38	72.7
220	-	-	-	1654	38	72.2

| Temp °C | Free | | | A_t $= A_o + A_d + A_f$ | A_o / A_t |
	υ cm^{-1}	$w_{1/2}$ cm^{-1}	A_f		
30	1679	26	4.0	89.0	0.34
50	1680	28	4.9	91.6	0.33
70	1681	28	5.0	92.3	0.32
90	1682	27	5.4	93.9	0.32
110	1683	26	5.1	94.1	0.30
130	1684	25	5.0	94.9	0.28
150	1684	25	5.3	95.7	0.27
170	1684	25	5.8	95.5	0.24
180	1685	24	6.3	94.6	0.23
190	1684	26	9.6	94.2	0.21
200	1684	26	17.1	90.9	-
210	1683	27	19.3	92.0	-
220	1683	27	19.8	92.0	-

The detailed results of the curve fitting of the amide I region throughout the temperature range 30-220°C are given in Table 5.1. One should always tabulate and carefully examine the results of curve-resolving in this manner, to determine whether they are consistent and make "physical" sense. In this example, the results exceeded our expectations and increased our confidence in the curve-fitting procedure we employed. The infrared band attributed to the ordered conformation is relatively narrow ($w_{1/2} \simeq 18$ cm^{-1}), is observed at 1636 cm^{-1}, and gradually shifts to 1641 cm^{-1} over the temperature range 30-190°C. This is precisely the relative breadth and temperature dependence we expected. At temperatures above the crystalline melting point of nylon 11, 190°C, the computer rejects any contribution close to this frequency. The disordered hydrogen bonded amide I band starts at 1645 cm^{-1} at 30°C and systematically shifts to 1654 cm^{-1} at 220°C. It is about twice as broad as the band attributed to the ordered amide I mode and almost identical in breadth with the analogous band in the completely amorphous nylon.[11] The "free" band similarly shifts from 1679 to 1683 cm^{-1} over the same temperature range. This band is inherently narrower than the disordered hydrogen bonded band, again an expected result. The areas of the bands contributing to the amide I mode are also given in Table 5.1. It should be emphasized that as the curve-fitting results were obtained on the *same* sample, the areas represent a quantitative but relative measure of the amount of the different species at the various temperatures.

There remains the question of the values of the absorption coefficients of the three different bands and how they change with the strength of the hydrogen bond. This we will address below. Although an interpretation of the results obtained in our study of nylon 11 is not our concern here, to bring this subject to a tidy close we present a graph of the respective areas of the three components to the amide I band as a function of temperature in figure 5.13. Also included is the total area of the amide I region and a thermogram of a nylon 11 sample prepared under the same conditions as that used for the infrared studies.

The area of the ordered band (A_o) steadily decreases to about two-thirds of its original value over a range 30-190°C, where-

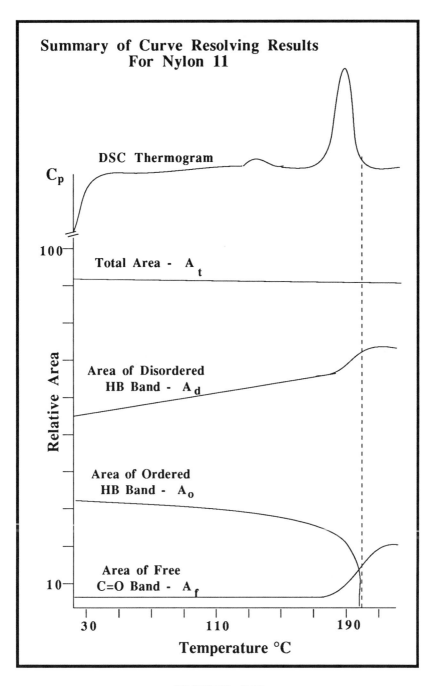

Summary of Curve Resolving Results For Nylon 11

DSC Thermogram

C_p

100 — Total Area - A_t

Relative Area

Area of Disordered HB Band - A_d

Area of Ordered HB Band - A_o

10 — Area of Free C=O Band - A_f

30   110   190

Temperature °C

FIGURE 5.13

upon a discontinuity is observed as it vanishes at higher temperatures. This correlates beautifully with the thermogram of nylon 11, which exhibits a melting transition close to 200°C.

Before we leave this section, we will show a few more characteristic examples of the infrared spectra of blends that we will return to in later discussions. In figure 5.14 we present spectra in the carbonyl stretching region recorded at 110°C, above the glass transition temperature, of an amorphous polyurethane (APU - see Chapter 2, page 133 for a schematic diagram of the chemical structure of this polymer) and blends of this polymer with an ethylene oxide-co-propylene oxide (EPO) copolymer. The spectrum of the pure APU is characterized by two broad overlapping bands at 1737 and 1711 cm^{-1}, which correspond to the free and hydrogen bonded carbonyl groups, respectively[12]. The APU - EPO blend system is miscible and with the addition of EPO to the mixture the former band increases at the expense of the latter. This is a consequence of urethane - ether interactions being formed (which we observe as free APU carbonyl groups - see figure 3.4) at the expense of urethane - urethane interactions (figure 3.1). Here, unlike the examples shown in figures 5.8 and 5.9, we do not see two separate bands that can be easily resolved, and we must employ the principles discussed above in order to use the least-squares curve fitting procedure with confidence. The interested reader is referred to reference 12 for further details and tabulated results.

Finally, we compare in figure 5.15 the spectra of 20:80 weight % APU blends with poly(ethylene oxide) (PEO) and poly(vinyl methyl ether) (PVME), recorded at 110°C. Both are polyethers, but the former blend is miscible while the latter is grossly immiscible. The difference between the spectra is striking. In the miscible PEO system the spectrum is dominated by the free band indicating the presence of a major contribution from urethane - ether interactions. Conversely, in the immiscible PVME system the spectrum is identical, to all extent and purposes, to that of pure APU (see figure 5.14). This indicates that APU essentially exists in separate phase from that of PVME and the equilibrium fraction of urethane - urethane interactions is the same as that found in the pure APU at the same temperature.

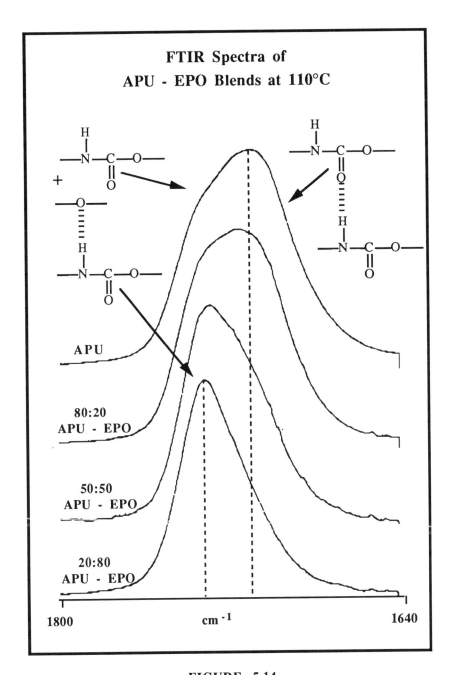

FIGURE 5.14

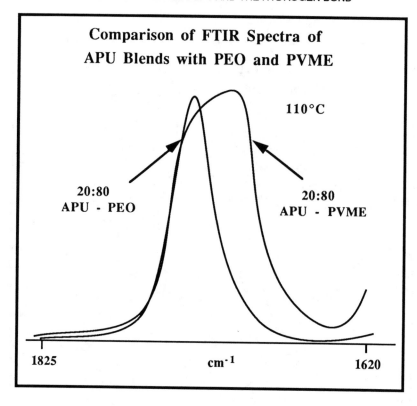

FIGURE 5.15

D. THE VIBRATIONAL SPECTRUM OF HYDROGEN BONDED SYSTEMS

Introduction

As we mentioned above, changes in the infrared spectrum of a material are considered diagnostic for the formation of hydrogen bonds. More specifically, the frequency shifts and intensity changes in the A-H stretching mode in the system A-H - - - B provide the fundamental parameters that have to be satisfied by any theoretical description. In this section we will be largely concerned with what happens to the A-H stretching mode, v_s, upon hydrogen bond formation. The low frequency A-H - - - B stretching mode, σ_s, is also diagnostic, but much more difficult to identify and observe (because of its location in the far infrared, coupling to other modes, etc.). In certain materials the bond connecting the atom B to the rest of its molecule also shows a degree of sensitivity. For example, if B is the oxygen of a carbonyl in a nylon we have:

$$\diagdown \!\!\! \diagup N-H ---O=C \diagup \!\!\! \diagdown$$

where the carbonyl stretching mode "loses" some of its double bond character upon hydrogen bonding and shifts to lower frequency, as illustrated for nylon 11 in the preceding section. Because the mass of the oxygen atom is much larger than a proton (and also in this example because the C=O is much stiffer than the N-H bond), the degree of sensitivity to perturbations, such as anharmonicity, is far less than the diagnostic N-H stretching mode. As we will see, this makes it extraordinarily useful, because we can readily identify "free" and "bonded" species in the spectrum, and it is the quantitative determination of the fraction of each of these species that will provide the link to thermodynamic quantities.

The in-plane and out-of-plane bending motions of the proton in the A-H - - B system are also sensitive to hydrogen bonding, but these motions are often coupled to other molecular vibrations. For example, in molecules containing amide groups, the in-plane N-H bend is coupled to C-N stretching and other modes, to give the so-called amide II and III vibrations. The amide II band (near 1530 cm^{-1}) shows a marked degree of sensitivity to hydrogen bonding, shifting significantly in frequency, but it is usually difficult to separately detect bands that can be assigned to "free" and bonded groups, so that these modes are not useful for our purposes and we will neglect them. We will confine most of our discussion to the stretching modes.

Potential Energy Diagrams for the A-H - - B System

In the preceding section we briefly considered the normal modes of the A-H - - B system. It is useful at this point to consider the form of the potential function describing the stretching modes, as this provides considerable insight into the spectroscopic changes we might expect as a function of the strength of the hydrogen bond, and provide a basis for the discussion of observed spectra.

There are a number of simple potential functions that could be applied to the A-H - - B system, but we do not need to consider these in any detail and for our purposes it is sufficient to consider the general shape of the curves that would correspond to the various species in the following equilibrium scheme:

$$
\begin{array}{c}
\qquad\qquad\qquad \text{(b)} \\
\text{AH} + \text{B} \rightleftharpoons \text{A-H- - -B} \\
\;\text{(a)} \qquad\qquad\qquad \updownarrow \\
\text{(A- - -H- - -B) (c)} \\
\updownarrow \\
\text{A- - -H-B} \rightleftharpoons \text{A}^{\ominus}\; {}^{\oplus}\text{HB} \\
\qquad\text{(d)} \qquad\qquad \text{(e)}
\end{array}
$$

These represent a non-hydrogen bonded system (a), hydrogen bonded systems of increasing strength (b and c) and the situation where proton transfer and disassociation occurs (d and e). Figure 5.16a shows the type of potential function that would describe a lone A-H group (cf section 5B). A Morse potential or the Schroeder-Lippincott semi-empirical function, formulated specifically for hydrogen bonded systems[13-15], fits the data well. Superimposed upon this (dotted lines) is the potential curve characteristic of a harmonic oscillator. For transitions between the ground and first excited state the two are close for *most vibrations*.

Anharmonic effects usually come into play in combination modes, where the curves start to diverge. A-H stretching modes of "free" or non-hydrogen bonded groups involve (relatively) large displacements, because of the small mass of the proton.

Potential Energy Diagrams
Diatomic Molecule A-H

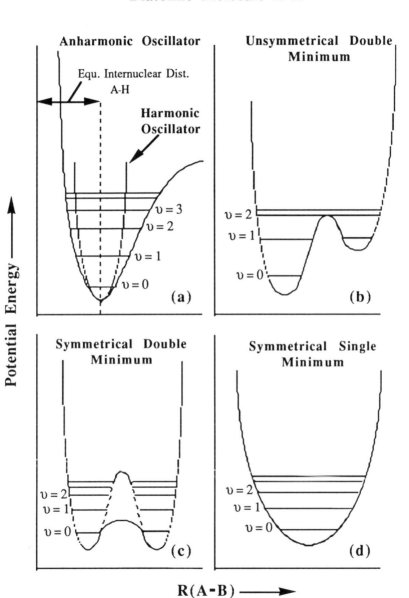

FIGURE 5.16

Presumably, this is partly responsible for an overtone band seen in the near infrared, described in the following section. Protons in hydrogen bonded groups sit in an oscillating field generated by the slower motions of the A and B atoms, and this apparently also leads to anharmonic effects.

Figure 5.16b displays the situation that occurs in weak or moderately strong hydrogen bonds. The hydrogen atom is located closer to one atom than the other (states (b) and (d) in the equilibrium scheme shown above). The potential curve is correspondingly asymmetric with a double minimum potential. The lower of the two minima is located near the donor atom. Note that in figure 5.16b the first overtone is predicted to be split, because the second potential energy minimum lies in the region of the $\upsilon=2$ vibrational energy level. This splitting has been observed in certain complexes (e.g., phenol with bases).

As the strength of the hydrogen bond increases and the length of the hydrogen bond decreases, potential functions of the type shown in figures 5.16c or 5.16d provide a better description. The hydrogen atom is not so clearly associated with one atom rather than the other. In certain special cases reliable crystallographic evidence has been obtained demonstrating the existence of truly symmetric hydrogen bonds. For strong hydrogen bonds we would therefore expect the potential function to possess two equal (or nearly equal) potential minima with a low barrier between the two (figure 5.16c), or be truly symmetric with one broad minimum. Strong hydrogen bonds lead to the appearance of extremely unusual spectra, and we will now consider the appearance of the A-H stretching mode in more detail.

The spectroscopic characteristics of various types of hydrogen bonds have been nicely summarized by Hadzi[16,17,] Odinokov, et al.[18] and in the special case of acid salts, Speakmen[19]. Essentially, we will consider spectra to be characteristic of four categories of hydrogen bonds; weak, medium, intermediate and strong. A key point is that strong hydrogen bonds do not necessarily represent an extrapolation of the characteristics of their weaker brethren; they can represent peculiar and challenging problems[16].

Table 5.2

Classification of Hydrogen Bonds

Type	IR Frequency Shift $\Delta\upsilon$ cm^{-1}	Enthalpy ΔH kcal. mol^{-1}	Examples
Weak	10 - 50	≈ 1	PVC - Polyesters
Medium	≈ 300	≈ 5	-OH; Amide, Urethane (Self-Association)
Intermediate	≈ 600	6 - 8	-COOH (Self-Association)
Strong	800 - 2000	> 8	Acid Salts (-COOH / NHP)

This approximate classification is summarized in Table 5.2, together with approximate frequency shifts and enthalpy changes for each category. (It should be kept in mind that a "weak" hydrogen bond is not the same as the "weak" interactions defined in Chapter 2, but can be considerably stronger.)

Having set the stage by setting up some (more or less) arbitrary criteria that allow us to place various hydrogen bonded systems in certain categories, we will now consider the spectroscopic characteristics of each type. It should be kept in mind that this picture is painted with a broad brush and certain systems show intermediate behavior (nature is awkward like that).

Weak hydrogen bonds are of the type that might occur between certain CH groups and halogens in, for example, poly(vinylidene fluoride) and PVC, and these same CH groups with ester carbonyls in various blends. Although the existence of similar hydrogen bonding interactions in polymers[20] have been questioned, they have been observed in low molecular

weight materials and the evidence is reviewed in the book by
Green[21]. Because of the weak nature of the interaction,
however, they are extremely difficult to detect unambiguously
using infrared spectroscopy and we will not consider them
further here.

Turning to medium strength hydrogen bonds anybody who
has dealt with the infrared spectra of nylons, urethanes,
phenolics etc. is familiar with the appearance of the N-H or O-H
stretching region of the spectrum. A fairly prominent, broad
band due to hydrogen bonded groups is observed near 3300
cm^{-1}, while a weak shoulder due to "free" N-H groups appear
near 3400 cm^{-1} or near 3500-3600 cm^{-1} for free OH groups.

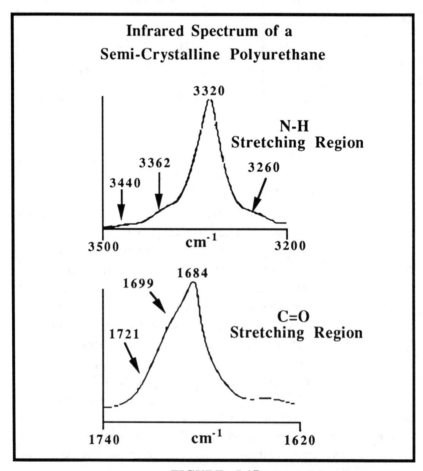

FIGURE 5.17

This region of the spectra of amides and urethanes is further complicated by anharmonic effects, which result in the appearance of overtones and combinations involving the carbonyl stretching mode. For example, scale expanded spectra of a semi-crystalline polyurethane in N-H and C=O stretching regions are shown in figure 5.17. In the N-H stretching region it is clear that the band envelope is composed of at least four bands. The modes at 3440 cm^{-1} and 3320 cm^{-1} are assigned to the NH stretch of the "free" and hydrogen bonded species, respectively[22]. Corresponding bands appear in the spectra of nylons[9]. The other two bands at about 3362 and 3260 cm^{-1} are assigned to overtones and combinations. Bands near 3200 cm^{-1} are also observed in the spectra of nylons, but the mode near 3362 cm^{-1} has no observable counterpart in spectra characteristic of the amide group. This is because the carbonyl stretching mode of urethanes occurs at a higher frequency, near 1680 cm^{-1} rather than 1640 cm^{-1}. The 3362 cm^{-1} band is probably the overtone of the fundamental near 1680 cm^{-1}. Overtone and combination modes are usually weak in intensity, but if their symmetry properties are appropriate and if they are close in energy (i.e., frequency) to a fundamental mode (in this case, the N-H stretch), they can "steal" intensity from it. The fundamental process involved is termed Fermi resonance and a discussion of this and other aspects of the role of anharmonicity in hydrogen bonding can be found in a review article by Sandorfy[23].

As an aside, anharmonicity is thought to be at least partly responsible for the unusual breadth or width at half-height of hydrogen bonded N-H and O-H vibrations. We have yet to discuss this aspect of the appearance of A-H stretching modes or a related topic, the extraordinary changes in intensity that can occur as a function of temperature. We will consider these separately when we consider quantitative measurements.

As we go from hydrogen bonds that we categorize to be of medium strength, to those we label intermediate, then the A-H stretching mode shifts to the 3100-2800 cm^{-1} range and so-called "satellite bands" can often be seen superimposed upon the broad fundamental profile. Polymers containing carboxylic acid groups hydrogen bonded in the form of characteristic pairs and the hydrogen bonds formed between phenol and pyridine are examples of this type. The spectra of poly(vinyl pyridine), poly(vinyl phenol) and a blend of these polymers is shown in figure 5.18[24].

It can be seen that the band near 3330 cm^{-1}, characteristic of O-H to O-H hydrogen bonds in pure poly(vinyl phenol) is replaced by a broad band whose peak position is difficult to identify precisely because it is centered beneath the CH stretching modes that lie between 2900 and 3100 cm^{-1}. Weak satellite bands near 2600 cm^{-1} can also be observed. These satellite bands could be due to overtones and combinations that are intensity enhanced by Fermi resonance.

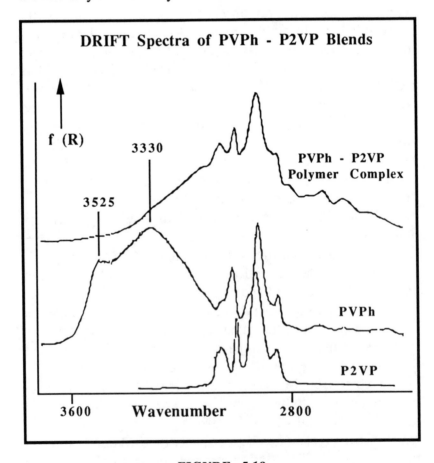

FIGURE 5.18

Because of the anharmonicity of the hydrogen bond, however, and the increasing effect of anharmonicity with hydrogen bond strength, there is another possibility, anharmonic coupling of the v_s (A-H stretching) and σ_s (H - - - B stretching) modes. This could give rise to satellite bands situated at wavenumbers $v_s \pm \sigma$ about the v_s peak. This is an appealing mechanism, but in a review Hadzi and Bratos[25] cite a considerable amount of evidence supporting the overtone-combination interpretation. In terms of the problem we wish to tackle this difference in interpretation doesn't matter much as we simply want to use the spectrum as a tool to estimate hydrogen bond strength and, more crucially, measure the number of hydrogen bonds present.

As we move from intermediate to strong hydrogen bonds, the spectra become increasingly peculiar, splitting into various components that are located throughout the mid-infrared range (for example, see the spectra reported by Odinokov et al.[18]). Spectra such as these have not often been obtained from polymers or polymer mixtures. One of the few examples is at of a polyamic acid model compound[26]:

complexed with N-methyl pyrrolidinone (NMP) whose spectra are shown in figure 5.19. In the spectrum of the pure material a strong N-H stretching mode, split into a doublet, appears near 3300 cm^{-1}, while a broad band centered near 3000 cm^{-1} can be assigned to hydrogen bonded carboxylic acid pairs. "Satellite" bands appear near 2600 cm^{-1}. Thus in this material we see bands characteristic of both medium and intermediate strength hydrogen bonds. NMP forms strong hydrogen bonds with carboxylic acid groups, however, and in the spectrum of the complex three characteristic broad bands centered near 2900, 2400 and 1900 cm^{-1}, together with various satellite bands, can be observed.

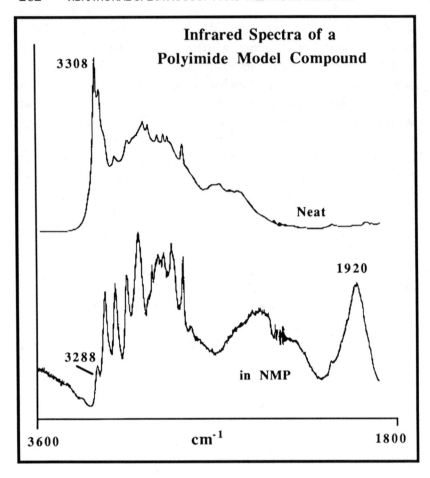

Infrared Spectra of a Polyimide Model Compound

3308

Neat

1920

3288

in NMP

3600   cm⁻¹   1800

FIGURE 5.19

These are characteristic of the "strong" hydrogen bonds identified by Hadzi[16,17] and Odinokov et al.[18] (and labeled "A", "B", "C" bands by Hadzi). Strong hydrogen bonds have also been observed in acid salts and thus could be an important characteristic of the structure of partially exchanged ionomers. There has been little work in this area up until now (January 1991), but this could change[27], so we will briefly outline the spectroscopic characteristics of such systems.

In a review article Speakman distinguishes between type A and type B acid salts[19].* In type A the proton is thought to be symmetrically placed (or in "pseudo" type A, almost symmetrically) and the hydrogen bond is very short. A broad band between 600 and 1200 cm[-1] is identified as the hydrogen bond stretching mode, which is now the vibration of a proton in a symmetric A -- H -- A type molecule (see preceding section for a discussion of potential functions). In type B structures the spectra of the free acid and the neutral salt seem to be superimposed, but the situation is confusing. Rather than observing spectra characteristic of COOH pairs, which you would anticipate from this description, the Hadzi A, B, C bands between 3000 and 1800 cm[-1] seem to appear i.e., an intermediate to strong but not symmetric H bond. In any event, we interpret type A to imply a truly symmetric (or close to) placement of the proton, while the type B is somewhat weaker with longer A - - A or A - - B distances and a corresponding non-symmetric placement. In dicarboxylic acids it is apparently much more common to find type A acid salts, with potassium hydrogen bis-phenylacetate, $(KH[_6H_5CH_2CO_2H]_2)$, being the classic example. (The structure of this complex has been determined by neutron diffraction.)

Bond Lengths, The Enthalpy of Hydrogen Bond Formation and Frequency Shifts

In this book we will be primarily concerned with the use of infrared spectroscopy to measure the number of free and hydrogen bonded groups. Although the variation of this balance with temperature is governed by the enthalpy of hydrogen bond formation, this latter quantity can be easily determined by measuring changes in the proportion of each species as a function of temperature and defining an appropriate equilibrium constant, as we discussed in the previous chapter on the

---

* The nomenclature in this field is a bit on the random side, with type A and type B structures (Speakman ref. 19), Type i and type ii (Hadzi, ref. 16) and Type I and type II (Odinokov, et al., ref. 18). Throw in A, B, C, and D and E bands and you've got potential confusion.

stoichiometry of hydrogen bonding. There are relationships between A-H frequency shifts and the enthalpy and bond lengths of hydrogen bonds, however, and for completeness we will briefly discuss them here (they have, on occasion, been applied to polymers). For low molecular weight materials there is a mass of data concerning the enthalpy of hydrogen bond formation and infrared frequency shifts.

The data for a large number of systems has been tabulated by Murthy and Rao[28] and (to a lesser extent) Pimentel and McClellan[29].

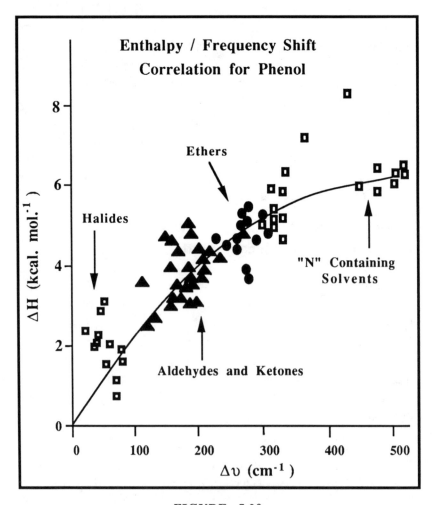

FIGURE 5.20

A number of enthalpy / infrared frequency shifts have been published and are reviewed (once more) by Pimentel an McClellan[29] and Dean Sherry[30]. A compilation of the data for phenols[31] is shown in figure 5.20. We will not review such correlations any further because, unsurprisingly, they vary from system to system. Nevertheless, one would intuitively expect a relationship between the depth of the potential curve and the vibrational energy levels involved in frequency shifts. It's just that such a relationship would also involve other variables, such as the other bonds in the system, the nature of the solvent, etc. As Hadzi and Bratos[25] point out, it is clear that the linear $\Delta H/\Delta \upsilon$ relationships established each involve only a narrow group of chemical types of molecules and well defined circumstances (i.e., phenols - esters will have a different relationship to phenols - ethers) [#].

The correlation between spectroscopic and crystallographic data has been reviewed by Novak[32]. For weak hydrogen bonds the correlation between υ_{A-H} and A - - B distance is excellent, particularly for O-H -- O hydrogen bonds. For stronger bonds there is the complication of identifying the precise location of the υ_{A-H} band, considering that a complex of modes appears in the spectrum (see figure 5.19). The frequency was determined by calculating the center of gravity of the bands (i.e., assuming that all intensity arises from the A-H stretching fundamental). The uncertainties created by band broadening in the spectra of weaker H-bonds were overcome by studying isotopically mixed crystals. Even so, one might still expect uncertainties in the spectroscopic and crystallographic data to produce major scatter in plots of υ_{A-H} vs $R_{A \ldots B}$. Not so, the correlations are excellent, and the interested reader is referred to Novak's article[32].

[#] The correlation indicated in figure 5.2 appears awful, but this is because data from a range of studies using many different techniques has been collected. Individual studies of particular groups of molecules often give good straight line correlations.

Band Widths and the Temperature Dependence of Absorption Coefficients

Given the sensitivity of the A-H stretching mode to the degree of hydrogen bonding and the strength of individual hydrogen bonds, quantitative measurements of the various A-H bands in a spectrum would seem to offer the best chance for the determination of the species present. Indeed, for low molecular weight materials measurements of the number of "free" groups, at least, are routinely made. Initial attempts to use the A-H stretching modes in the infrared spectra of polymers, however, often produced results that made no sense. For example, values of the enthalpy of hydrogen bond formation between amide groups ranging from 8 to 14 kcal. mol[-1] were obtained[33,34], which can be compared to values of about 5 kcal. mol[-1] for low molecular weight amides. The main problems are:

1. The band due to "free" A-H groups is usually extremely weak and overlaps the strong band due to hydrogen bonded species. Its intensity is thus difficult to measure with the required degree of accuracy.

2. The hydrogen bonded A-H band is very broad and its absorption coefficient is a very strong function of temperature.

These problems are illustrated in Figures 5.21(a) and (b), which show the N-H stretching region of the infrared spectrum of an amorphous polyamide and polyurethane, respectively. With increasing temperature the broad band due to the hydrogen bonded group shifts to higher frequency and decreases in intensity, dramatically so in the case of the amorphous polyurethane. The weak band due to free N-H, near 3444 cm[-1], apparently increases in intensity, but it is possible to show that much of this change is due to the increase in overlap with the band due to the hydrogen bonded species as it shifts to higher frequency[11,22]. Why is the A-H band so broad and why does its intensity depend so strongly on temperature?

Sandorfy[23] has reviewed the question of the unusual breadth of the hydrogen bonded A-H stretching mode and essentially concludes that it is due to both anharmonicity and disorder.

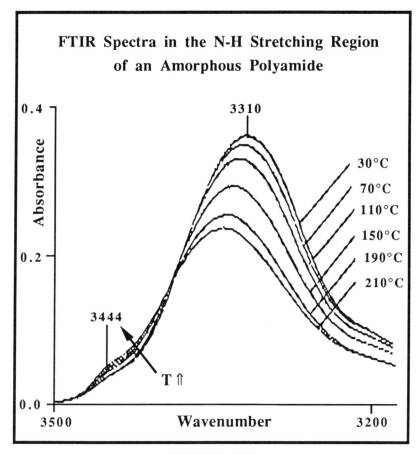

FIGURE 5.21

As we noted above, bands due to overtones and combinations often appear in this region of the spectrum and significantly overlap the hydrogen bonded N-H fundamental. It is also possible that the A-H stretching mode is anharmonically coupled to oscillations of the A - - B atoms, although this is far more likely in systems where strong hydrogen bonds occur. Even allowing for overtones, combinations, (e.g., by careful curve resolving or choice of system), etc. the A-H stretching mode remains unusually broad and in our opinion the most significant cause is disorder. Motions of the proton are fast relative to slow A - - B oscillations, so that A-H stretching vibrations occur in a slowly oscillating field. The distribution of

A - - B distances is then further broadened by the inherent disorder found in liquids and amorphous polymers, where one would anticipate that there would be a distribution of equilibrium A - - B distances.

We have used this concept of disorder to interpret the changes in intensity that occur as a function of temperature[11]. We assume that at a particular temperature the distribution of the "strengths" or bond lengths of hydrogen bonded N-H groups is Gaussian, centered about a mean position, v_1, which corresponds to the average strength of the intermolecular interaction. At some elevated temperature, T_2, the average strength of the hydrogen-bonded N-H groups will diminish, as the volume of the polymer and the average intermolecular distance between chains increases with increased thermal motion. We will also assume that the form of the distribution of the strengths of the hydrogen-bonded N-H groups at T_2 is identical to that occurring at T_1, the only difference being the mean position, v_2, where the frequency shift v_2-v_1 reflects the reduction of the average strength of the hydrogen bonds on raising the temperature.

If the broad hydrogen-bonded N-H band observed in the experimental spectrum reflects only the normalized weight fraction distribution of hydrogen-bonded species, then we would anticipate identical curves of the same area centered at different frequencies. This is not what is observed and we will assume that the reduction in area observed experimentally upon increasing the temperature from T_1 to T_2 is solely due to a variation of absorption coefficient with the hydrogen bond strength and hence frequency, $a(v)$. This naturally assumes that the concentration of "free" N-H groups changes only a small amount in the temperature range considered, a result that has been obtained experimentally[11].

The intensity of the hydrogen-bonded N-H band at T_1 may thus be described by a function of the form:

$$I_1(v) = a(v) \frac{1}{\sigma (2\pi)^{0.5}} \exp\left[-\frac{1}{2}\left(\frac{v - v_1}{\sigma}\right)^2 \right] \quad (5.20)$$

This equation is made up to two contributions, namely a term that accounts for the variation of absorption coefficient with frequency and a term that describes the Gaussian distribution of species present. Similarly, at T_2:

$$I_2(\upsilon) = a(\upsilon) \frac{1}{\sigma (2\pi)^{0.5}} \exp\left[-\frac{1}{2}\left(\frac{\upsilon - \upsilon_2}{\sigma}\right)^2\right] \quad (5.21)$$

Dividing these two equations and taking the natural logarithm of both sides yields:

$$\ln\frac{I_1(\upsilon)}{I_2(\upsilon)} = -\frac{1}{2}\left(\frac{\upsilon - \upsilon_1}{\sigma}\right)^2 + \frac{1}{2}\left(\frac{\upsilon - \upsilon_2}{\sigma}\right)^2 \quad (5.22)$$

Substitution of $\upsilon_2 = \upsilon_1 + \Delta\upsilon$, expanding, rearranging, and collecting terms allows the equation to be cast in the form:

$$\ln\frac{I_1(\upsilon)}{I_2(\upsilon)} = -\left(\frac{\Delta\upsilon}{\sigma^2}\right)\upsilon + \frac{\Delta\upsilon}{\sigma^2}\left(\upsilon_1 + \frac{\Delta\upsilon}{2}\right) \quad (5.23)$$

This is a linear equation and a plot of the left-hand side of equation 5.23 vs. frequency, υ, should yield a straight line with a slop of $-(\Delta\upsilon/\sigma^2)$ and an intercept of $(\Delta\upsilon/\sigma^2)$ $(\upsilon_1 + \Delta\upsilon/2)$. From this the unknown parameters υ_1 and $\Delta\upsilon$ can be determined. The only other unknown parameter is the standard deviation, σ. The width at half-height of the hydrogen-bonded N-H stretching band is remarkably constant, having values lying between 130 ± 2 cm^{-1} over the range 30-210°C.[11] From a table of the ordinates of a standard normal curve, the standard deviation may be found by dividing the width at half-height by the corresponding number of standard deviations, in this case 2.34. Accordingly, we have assumed a width at half-height of 130 cm^{-1}, which yields a value of 55.6 for σ. It did not escape our attention that this assumption could lead to significant errors in the subsequent calculation of υ_1 and $\Delta\upsilon$. Changes in the value of σ, however, do not materially affect the overall form of the results and the arguments we wish to make. They only affect the absolute numbers.

Returning to equation 5.23, we may obtain the values of $I_1(\upsilon)$ and $I_2(\upsilon)$ from a pair of experimental spectra recorded at two different temperatures, T_1 and T_2, and the values of the slope and intercept calculated from a least-squares regression. For example, using the spectral data obtained at 30 and 210°C, we calculated $\Delta\upsilon$ to be 22.7 cm^{-1} and a value υ_1 equal to 3367 cm^{-1}, with a correlation coefficient of better than 0.99. This suggests that the peak maximum representing the average position of the distribution of the strengths of the hydrogen-bonded N-H groups at 30°C is at 3367 cm^{-1}. The experimentally observed peak maximum is seen at 3310 cm^{-1}. This, in turn, means that the experimentally observed peak maximum occurs at a lower frequency than that of the "true" peak maximum representing the distribution of the strengths of the hydrogen bonds, because it has been multiplied by a non-linear function describing the variation of absorptivity coefficient with frequency. (Incidentally, if the above hypothesis is correct, it may be one of the reasons why it has not been possible to *generally* scale the frequency difference between "free" and hydrogen-bonded N-H or O-H bands with the enthalpy of dissociation.)

The average value of υ_1 was obtained from the spectral data recorded at various temperatures with respect to the reference spectrum recorded at 30°C. To our delight, an average value for υ_1 of 3367 ± 3 cm^{-1} was determined. (Naturally, the values of $\Delta\upsilon$ varied, but were entirely consistent with the experimental observations). One can now substitute back into equation. 5.20 to obtain the absorptivity coefficient as a function of frequency.

The results of this calculation are displayed graphically in figure 5.22. A very strong dependence of the absorption coefficient on frequency is immediately apparent. In fact, the results indicate that the absorption coefficient effectively doubles over the 40 cm^{-1} range from 3320 to 3360 cm^{-1}.

Although the above model appears intuitively satisfying, we would be the first to admit that there are elements of a "circular argument" present. The reduction in area of the hydrogen -

Absorption Coefficient vs. Frequency

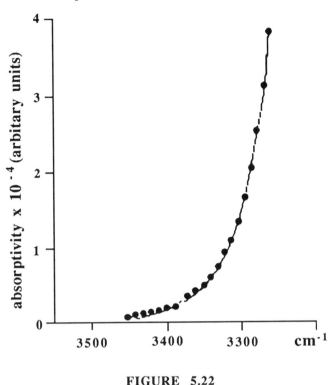

FIGURE 5.22

bonded N-H band with temperature has been assumed to be solely due to the change in absorption coefficient with frequency and we employed the experimental data to "back calculate" the $a(\upsilon)$ function. In fact, for the specific case of the amorphous polyamide presented here, where both the theoretical distribution of hydrogen-bonded strengths and the experimentally observed hydrogen-bonded N-H band are assumed to be Gaussian with the same standard deviations, it can be demonstrated mathematically that the $a(\upsilon)$ function is exponential in form. There is, however, no *a priori* assumption of the form of the $I(\upsilon)$ function in the model. Nevertheless, it is clear that there is a very strong dependence of the absorption coefficient upon hydrogen-bonding strength and it can be shown that this strong variation of absorption coefficient with frequency has some basis in fundamental theory. The intensity, I_i, of a given normal mode is proportional to the square of the dipole moment change with normal coordinate. For the purposes of this discussion it is only necessary to consider an isolated N-H bond:

$$I_i = C \left(\frac{\delta \mu}{\delta Q} \right)^2 \qquad (5.24)$$

The dipole moment, μ, is related to the bond distance, r, and the effective charge, q, by:

$$\mu = q r \qquad (5.25)$$

Differentiation with respect to the normal coordinate (stretching mode) leads to:

$$\frac{\delta \mu}{\delta Q} = q \frac{\delta r}{\delta Q} + r \frac{\delta q}{\delta Q} \qquad (5.25)$$

It is well established that the N-H bond length increases upon hydrogen bonding and the effective charge also changes with hydrogen bond strength. (see preceding section). Accordingly, changes in the parameters given in equation 5.26 would lead to a change in $\delta \mu / \delta Q$. As intensity is proportional to the square of the dipole moment change with normal coordinate (equation 5.24), the absorption coefficient should be a strong function of frequency.

E. QUANTITATIVE INFRARED SPECTROSCOPIC MEASUREMENTS OF THE FRACTION OF HYDROGEN BONDED GROUPS

Having considered various fundamental aspects of the vibrational spectra of hydrogen bonded systems we are now in a position to discuss the measurement of the fraction of hydrogen bonded (f_{HB}) and free ($f_F = 1-f_{HB}$) species. As we saw in

Chapter 4, these quantities can be related to equilibrium constants that (in principle) describe the stoichiometry of hydrogen bonding. We will demonstrate in this section that the distribution of hydrogen bonded species is described extremely well by association models and that we can indeed use "transferable" equilibrium constants in certain systems. We will divide our discussion into two principle parts, the measurement of self-association equilibrium constants and the determination of inter-association constants.

Self-Association Equilibrium Constants

In describing self-association equilibrium constants in polymer materials, one can further differentiate between two classes of polymers; those that contain carbonyl groups and those that do not. It is usually a fairly straightforward matter to determine the fraction of hydrogen bonded species when carbonyl groups are present, as the sub-title to this chapter might suggest, but can be frustratingly difficult when they are not. We will start with the less difficult problem, polymers containing carbonyl groups.

Earlier in this chapter (section C) we presented infrared spectra of nylon 11 recorded as a function of temperature (figure 5.10). At temperatures above the melting point of this polymer the amide I region of the spectrum can be curve resolved into two bands, one due to hydrogen bonded groups and one due to free groups, the former having the frequency and band shape characteristic of the amorphous phase. In early work on the vibrational spectra of hydrogen bonded systems, it was often assumed that the absorption coefficients of the free and bonded bands were identical, but we now know that this is not so (in some of our early work we also committed this original sin). We would not anticipate these differences to be as large as those observed between free and bonded A-H groups, however, as the mass of the oxygen atom and its connection to the carbon atom by a stiff double bond make it less susceptible to perturbations. There is certainly a degree of sensitivity, but C=O band shifts are often of the order of 20-30 cm^{-1} upon hydrogen bonding rather than the 200-300 cm^{-1} observed for the O-H or N-H mode. Similarly, the absorption coefficient of the hydrogen bonded band is somewhat greater than that due to free groups, but is far less sensitive to temperature[9,11,22]. That suggested to us[35] a method for determining the *ratio* of the absorption coefficients for the free and bonded bands, which is all we require for quantitative measurements of the relative proportions of these species and, additionally, has the immense advantage of not requiring a measurement of film thickness. Essentially, if in a sample at temperature T_1 the areas of the hydrogen bonded and free bands are A_{HB} and A_F, respectively, then using equation 5.18, the total concentration of carbonyl groups is given by:

$$c = \frac{(A_{HB})_{T_1}}{b\,a_{HB}} + \frac{(A_F)_{T_1}}{b\,a_F} \tag{5.27}$$

At temperature T_2 the total concentration and film thickness are not changed (one must check for example, that flow has not occurred by measuring the intensity of a band that is insensitive to hydrogen bonding, such as a CH stretching mode), so that we can write:

$$c = \frac{(A_{HB})_{T_2}}{b\,a_{HB}} + \frac{(A_F)_{T_2}}{b\,a_F} \tag{5.28}$$

where it is assumed a_{HB} and a_F are also unaltered. This seems to be a reasonable assumption providing that T_2-T_1 is of the order of 100°C or less (changes in a_{HB} appear to be less than 5%)[35,36]. Subtracting:

$$\frac{a_{HB}}{a_F} = \frac{(A_{HB})_{T_2} - (A_{HB})_{T_1}}{(A_F)_{T_1} - (A_F)_{T_2}} \tag{5.29}$$

The fraction of hydrogen bonded or free groups is then given by:

$$f_{HB} = (1 - f_F) = \frac{A_{HB}}{A_{HB} + A_F\left(\dfrac{a_{HB}}{a_F}\right)} \tag{5.30}$$

Table 5.3
Absorptivity Ratios

Carbonyl Type	Type of Association	a_{HB} / a_F	Reference
Amide	Inter-association in aromatic nylon - polyether blends	1.2	38
Ester	Inter-association in phenoxy blends	1.23	37
Acetate	Inter-association in PVPh blends	1.5	35
Carboxylic Acid	Self-association and Inter-association in EMAA - polyether blends	1.6	7
Urethane	Self-association in aliphatic polyurethane	1.7	22

A similar procedure can be applied to carbonyl bands engaged in inter-association (e.g., esters) and typical values of a_{HB}/a_F are summarized in Table 5.3, falling in the range 1.2 - 1.7.

If self-association can be described by a single equilibrium constant K_B, and this is certainly true of polymers containing carboxylic acid groups when association consists of the formation of hydrogen bonded pairs, and is *apparently* also a good assumption for polymers containing amide and urethane functional groups, then K_B is simply calculated from the equations given in Chapter 4; for self association in chains:

$$K_B = \frac{(1 - f_F^o)}{\left[f_F^o \right]^2}$$

(5.31)

for self-association as cyclic dimers:

$$K_B = \frac{(1 - f_F^o)}{2\left[f_F^o\right]^2} \tag{5.32}$$

In calculating phase behavior we will require a knowledge of the variation of the equilibrium constants with temperature, which can be obtained from the usual van't Hoff relationship (see Chapter 4, equation 4.70). We will present examples later in this section.

We now turn our attention to the measurement of self-association equilibrium constants in polymers that do not have carbonyl groups. As we demonstrated in the preceding section, the absorption coefficient of the *hydrogen bonded* N-H band varies considerably with temperature, so that the procedure described above, based on measuring the ratio of the free and bonded bands at different temperatures, cannot be used. Furthermore, the intensity of the free band is (in most homopolymer spectra) very weak and significantly overlapped by the broad band due to hydrogen bonded species (see figure 5.17), so that measurements of peak height or area are subject to considerable error, even with the most carefully and rigorously applied curve-resolving procedures. In amides and urethanes we fortunately have an alternative to measurements in the N-H stretching region, but in polymers containing O-H groups, such as poly(vinyl phenol) (PVPh), and those containing certain types of N-H groups, we do not.

The absorption coefficient of the *free* O-H or N-H band is apparently not particularly sensitive to temperature and in low molecular weight materials, such as phenol, equilibrium constants can be obtained by solution studies in cells of known thickness. The solvent employed is usually "inert," i.e., non-hydrogen bonding. Unfortunately, we cannot apply the same procedure to polymers such as PVPh, because they usually will not dissolve in such solvents. One alternative is to synthesize a series of random copolymers, for example vinyl phenol-co-styrene, of systematically varying composition. The number of phenol O-H groups decreases with increasing styrene content, which in turn, increases the number of free O-H groups, as illustrated in figure 5.23. By establishing an "internal standard" band, quantitative studies can be made. This is clearly a time consuming business, however, but in certain cases may prove to be a necessary one. At the time of writing this book, such detailed studies have yet to be made. Instead, we have made the assumption that we can use equilibrium constants derived from low molecular weight materials to describe the associations in polymers containing the same types of functional groups. As we saw in Chapter 4, this follows directly from the use of a Flory lattice model and simply requires that we account for the difference in the molar volume of the species. For PVPh, therefore, we could obtain equilibrium constants using:

$$K_B^{Ph} V_B^{Ph} = K_B^{PVPh} V_B^{PVPh} \qquad (5.33)$$

where V_B^{Ph} is the molar volume of the phenol molecule and V_B^{PVPh} the chemical repeat unit of poly(vinyl phenol), respectively.

The validity of this approach depends upon a number of things, but the most crucial factors appear to be:

1. We do not introduce unusual steric or electronic effects on going from the low molecular weight model compound to the polymer.

2. The equilibrium constant is independent of the molecular weight of the polymer chain.

3. The polymer chain is sufficiently flexible that hydrogen bonds can form according to their intrinsic proclivities.

Problem (1) is straightforward enough; we would not expect the substitution of an alkyl group in the para position relative to the O-H group to significantly affect equilibrium constant of phenol relative to PVPh, for example. If it did, we could alternatively use p-cresol as our model system. The equilibrium constant is changed due to the diluting effect of the volume occupied by the alkyl group, but this is accounted for by equation 5.33. Problems (2) and (3) are much more significant, however.

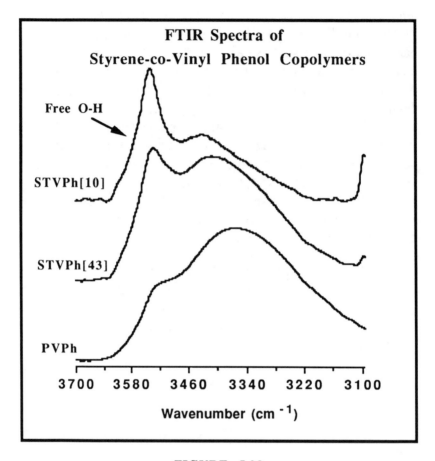

FIGURE 5.23

Stadler[39] has argued that association constants depend upon chain length. We do not agree, for reasons that we will present in the following chapter. Instead, we think the crucial factor will prove to be (3), chain stiffness. The Flory lattice model that we employed in Chapter 4 assumes that the chains are perfectly flexible. The chains that we referred to in that chapter were small molecules, such as phenol, linked by hydrogen bonds. In polymers that hydrogen bond we have two sets of chains, however, one covalent and the other hydrogen bonded, and it is possible (or even likely) that the *covalent* chain stiffness can affect the number of hydrogen bonds that can form and so alter the equilibrium constants. *This should not be a problem in polymer systems where the equilibrium constants can be measured directly by experiment, in that these quantities should then reflect such factors,* but could be a problem when we transfer equilibrium constants from small molecules. Fortunately, in our blend work we have considered mixtures of PVPh with ester containing polymers, so that the measurement of the number of free and bonded groups in these blends can serve as a check on the degree of validity of our assumption of transferability for the system phenol / PVPh. We will consider these results shortly. First, we wish to discuss the measurement of equilibrium constants in low molecular weight materials, as there are various problems that must be considered and not all reported equilibrium constants can be transferred to polymer systems.

Measurement of Self-Association Constants by Solution Studies of Small Molecules

We will use phenol to illustrate the determination of equilibrium constants from the measurement of the free O-H (or N-H) band. The self-association of this molecule has been studied by Whetsel and Lady.[40] This splendid paper has the sometimes neglected virtue of tabulating all data obtained, in this case band intensities in both the near and mid-infrared regions of the spectrum, so that we were able to reconsider the use of various association models and perform our own calculations. We reached the same conclusions as these authors, however, and present their results and conclusions here.

Whetsel and Lady considered a general association scheme where the equilibrium constants are initially expressed in terms of concentration, c (mol. l^{-1}), in the following form:

$$h \, c_{B_1} \underset{\longleftarrow}{\overset{K_h}{\longrightarrow}} c_{B_h} \qquad K_h = \frac{\left[c_{B_h} \right]}{\left[c_{B_1} \right]^h}$$

$$(5.34)$$

We continue to use the subscript B to represent a self-associating component, so that c_{B_h} represents the concentration of species consisting of h hydrogen bonded phenol molecules. If we now consider the infrared absorption due to "free" OH groups, this consists of the contributions from "monomers," by which we mean phenol molecules that have no hydrogen bonded partners at all, plus the contributions from the end groups of dimers, O-H- - -O-H, trimers, and so on. Using equation 5.34 together with the equation relating the absorbance (A) of a band to the cell thickness (b), absorption coefficients (a_i) and concentrations (c_i) of the species in a multicomponent system (equation 5.18), we can write:

$$\frac{A}{b} = a_1 c_{B_1} + a_2 c_{B_2} + a_3 c_{B_3} + \text{---------}$$

$$= a_1 c_{B_1} + a_2 K_2' c_{B_1}^2 + a_3 K_3' c_{B_1}^3 + \text{----} \tag{5.35}$$

where a_1 is the absorption coefficient due to a "monomer," etc. We also have the equation describing the stoichiometry of the system, given by:

$$c_B = c_{B_1} + 2 c_{B_2} + 3 c_{B_3} + \text{---------}$$

$$= c_{B_1} + 2 K_2' c_{B_1}^2 + 3 K_3' c_{B_1}^3 + \text{----} \tag{5.36}$$

where c_B is the concentration of phenol in the solution being considered.

Clearly, there are an infinite number of variables (a_i and K_i') in equations 5.35 and 5.36 and in order to obtain solutions certain assumptions must be made. One such assumption is that the absorption coefficient of free O-H groups is independent of h, so that $a_1 = a_2 = a_3$, etc. The absorption coefficient of monomers is probably different to subsequent h-mers[40] but this can be accounted for if data of sufficient accuracy is obtained. It clearly becomes a less critical problem if fairly concentrated solutions are studied, where c_{B_1} is small. The second set of assumptions concerns the type of association models that are employed. For this purpose it is useful to consider equilibrium constants that describe the stepwise formation of species (as in Chapter 4):

$$B_1 + B_{h-1} \;\;\overset{K}{\rightleftharpoons}\;\; B_h \tag{5.37}$$

If K is assumed independent of h, then in terms of molar concentrations, an equilibrium constant, K_c can be defined (cf equation 4.16) as:

$$K_c = \frac{\left[c_{B_h} \right]}{\left[c_{B_1} \right]\left[c_{B_{h-1}} \right]} \tag{5.38}$$

and the following substitution can be made in equation 5.35:

$$K_h^{'} = K_c^{h-1} \tag{5.39}$$

If dimers have different association constant (K_{2_c}) to subsequent h-mers (k_c), then:

$$K_h^{'} = K_{c_2} K_c^{h-2} \tag{5.40}$$

In a similar fashion, other independent constants can be defined (e.g., K_{c_3}, K_{c_4}, etc.) and the relationships appropriate to the model being considered substituted into equations 5.35 and 5.36, which are then solved iteratively.

Examples of the infrared spectra of phenol - CCl_4 solutions obtained in both the near and mid-infrared are shown in figure 5.24 and it can be seen that the free O-H band is readily measured, particularly in the near infrared where the band due to the hydrogen bonded O-H group is extremely weak (this is possibly because the large displacements allowed the light proton in the "free" state, relative to when it is hydrogen bonded, result in a greater breakdown of the harmonic oscillator approximation and hence, the selection rules that in principle do not permit the appearance of overtones and combinations).

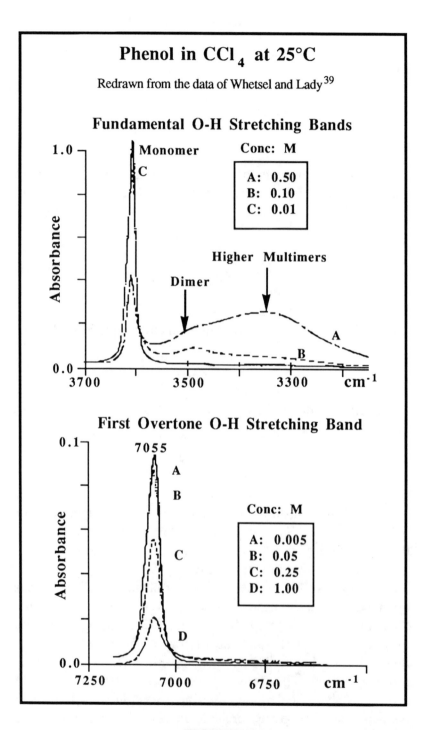

Phenol in CCl$_4$ at 25°C

Redrawn from the data of Whetsel and Lady[39]

Fundamental O-H Stretching Bands

Monomer

Conc: M

A: 0.50
B: 0.10
C: 0.01

Dimer

Higher Multimers

1.0

0.0

Absorbance

3700   3500   3300   cm^{-1}

First Overtone O-H Stretching Band

0.1

7055

A

B

C

D

Conc: M

A: 0.005
B: 0.05
C: 0.25
D: 1.00

0.0

Absorbance

7250   7000   6750   cm^{-1}

FIGURE 5.24

Whetsel and Lady[40] examined the fit of various association models to the data obtained over a range of concentrations and temperatures. Certain models yielded negative formation constants and could be discarded. A model where the formation of dimers is described by a different equilibrium constant to subsequent h-mers gave a good fit to the data over the widest range of concentrations and is the one we have subsequently employed. Before considering the values of these equilibrium constants one additional result of this study needs to be discussed, as it has a general impact on the use of equilibrium constants derived from solution studies.

Many infrared spectroscopic studies of hydrogen bonding in small molecules have used carbon tetrachloride as a solvent as it is usually considered to be inert and non-polar. As Marcus[41] points out, however, there is considerable evidence that CCl_4 (and aromatic hydrocarbons) interact with molecules containing OH groups, possibly through the formation of (very) weak hydrogen bonds.

Effects of Solvent and Concentration on Intensity of First Overtone O-H Stretching Band of Phenol

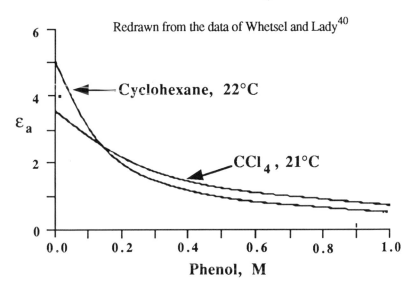

FIGURE 5.25

van't Hoff Plot

Redrawn from data of Whetsel and Lady [40]

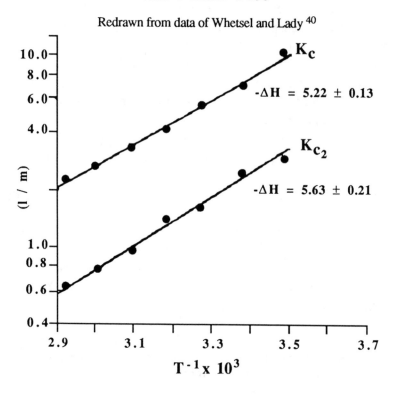

FIGURE   5.26

There certainly appears to be some sort of specific solvation effect and Whetsel and Lady[40] observed that in dilute solutions the apparent absorptivity of the first overtone free O-H band of phenol is approximately 50% larger in cyclohexane than in CCl_4, as shown in figure 5.25. At concentrations above 0.15 M, the values are more comparable. Clearly, there could be major errors in using equilibrium constants obtained from dilute solution studies in CCl_4, and in our work we used the values of equilibrium constants obtained from studies of cyclohexane solutions.

For the formation of dimers the equilibrium constant was reported to have the value:

$$K_{c_2} = 2.10 \text{ l. mol}^{-1}$$

and for higher multimers:

$$K_c = 6.68 \text{ l. mol}^{-1}$$

In our work we use Flory's equilibrium constants, expressed in terms of volume fractions, which can be obtained from the above values using (see section 4B):

$$K_2 = \frac{\Phi_{B_2}}{\Phi_{B_1}^2} \cdot \frac{1}{2} = K_{c_2} \cdot \frac{1000}{V_B} \qquad (5.41)$$

$$K_B = \frac{\Phi_{B_{h+1}}}{\Phi_{B_h} \Phi_{B_1}} \frac{h}{h+1} = K_c \cdot \frac{1000}{V_B} \qquad (5.42)$$

where V_B is the molar volume of phenol in $cm^3 \text{ mol}^{-1}$. As Flory noted, either is an acceptable equilibrium constant (see Chapter 4). The enthalpy of hydrogen bond formation can be obtained from the usual van't Hoff plot of ln K vs. 1/T. The data of Whetsel and Lady[40] is shown in figure 5.26 and the values of h_2 = 5.63 kcal. mol^{-1} and h_B = 5.22 kcal. mol^{-1} are those used in the work reported later in this book.

Inter-Association Equilibrium Constants

In discussing the measurement of the number of inter-association hydrogen bonds and the equilibrium constants describing these interactions we can distinguish between three different situations. First, recall that we are most often dealing with blend systems where one component self-associates while the second does not, but has a functional group capable of hydrogen bonding to segments of the first. Given such mixtures, we can then come across the following types of systems:

1. Where the self-associating polymer has a carbonyl group that can be used as a probe of the distribution of hydrogen bonded species (e.g. amide or urethane).

2. Where the non self-associating polymer has a carbonyl group (e.g. ester).

3. Where neither component has a carbonyl group and equilibrium constants must be transferred from studies of low molecular weight analogues (e.g., PVPh mixed with polyethers).

A blend of an amorphous polyurethane (APU) with an ethylene oxide-co-propylene oxide (EPO) copolymer is an example of type (1) systems. The type of interactions that can occur are illustrated in figure 5.27 and the infrared spectra of various mixtures of these polymers were presented earlier in figure 5.14.

The intensity of the free carbonyl band increases with the concentration of ether functional groups in the mixture, as urethane N-H to ether oxygen hydrogen bonds are formed at the expense of self-association interactions. As we showed in Chapter 4 (equation 4.66), the fraction of free carbonyl groups is given by:

$$f_F^I = (1 - K_B \Phi_{B_1}) \qquad (5.43)$$

where in this system it is assumed that self-association is described by a single equilibrium constant, K_B, which is obtained by the measurement of the fraction of free groups in films of the pure polyurethane, as described above. The volume fraction of non hydrogen bonded urethane segments, Φ_{B_1}, and the volume fraction of non hydrogen bonded ether segments, Φ_{A_1}, are related to the stoichiometry of hydrogen bonding, given earlier in Chapter 4 but reproduced here for convenience.

$$\Phi_B = \frac{\Phi_{B_1}(1 + K_A \Phi_{A_1}/r)}{\left[1 - K_B \Phi_{B_1}\right]^2} \qquad (5.44)$$

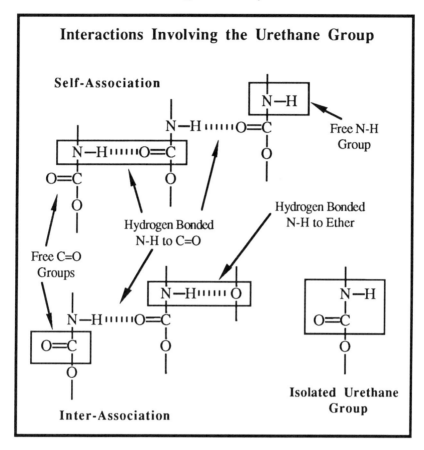

FIGURE 5.27

$$\Phi_A = \Phi_{A_1} \left[1 + \frac{K_A \Phi_{B_1}}{(1 - K_B \Phi_{B_1})} \right] \qquad (5.45)$$

The factor r is a known quantity (the ratio of molar volumes V_A/V_B), so that there are four unknowns, K_A, K_B, Φ_{B_1}, and Φ_{A_1}. As we have just stated, K_B is obtained from the spectrum of the pure polyurethane, while Φ_{B_1}, and Φ_{A_1} depend upon concentration according to equations 5.44 and 5.45 above. Hence, in principle, it is possible to measure the fraction of free carbonyls in a single blend of known composition (Φ_A, Φ_B) and use equations 5.43, 5.44 and 5.45 to determine K_A.

This value of K_A can then be used to calculate the fraction of free carbonyl groups at any other composition, using the same equations to determine Φ_{A_1}, Φ_{B_1} and hence, f_F, and such calculations would serve to demonstrate how well this association model describes the stoichiometry of hydrogen bonding. The equations fit the experimental data obtained on this system extremely well[12], as shown in figure 5.28. In initial work we did indeed calculate K_A from data from a single mixture composition, but once we saw the general excellence of the agreement we then obtained K_A by a least squares fit to the data set as a whole, in order to increase the accuracy of our determination of this parameter for use in subsequent calculations of phase behavior.

Comparison of the Experimental and Theoretical Determined Fraction of Free Urethane Carbonyl Groups in APU - Polyether Mixtures

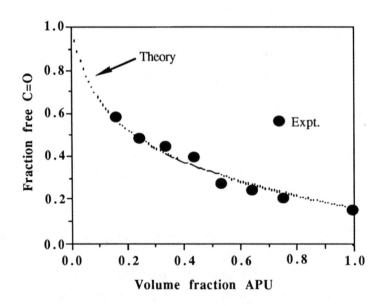

FIGURE 5.28

A typical example of type (2) systems is a blend of PVPh with a polymer containing an ester group, such as poly(ethyl methacrylate) (PEMA)[42]. The procedure is essentially the same, except that for this system we cannot, at the present time, obtain self-association equilibrium constants by infrared studies of PVPh alone, for reasons outlined above, and instead "transferred" values from phenol by simply accounting for the difference in molar volume using equation 5.33. Because two equilibrium constants are necessary to describe self-association in these types of molecules, we employed the more complicated equations for the stoichiometry of hydrogen bonding given by equations 4.89 and 4.92. Here, we choose to work with the fraction of hydrogen bonded ester group, f_{HB}, rather than the fraction free, where the carbonyl band we are measuring is the ester group in the non self-associating component (actually, the graduate student who did this work made this choice and once he'd produced the plots we did not have the heart to make him change them). As we showed in Chapter 4 (see equations 4.69 and 4.99), f_{HB} is simply given by:

$$f_{HB} = 1 - f_F = 1 - \frac{\Phi_{A_1}}{\Phi_A} \qquad (5.46)$$

Typical spectra obtained from an 80:20 PVPh - PEMA blend as a function of temperature are shown in figure 5.29. The values of f_{HB} obtained from spectra obtained as a function of composition are then again used to obtain K_A, using a least squares fit to equation 5.46 and the equations describing the stoichiometry of hydrogen bonding. This value of K_A gives an impressive fit to all the data, as shown in figure 5.30A.

The fact that the values of K_A successfully reproduce the stoichiometry of hydrogen bonding across the composition range is encouraging, but not proof of the validity of this approach. More convincing evidence is obtained by considering the values of the equilibrium constants obtained from studies of other methacrylate polymers (and also some copolymer studies, which we will discuss in the following section).

FTIR Spectra of PVPh - PEMA Blends

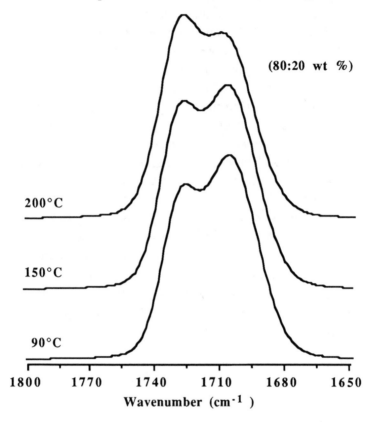

FIGURE 5.29

As we noted at the end of Chapter 4 (p. 219), when we are considering mixing a specific self-associating polymer, in the case PVPh, with a range of other polymers containing the same functional group, in this case the ester side chain of various methacrylates, we can in principle use the *same* value of K_A. The difference in molar volume of the chemical repeat units of these polymers is accounted for by the factor r (equal to the ratio of the molar volumes, V_A/V_B) in the equations describing the stoichiometry of hydrogen bonding. Accordingly, values of K_A calculated from spectroscopic studies of poly(propyl methacrylate), PPMA, for example, should be in good agreement.

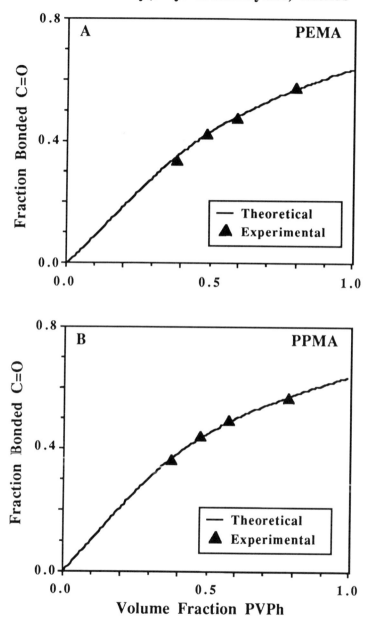

PVPh - Poly(alkyl methacrylate) Blends

A   **PEMA**

— Theoretical
▲ Experimental

B   **PPMA**

— Theoretical
▲ Experimental

Volume Fraction PVPh

FIGURE 5.30

Table 5.4
Determination of K_A

Temperature °C	K_A	
	PEMA	PPMA
150	5.62	5.76
160	5.02	5.18
170	4.56	4.71
180	4.15	4.28
190	3.80	3.92
200	3.47	3.60

Figure 5.30B shows a comparison of calculated and experimental values of f_{HB} for PPMA and the values of K_A obtained as a function of temperature for both PPMA and PEMA are compared in Table 5.4. Given the errors involved in curve resolving the agreement is excellent. Furthermore, the values of the molar volumes of the chemical repeat units of PEMA and PPMA were calculated from group contributions and errors here would lead to a systematic deviation.

Plots of ln K vs 1/T can be used to obtain h_A, the enthalpy of phenolic O-H - ester hydrogen bonds. Such plots are shown in figure 5.31 and again the data from PEMA and PPMA are in good agreement, giving values of h_A equal to 3.8 and 3.75 kcal. mol.$^{-1}$, respectively[42].

It might occur to the reader that an unbiased estimation of the three equilibrium constants could be obtained by permitting them all to vary in a least-squares best fit of all the experimental data, thus removing the need to transfer values of K_2 and K_B from solution measurements. Unfortunately, this is not feasible as with three adjustable variables the equations are classically ill conditioned. That is, we can obtain numerous different solutions yielding acceptable best fits of the data within the limits of the experimental error. Values of the self-association constants need to be independently determined.

van't Hoff Plot for PVPh Blends

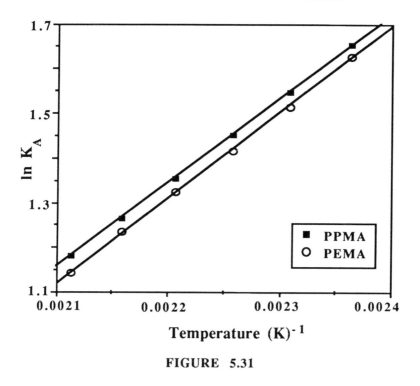

FIGURE 5.31

Finally, there is the situation where at the present time we are not able to obtain either the self-association or inter-association constants from spectroscopic measurements because there is no suitable functional group involved. An example is blends of PVPh with polyethers. We have studied such systems[43] and here the validity of the assumptions must be tested by the ability of the model to successfully predict phase behavior, or trends in phase behavior for a series of similar polymers.

Copolymer Blends and the Transferability of Spectroscopically Determined Equilibrium Constants

In the preceding paragraphs and in Chapter 4, we have discussed hydrogen bonding in homopolymer systems where equilibrium constants are defined in terms of chemical repeat units. Here, we will consider the effect of random copolymerization (block copolymers lie outside the scope of this book), as some of our studies also demonstrate that our assumption of the transferability of equilibrium constants works well for certain systems. We will confine our discussion to copolymers where one of the components hydrogen bonds and the second is an "inert" diluent (non hydrogen bonding). We start by considering styrene-co-vinyl phenol (STVPh) copolymers. The self-association of the phenolic O-H groups in these copolymers can be described in two equivalent ways. First, we can simply apply the equilibrium constants obtained from phenol, as described above, but when we calculate the stoichiometry of hydrogen bonding we require that Φ_B be equal to the volume fraction of vinyl phenol chemical repeat units in the mixture, equal to the volume fraction of STVPh copolymer in the blend multiplied by the volume fraction of vinyl phenol in the copolymer. Alternatively, an average chemical repeat of the copolymer may be defined in such a manner that it contains one vinyl phenol unit, as illustrated below:

The equilibrium constants K_2 and K_B are then modified according to the difference in molar volume between this average repeat unit and a vinyl phenol segment using equation 5.33. The quantity Φ_B is then the volume fraction of the copolymer. To repeat, the methods are equivalent and give identical results. We find it convenient for computational purposes to use the second approach, however, and this is built into our programs.

The synthesis of various styrene-co-vinyl phenol copolymers has allowed an additional check on the transferability of equilibrium constants[44]. As we noted at the end of Chapter 4, if we change the molar volume of the repeat unit of the self-associating component, K_A should vary in proportion. Using various styrene-co-vinyl phenol copolymers values of K_A were determined spectroscopically using the least-squares procedure described above for miscible blends with poly(n-butyl methacrylate) (PBMA) and poly(n-hexyl methacrylate) (PHMA). The expected linear dependence of K_A on the vinyl phenol content of the copolymer was obtained, as shown for the STVPh - PBMA blends in figure 5.32. Alternatively, the initial value of K_A obtained from the studies of PVPh - PEMA homopolymer blends, reported above, can be used to calculate the fraction of hydrogen bonded groups and the results compared to the values obtained experimentally[45]. Such a comparison is presented in Table 5.5 and it can be seen that the agreement is excellent, indicating that at least in this system our assumption of equilibrium constant transferability appears to be reasonable.

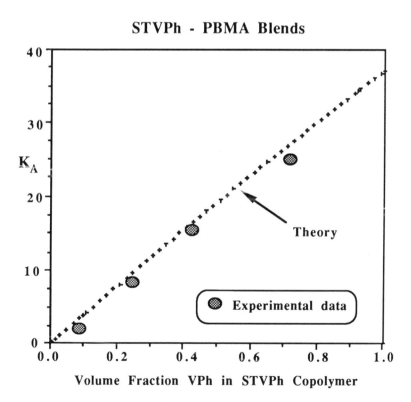

FIGURE 5.32

A final note concerning these calculations is in order. Figure 5.32 shows K_A varying with the size of the "average repeat unit" of the STVPh copolymer, rather than the size of the methacrylate unit. This was explained in the final section of Chapter 4, but can seem counter-intuitive. Essentially, we use V_B, the molar volume of the self-associating component, to define the lattice cell size, so that *both* K_A and K_B scales with V_B according to equation 5.33. Clearly, the size of the inter-associating unit, V_A, must affect the value of K_A, but this is brought into the equations through the factor $r = V_A/V_B$, hence, the appearance of K_A/r terms in the stoichiometric equation for Φ_B.

Table 5.5

Comparison of Experimentally Determined and Theoretically Calculated Fraction of Hydrogen Bonded Carbonyl Groups in Poly(n-butyl methacrylate) - Styrene-co-Vinyl Phenol Blends

Polymer Blend	Fraction Hydrogen Bonded Carbonyls	
PBMA:STVPh[x]	Experiment	Calculated
PBMA:STVPh[8]	0.20	0.22
20:80	0.09	0.09
50:50	0.04	0.03
80:20		
PBMA:STVPh[25]		
20:80	0.38	0.41
50:50	0.24	0.24
80:20	0.10	0.08
PBMA:STVPh[43]		
20:80	0.47	0.48
50:50	0.30	0.32
80:20	0.14	0.13
PBMA:STVPh[75]		
20:80	0.52	0.54
50:50	0.38	0.39
80:20	0.18	0.19

F. MAPPING PHASE DIAGRAMS USING INFRARED SPECTROSCOPY

As we have seen in the preceding paragraphs, the association model we have employed allows the calculation of the number of each type of hydrogen bonded species present at a particular temperature and composition and also apparently allows equilibrium constants determined in one system to be transferred to a similar one by simply accounting for the difference in molar volume of the species. We would not expect these relationships to hold upon phase separation, even if the phase separated domains are each mixtures of the two species; the composition and hence, stoichiometric relationships in each phase would be different to that of a completely miscible system.

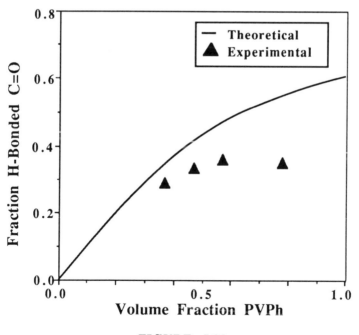

Comparison of the Theoretical and Experimental Fraction of Hydrogen Bonded Carbonyl Groups for PVPh - PBMA Blends at 200° C

FIGURE 5.33

**Comparison of the Theoretical and Experimental
Fraction of Hydrogen Bonded Carbonyl Groups
for a 50:50 PVPh - PBMA Blend**

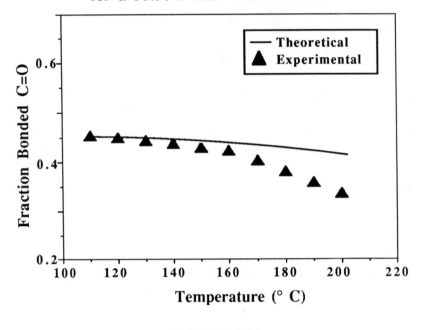

FIGURE 5.34

This can be illustrated by a consideration of PVPh - poly(n-butyl methacrylate) (PBMA) blends. Using the equilibrium constants described in the preceding section, we have calculated phase diagrams that predict that this polymer should be single phase at room temperature, but display an LCST near 80°C[42]. The position of this predicted LCST will naturally be affected by errors in the various parameters we employ, as we will discuss later in this book.

Of interest to us here is that cloud point measurements demonstrate phase separation near 160°C, as illustrated previously in chapter 1 (figure 1.14), and the fraction of hydrogen bonded carbonyl bands determined at 200°C indeed deviates significantly from that predicted using the association model as shown in figure 5.33. The cloud point measurements reported in this study are crude and probably do not accurately reflect the position of the LCST. (The Tg of PVPh used in this study is about 190°C, so that once appreciable phase separation occurs (~160°C), remixing cannot be induced by lowering the temperature.) It therefore occurred to us that *if we can assume that the equilibrium constants we employ give an accurate representation of the stoichiometry of hydrogen bonding*, then the temperature at which the measured fraction of hydrogen bonded species deviates from that predicted by the model is a measure of phase separation, *at the probe size of the functional group*. This is illustrated for the PVPh - PBMA system in figure 5.34, where for 50:50 blends, for example, the infrared measurements suggest an LCST near 140°C[42].

In recent work we have examined in more detail how we might use such observations to "map" phase diagrams[38]. Consider the phase diagram illustrated in figure 5.35. At the temperature denoted T and a blend composition Φ_B, the system is in a two phase region and will separate into phases of composition Φ_B^x and Φ_B^y, defined by the binodal (as always, the suffix B refers to a self-associating component). Employing the lever rule, the fraction of B present in the two phases is, respectively:

$$\frac{\Phi_B^y - \Phi_B}{\Phi_B^y - \Phi_B^x}\left[\frac{\Phi_B^x}{\Phi_B}\right] \quad \text{and} \quad \frac{\Phi_B - \Phi_B^x}{\Phi_B^y - \Phi_B^x}\left[\frac{\Phi_B^y}{\Phi_B}\right] \qquad (5.47)$$

Now, if we define f_F^x and f_F^y, as the fraction of "free" carbonyl groups at compositions corresponding to Φ_B^x and Φ_B^y, then the fraction of "free" carbonyl groups, f_F, that will be observed at the blend composition Φ_B is given, after some rearrangement of the equations, by:

$$\Phi_B f_F (\Phi_B^y - \Phi_B^x) = \Phi_B^y \Phi_B^x (f_F^x - f_F^y) + \Phi_B (\Phi_B^y f_F^y - \Phi_B^x f_F^x)$$

$$(5.48)$$

Accordingly, at a given temperature within a two phase region at equilibrium, a plot of $\Phi_B \cdot f_F$ versus Φ should yield a straight line with a slope of:

$$\frac{\Phi^y_B f^y_F - \Phi^x_B f^x_F}{\Phi^y_B - \Phi^x_B}$$

and an intercept of:

$$\frac{\Phi^y_B \Phi^x_B (f^y_F - f^x_F)}{\Phi^y_B - \Phi^x_B}$$

Using data obtained at 70 and 150°C from a blend of an aromatic polyamide (MPD6) with polyethylene oxide (PEO)[38], we obtained the plots shown in figure 5.36. All the 150°C data satisfactorily fits a straight line (y = 0.257x + 0.0127 : R^2 = 0.995) strongly suggesting a two phase system over a very wide composition range.

Schematic Phase Diagram

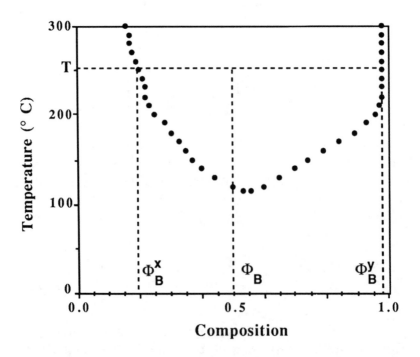

FIGURE 5.35

MPD6 - PEO BLENDS

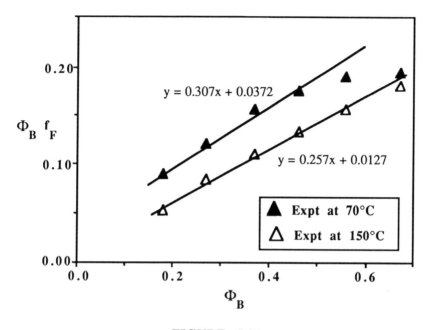

FIGURE 5.36

On the other hand, at 70°C, the two data points corresponding to the higher concentrations of MPD6 deviate significantly from the straight line ($y = 0.307x + 0.0372$: $R^2 = 0.990$) that satisfactorily fits the remaining data. Assuming equilibrium has been attained, this implies that at these two compositions ($\Phi_B = 0.56$ and 0.67) the MPD6-PEO blends exist in a single phase. Having obtained values of K_B and K_A for this system[38], we calculated the theoretical fraction of "free" carbonyl bands as a function of blend composition for *hypothetically miscible* MPD6 - PEO blends at 90, 110, 130 and 150°C, and compared the results to the experimental data, as illustrated in figure 5.37. With increasing temperature the agreement between the experimental data and the theoretical fraction of "free" carbonyl bands for a single phase becomes increasingly poor. This is to be expected, of course, if the phase diagram resembles that shown above in figure 5.35.

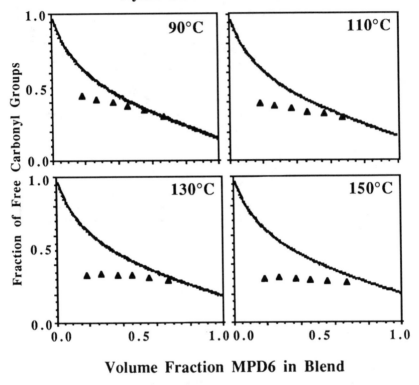

**Comparison of Experimental Values to the
Theoretical Fraction of Free Carbonyls
Assuming a Single Phase
Nylon MPD6 - PEO Blends**

Volume Fraction MPD6 in Blend

FIGURE   5.37

However, because we have obtained the slope and intercept
from the linear plot of $\Phi_B \cdot f_F$ versus Φ_B (equation 5.48, figure
5.36) we can now calculate a curve of the theoretical fraction of
"free" carbonyl bands as a function of blend composition for a
two phase system at equilibrium using:

$$f_F = slope + \frac{intercept}{\Phi_B} \qquad (5.49)$$

Figure 5.38 shows the experimental data obtained at 70 and
150°C, together with the theoretical curves for both the single
and two phase systems.

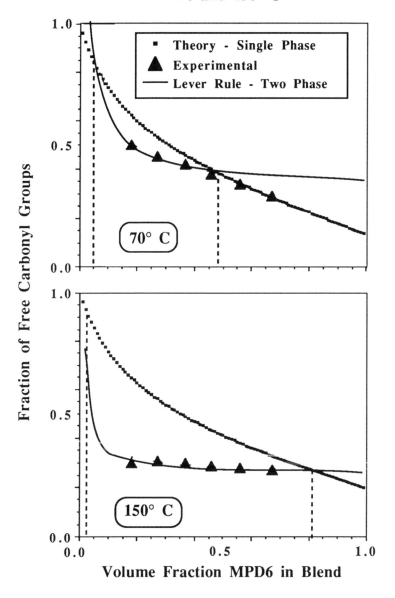

Nylon MPD6 - PEO Blends at 70 and 150° C

FIGURE 5.38

Phase Diagram for Nylon MPD6 - PEO Blends
Determined from Infrared Spectroscopy

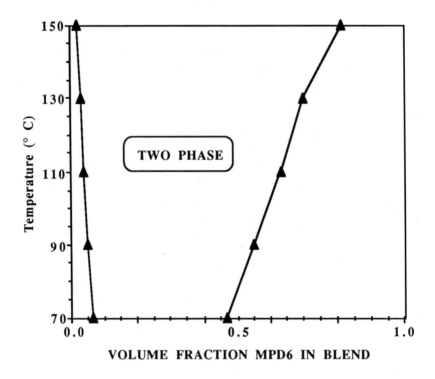

FIGURE 5.39

At the former temperature, two of the data lie on the single phase curve ($\Phi_B = 0.56$ and 0.67), three on the two phase curve ($\Phi_B = 0.18$, 0.27 and 0.37) and one within experimental error could fit on either curve ($\Phi_B = 0.46$). At 150°C, all the experimental data fall upon the two phase curve. At a given temperature, the intercept of the two theoretical curves yields the composition limits of the two phase region of the phase diagram (binodal). Accordingly, we can now map a phase diagram that has been derived from experimental infrared measurements. This is illustrated in figure 5.39.

It is important to note that throughout the above discussion we have made the implicit assumption that the phases formed in figure 5.35 of composition Φ_B^x and Φ_B^y are above their respective Tgs at the temperature of measurement (between 70 and 150°C). This appears to be a reasonable assumption for the MPD6 and PEO system[38] but probably would not be true for PVPh - PBMA blends.

G. REFERENCES

1. Wilson, Jr., E. B., Decius, J. C. and Cross, P. C. *Molecular Vibrations*, McGraw-Hill, New York, 1955.

2. Painter, P. C., Coleman, M. M. and Koenig, J. L. *The Theory of Vibrational Spectroscopy and its Application to Polymeric Materials*, John Wiley and Sons, 1982.

3. Krimm, S., *Advances in Polymer Science*, 1960, **2**, 51-172.

4. Griffiths, P. R., *Chemical Infrared Fourier Transform Spectroscopy*, Wiley, New York, 1975.

5. Griffiths, P. R., and Haseth, J. A., *Fourier Transform Infrared Spectroscopy*, Wiley, New York, 1986.

6. Koenig, J. L., *Advances in Polymer Science*, 1983, **54**, 87.

7. Lee, J. Y., Painter, P. C. and Coleman, M. M., *Macromolecules*, 1988, **21**, 346.

8. Xu, Y., Painter, P. C. and Coleman, M. M., *Makromol. Chemie. Macromol. Symp.*, (in press)

9. Skrovanek, D. J., Painter, P. C. and Coleman, M. M., *Macromolecules*, 1986, **19**, 699.

10. Maddams, W. F., *Applied Spectroscopy*, 1980, **34**, 245.

11. Skrovanek, D. J., Howe, S. E., Painter, P. C. and Coleman, M. M., *Macromolecules*, 1985, **18**, 1676.

12. Coleman, M. M., Skrovanek, D. J., Hu, J. and Painter, P. C., *Macromolecules*, 1988, **21**, 59.

13. Lippincott, E. R. and Schroeder, R. J., *Chem. Phys.*, 1955, **23**, 1099.

14. Lippincott, E. R. and Schroeder, R. J., *Phys. Chem.*, 1957, **61**, 921.

15. Reid, C. R., *Chem. Phys.*, 1959, **30**, 182.

16. Hadzi, D., *Chimia*, 1972, **26**, 7.

17. Hadzi, D., *Pure and Applied Chem.*, 1965, **11**, 435.

18. Odinokov, S. E., Mashkovsky, A. A., Glazunov, V. P., Iogansen, A. V. and Rassadin, B. V., *Spectrochim Acta*, 1976, **32A**, 1355.
19. Speakman, J. C., *Structure and Bonding*, 1972, **12**, 141.
20. Vorenkamp, E. J. and Challa, G., Polymer, 1988, **29**, 86.
21. Green, R. D., *Hydrogen Bonding by C-H Groups*, Wiley, New York, 1974.
22. Coleman, M. M., Lee, K. H., Skrovanek, D. J. and Painter, P. C., *Macromolecules*, 1986, **19**, 2149.
23. Sandorfy, C. in *The Hydrogen Bond, II*, Schuster, P., Zundel, G., and Sandorfy, C., Editors, North Holland Publishing Co., 1976.
24. Lee, J. Y., Moskala, E. J., Painter, P. C. and Coleman, M. M., *Applied Spectroscopy*, 1986, **7**, 991.
25. Hadzi, D. and Bratos, S. *Vibrational Spectroscopy of the Hydrogen Bond*. in *The Hydrogen Bond II* (Schuster, P., Zundel, G. and Sandorfy, C., Editors) North Holland Publishing Co. (1976) Chapter 12.
26. Thomson, B., Park, Y., Painter, P. C. and Snyder, R. W., *Macromolecules*, 1989, **22**, 4159.
27. Coleman, M. M., Lee, J. Y. and Painter, P. C., *Macromolecules* (in press).
28. Murthy, A. S. N. and Rao, C. N. R., *Appl. Spect. Rev.* 1968, **2**, 69.
29. Pimentel, G. C. and McCellan, A. L., *Ann. Rev. Phys. Chem.*, 1971, **22**, 3247.
30. Dean Sherry, A., in *The Hydrogen Bond, Vol. III* (Schuster, P., Zundel, G. and Sandorfy, C. Editors) North Holland Publishing Co. (1976) Chapter 25.
31. Park, Y. and Painter, P. C., unpublished results.
32. Novak, A. *Structure and Bonding*, 1974, **18**, 177.
33. Schroeder, L. R. and Cooper, S. L., *J. Appl. Phys.*, 1976, **47**, 4310.
34. Garcia, D. and Starkweather, Jr, H. W., *J. Polym. Sci., Polym. Phys. Ed.*, 1985, **23**, 537.
35. Moskala, E. J., Howe, S. E., Painter, P. C., and Coleman, M. M., *Macromolecules*, 1984, **17**, 1671.
36. Hu, J., Painter, P. C. and Coleman, M. M., Unpublished results.
37. Garton, A., *Polym. Eng. Sci.* 1984, **24**, 112.
38. Bhagwagar, D. E., Painter, P. C., Coleman, M. M. and Krizan, T. D., *J. Polym. Sci. - Phys. Ed.*, accepted.
39. Stadler, R., *Macromolecules*, 1988, **21**, 121.
40. Whetsel, K. B. and Lady, J. H., in *Spectrometry of Fuels*, Friedel, H. Ed., Plenum, London 1979, p259.

41. Marcus, Y., *Introduction to Liquid State Chemistry*, John Wiley and Sons, New York, 1977, p176.
42. Serman, C. J., Painter, P. C. and Coleman, M. M., *Polymer*, (in press).
43. Serman, C. J., Xu, Y., Painter, P. C. and Coleman, M. M., *Polymer*, (in press).
44. Xu, Y., Painter, P. C. and Coleman, M. M., Unpublished results.
45. Xu, Y., Painter, P. C. and Coleman, M. M., *Makromol. Chemie. Macromol. Symp.*, (in press).

## CHAPTER 6

Association Models and The Thermodynamics of Mixing Molecules with Strong Specific Interactions

A. INTRODUCTION–WHY USE ASSOCIATION MODELS ?

We now turn our attention to the thermodynamics of mixing molecules that have strong specific interactions. As we noted in Chapter 1, there are various types of "chemical" interactions that can be regarded as strong and specific, but hydrogen bonds are a particularly important category and presently the most amenable to analysis. We will therefore present our discussion in terms of these interactions, but the *general* points we will make apply to mixtures where *identifiable* associated species form through any sort of specific interaction.

The mixing of molecules that hydrogen bond is fundamentally a polymer problem and has received attention from some of the great scientists who laid the foundations of polymer solution theory, notably Huggins[1], Flory[2,3] and Prigogine[4,5]. Huggins published a review of the structural role of what were then called hydrogen bridges in 1936, but of more use to us here is the work of Flory and Prigogine, as this provides the basis for a description of the thermodynamics of hydrogen bonding through the use of so-called association models.

More than thirty years ago Prigogine[4,5] pointed out that there is no satisfactory theory of the strong orientational effects that occur in mixtures involving strong, specific interactions, principally because the rotational partition function is no longer independent of the translational partition function (see Section 1C). Prigogine proposed that the formation of a complex be treated by using the assumption of a chemical equilibrium between the monomolecules of the associated species. Such a "chemical" theory of solutions was first proposed by Dolezalek[6] around the turn of the century and in its original form interpreted all deviations from ideality as a consequence of the formation of new chemical species (aggregates). In this view, the problem of solution thermodynamics becomes the identification of the "true" species present. The early history of solution theory is enlivened by the bitter polemics between the "physical" and "chemical" schools of thought[7-9], but as Hildebrand has pointed out in the 1950 edition of his book[10], the former view prevailed in the form of regular solution theory, in part because "chemical" treatments were badly misapplied to inappropriate systems. Nevertheless, the "chemical" approach can be applied to systems where the interactions are weak as, for example, in the quasi-chemical approximation described by Guggenheim[11]. This model seeks to account, at least in part, for the non-random distribution of contacts in systems where the exchange energy is not zero through equations that have the same form as the law of mass action (equilibrium constant) expressions for chemical reactions. Amongst other difficulties, however, it requires the assumption of arbitrary co-ordination numbers and involves the difficult problem (in these systems) of counting species rather than intermolecular contacts. Hildebrand and Scott[10] estimated that for weak interactions corrections for non-random contacts are relatively small and can be neglected, particularly when other approximations (e.g., zero volume change upon mixing) inherent in regular solution theory are considered. Accordingly, it is usually considered unsatisfactory to treat "physical" interactions using a "chemical" model.

In the same manner, it is often difficult and sometimes inappropriate to treat strong specific interactions using "physical" models[4,10,12-14], by which we mean those that do not recognize the formation of a complex or associate. In this category is the contact point approach described by Barker[15,16] and Tompa[17] which divides the surface of a molecule into several contact points so that the energy of nearest-neighbor interactions depend upon the relative orientations of the molecules concerned. This approach can become algebraically complicated and introduces a number of parameters that cannot be determined independently, but only estimated by a fit to thermodynamic data. As Tompa[17] pointed out, it is uncertain how much physical significance can be attached to parameters so obtained. A similar approach was used by ten Brinke and Karasz[18] to describe specific interactions in polymers, but the problem was made more tractable by considering segments that have a single site that is capable of engaging in a specific interaction. Separate parameters defining the energy of non-specific interactions and the enthalpic and entropic components of specific interactions were then obtained. This model was later generalized to account for compressibility or free volume effects by Sanchez and Balazs[19]. The random contact assumption of the Flory Huggins model was maintained, however, and specific interactions between like molecules were neglected. As we will show later in this chapter, these factors play a major role in the thermodynamic properties of mixtures where interactions are "strong" (according to our definition - Table 1.1). For systems where interactions are relatively "weak," however, as in poly(styrene)/poly(vinyl methyl ether) blends (an overall χ of the order of -1×10^{-4})[20], this approach has worked remarkably well.[19]

If definite aggregates are formed, the "physical" approach often cannot account for various properties. Marsh and Kohler[14] cite the example of a chemical equilibrium between monomers and dimers in a low density gas where the dimerization constant is large. The main deviation from ideal behavior is not linear with pressure and the gas cannot be characterized by a very large second virial coefficient. Similarly, the dielectric properties of amides and alcohols can only be explained by the presence of chains of hydrogen bonded molecules[13]. Following Kohler[12,14], we will therefore define an associated species as one which has a lifetime that is at least an order of magnitude longer than that of an intermolecular vibration ($\sim 10^{-13}$ secs), so that it would be reasonable to assume that the complex is a distinguishable species for which a partition function can be constructed. Kohler[12] cites hydrogen bond lifetimes ranging from 10^{-11} to 10^{-5} secs, the shorter lifetime corresponding to hydrogen bonds between small, approximately spherical, molecules such as water, for which association models may therefore be inappropriate. The longer lifetimes correspond to the formation of cyclic dimers between anisotropic molecules, as in the carboxylic acids. Polymers containing amide, urethane, hydroxyl, etc., functional groups should lie between these extremes and we will therefore assume that association models are appropriate for these materials.

The great advantage of association models is the relatively simple way in which partition functions can be constructed. In contrast to the application of quasi-chemical models to the description of weak interactions, where a coordination number has to be assumed in order to count all contacts, in applying an association model to hydrogen bonds the number of specific interaction contacts is usually known (e.g., an amide group can hydrogen bond to up to two neighbors). Non-specific interactions can be handled using a χ parameter in the usual way (see below). If we were to use a contact point approach to account for all interactions, however, we would end up with algebraically complex expressions and parameters that could not readily be determined from experimental measurements. It would be necessary to account for the enthalpic component of specific interactions corresponding to both self and inter-association, the entropic changes corresponding to loss of rotational freedom, interactions associated with non-specific contacts, and so on. Finding the maximum term, corresponding to the equilibrium distribution of species, would not be a trivial task. Association models, however, start with the assumption of a *prior knowledge* of the equilibrium distribution of associated species present at a given temperature and composition. For molecules that interact through the formation of hydrogen bonds we saw in Chapters 4 and 5 that we could describe this distribution through appropriately defined equilibrium constants that (in many mixtures) can be determined experimentally by infrared spectroscopy. In mixtures of simple (non-polymeric) amides and ethers, for example, we have chains of reversibly associated molecules that have (in principle) a known distribution of chain lengths. If we treat each of these species (amide "monomers," amide "dimers," amide "trimers," etc., and equivalent amide-ether chains) as distinguishable (according to chain length), we can apply Flory's[2] treatment for the mixing of heterogeneous polymers. Essentially, these distinguishable species are initially considered as if they were separate and oriented. The free energy of disorienting and mixing the monomers, dimers, etc., with one another is then determined using Flory's equations, thus obtaining a free energy relative to the separate and oriented reference state. The same procedure is applied to the distribution of species found in the initially pure amide and ether components, so that a free energy is again found relative to the same reference state. The difference in these two equations gives a free energy that includes a combinatorial entropy of mixing and the free energy changes that occur due to the change in the distribution of hydrogen bonded species. Physical interactions can then be handled in the usual manner by defining a χ parameter, thus assuming random contacts between the associated molecules (equivalent to the assumption of random mixing of covalent polymer molecules).

We will proceed with a description of this methodology shortly, but finish here with one or two additional points. First, the form of the equations describing the free energy changes due to hydrogen bonding corresponds to Flory's expression for the entropy of mixing heterogeneous polymers. This is now a free energy term, however, because the enthalpy and entropy components are accounted for *together* by the equilibrium constants that define the distribution of species as a function of concentration and that are a measure of the entropic and enthalpic contributions to the free energy per hydrogen bond. Secondly, the various entropic contributions (combinatorial, rotational, etc.) to the free energy are accounted for simultaneously. In our amide/ether example there will be fewer amide/amide hydrogen bonds as the mixture is diluted with ether molecules and thus an increase in entropy due to the gain in rotational freedom of the "liberated" amides. Similarly, there will be a contribution that decreases the entropy of mixing due to the formation of amide/ether hydrogen bonds. In an association model these entropic contributions are accounted for simply in terms of the change in the number of species with concentration. When dealing with hydrogen bonding in polymers we will need an excess function that does not include the combinatorial component, however, and as we will see, this will require a careful consideration of reference states.

Following the arguments we made in Chapter 3, we will assume that hydrogen bonds involving most functional groups results in the formation of chains of associated species, rather than cyclic complexes, and we will consider these first. Mixtures involving molecules with functional groups where cyclic complexes have been unambiguously identified (e.g., carboxylic acids) can be handled in an equivalent manner, but there are slight differences in the final form of the equations. This will be described in an appendix to this chapter.

B.   HYDROGEN BONDING IN SMALL MOLECULES–OPEN CHAIN ASSOCIATION IN MIXTURES WHERE ONE COMPONENT SELF-ASSOCIATES

The use of association models to account for hydrogen bonding in small molecules has been described by a number of authors (for example, references 21-26). In adapting this approach to polymers[27-29], we initially followed Nagata's description[23], but in more recent work, we considered a modification of Flory's lattice filling procedure, which we will describe here. We will confine our analysis to the specific but commonly encountered situation where one component self-associates, while the second does not, but may have a functional group that can hydrogen bond to segments of the first (e.g., amides mixed with ethers). As in Chapter 4, we consider a "small" molecule to be one with a relatively low molecular weight (compared to a polymer) and only one functional group capable of forming a hydrogen bond, although for the self-associating component, this can hydrogen bond with up to two neighbors, thus allowing the formation of chains. We will again initially neglect physical interactions, because as long as the hydrogen bonded chains are randomly mixed with one another then these can be accounted for separately by the simple addition of a χ parameter.

Our purpose in Chapter 4 was to define appropriate equilibrium constants and to discuss their "transferability". To accomplish this we divided the molecules into segments on the basis of an arbitrarily defined "interacting unit." We could continue that approach here, but it is unnecessary and we will employ a model where the molar volume of the self-associating molecule is used to define the lattice cell size. For self-associating small molecules mixed with an *inert* (non-hydrogen bonding) diluent an expression for the number of configuration available to the chain is then given by (see Chapter 4):

$$
\Omega_H = \frac{1}{\prod_{h=1}^{\infty} n_{B_h}!} \left\{ \frac{(z-1)^{\sum n_{B_h}(h-1)}}{\sigma^{\sum n_{B_h}}} \right\} \left\{ \frac{n_0!}{\left(n_0 - \sum n_{B_h}h\right)!} \cdot \frac{1}{n_0^{\sum n_{B_h}(h-1)}} \right\}
$$

$$
= \left\{ \frac{(z-1)^{\sum n_{B_h}(h-1)}}{\sigma^{\sum n_{B_h}} n_0^{\sum n_{B_h}(h-1)}} \right\} \left\{ \frac{n_0!}{n_A! \prod_{h=1}^{\infty} n_{B_h}!} \right\} \tag{6.1}
$$

where n_0 is equal to the total number of lattice sites, $(n_A + \sum n_{B_h}h)$, σ is a symmetry number for the "polymer" molecules and z is the coordination number of the lattice (our nomenclature is the same as that previously employed and for convenience is summarized at the beginning of this book). This is with respect to Flory's reference state, where the B_h-mers are treated as distinguishable according to their length and are initially separate and oriented. The number of configurations available to the $\sum n_{B_h}^0$ molecules in pure B can also be directly written down and an expression for the free energy of mixing obtained from these two equations in the usual manner by taking logarithms and using Stirling's approximation. An expression for the weak dispersive interactions between segments can also be added at this point.

For the more complicated case where the A segments can hydrogen bond to B segments, it is convenient to define three types of species; Σn_{B_h} chains of B units alone; Σn_{B_hA} chains of B units that have an A unit hydrogen bonded to the appropriate end (e.g., an ether functional group attached to the N-H end of the chain shown in figure 3.4); and n_{A_1} molecules of type A that are not hydrogen bonded at all. Using Flory's[2] lattice filling methodology we obtain:

$$
\Omega_H = \frac{1}{\prod_{h=1}^{\infty} n_{B_h}! \prod_{h=1}^{\infty} n_{B_hA}! \ n_{A_1}!} \left\{ \frac{(z-1)^{\left(\Sigma n_{B_h}(h-1) + \Sigma n_{B_hA} h\right)}}{\sigma^{\left(\Sigma n_{B_h} + \Sigma n_{B_hA}\right)}} \right\}
$$

$$
\times \left\{ \frac{n_0!}{n_0^{\Sigma n_{B_h}(h-1)} \ n_0^{\Sigma n_{B_hA} h}} \right\}
$$

(6.2)

For simplicity of presentation we have assumed that the molar volume of the A and B molecules are the same. The more usual case where there are r segments ($= V_A/V_B$) in each A molecule is easily handled (see Flory) and will be included in our final results.

In order to demonstrate the validity of a methodology we will later apply to polymers we now proceed to derive equation 6.2 in a different manner, by considering the probability that a random mixture of A and B molecules would spontaneously occur in a configuration equivalent to that of the hydrogen bonded chains. To do this we first write the usual expression for the number of configurations available to non hydrogen bonded A and B molecules:

$$
\Omega_{A,B} = \frac{n_0!}{n_A! \ n_B!}
$$

(6.3)

The result we wish to obtain, Ω_H given by equation 6.2, can be found from:

$$\Omega_H = \Omega_{A,B} \cdot \Omega_{conn} \qquad (6.4)$$

where Ω_{conn}, in effect, determines that fraction of configurations in $\Omega_{A,B}$ that correspond to those found in the hydrogen bonded system. It is obtained from the probability of spontaneously finding the appropriate molecules being adjacent to one another in such a fashion that they could be "connected" by hydrogen bonds. If we first consider the number of ways that the B units can be connected to form B_h mers, by analogy with Flory's lattice filling methodology we determine the number of ways of forming the jth chain by "connecting" adjacent units after $j - 1$ chains have been so formed. The number of "unconnected" B units remaining is equal to:

$$\left(n_B - \sum_{i=1}^{j-1} h_i \right)$$

where h_i is the number of B molecules in the ith chain. This is, of course, equal to the number of sites available for the first unit of the jth chain. We now need the probability that this molecule is adjacent to $(h_j - 1)$ other B units and the probability that each of these molecules are as yet unconnected. Using the usual Flory approximations and assumptions[2,11], these terms are, respectively:

$$\left\{ \frac{n_B}{n_0} \right\}^{h_j - 1} \quad \text{and} \quad \frac{n_0 - \sum_{i=1}^{j-1} h_i - (\zeta - 1)}{n_B}$$

where the latter term represents the number of unconnected or available molecules for the ζth segment of the jth chain. Following Flory[2], we thus obtain the number of configurations (or "connections") available to the jth chain, υ_j as:

$$\upsilon_j = \frac{(z-1)^{(h_j-1)}}{\sigma} \frac{\left(n_B - \sum_{i=1}^{j-1} h_i \right)!}{\left(n_B - \sum_{i=1}^{j} h_i \right)!} \left(\frac{1}{n_B} \frac{n_B}{n_0} \right)^{h_j-1} \qquad (6.5)$$

Hence, the number of configurations available to the B_h-mers is given by:

$$\Omega_{B_h} = \frac{1}{\prod\limits_{h=1}^{\infty} n_{B_h}!} \frac{(z-1)^{\sum n_{B_h}(h-1)}}{\sigma^{\sum n_{B_h}}} \frac{n_B!}{\left(n_B - \sum n_{B_h}h\right)!} \left(\frac{1}{n_0}\right)^{\sum n_{B_h}(h-1)} \tag{6.6}$$

where we use:

$$\sum_{i=1}^{\sum n_{B_h}} h_i = \sum_{h=1}^{\infty} n_{B_h} h = \sum n_{B_h} h \tag{6.7}$$

Similarly, for the B_hA-mers we obtain:

$$\Omega_{B_hA} = \frac{1}{\prod\limits_{h=1}^{\infty} n_{B_h}!} \left\{ \frac{(z-1)^{\sum n_{B_hA}h}}{\sigma^{\sum n_{B_hA}}} \right\} \left\{ \frac{\left(n_B - \sum n_{B_h}h\right)!}{\left(n_B - \sum n_{B_h}h - \sum n_{B_hA}h\right)!} \right\}$$

$$x \left\{ \frac{1}{n_B} \frac{n_B}{n_0} \right\}^{\sum n_{B_hA}(h-1)} \left\{ \frac{n_A!}{\left(n_A - \sum n_{B_hA}\right)!} \right\} \left\{ \frac{1}{n_A} \frac{n_A}{n_0} \right\}^{\sum n_{B_hA}} \tag{6.8}$$

Hence using $\Omega_{conn} = \Omega_{B_h} \Omega_{B_hA}$ and equation (6.4):

$$\Omega_H = \left\{ \frac{n_0!}{n_B! \, n_A!} \right\}$$

$$x \left\{ \frac{1}{\prod\limits_{h=1}^{\infty} n_{B_h}! \prod\limits_{h=1}^{\infty} n_{B_hA}! \, n_{A_1}} \frac{(z-1)^{\left(\sum n_{B_h}(h-1) + \sum n_{B_hA}h\right)}}{\sigma^{\left(\sum n_{B_h} + \sum n_{B_hA}\right)}} \right\}$$

$$x \left\{ \frac{n_A! \, n_B!}{n_0^{\sum n_{B_h}(h-1)} \, n_0^{\sum n_{B_hA}(h-1)} \, n_0^{\sum n_{B_hA}}} \right\} \tag{6.9}$$

which is the same as equation 6.2. This demonstrates that we can find the number of configurations available to chains of B and A units by determining the probability that a mixture of the monomers would spontaneously occur in a configuration equivalent to that of the hydrogen bonded chains, a result we will put to good use later.

To obtain the free energy of mixing we also need the number of configurations available to the hydrogen bonded polymers in pure B, given by:

$$
\Omega_H^\circ = \left\{ \frac{1}{\prod n_{B_h}^\circ !} \right\} \left\{ \frac{n_B !}{\dfrac{\sum n_{B_h}^\circ (h-1)}{n_B}} \right\} \left\{ \frac{(z-1)^{\sum_h n_{B_h}^\circ (h-1)}}{\sigma^{\sum n_{B_h}^\circ}} \right\} \tag{6.10}
$$

Taking logarithms, using Stirling's approximation and expressing the n's as moles, we obtain for the free energy of mixing:

$$
\begin{aligned}
\frac{\Delta G_m}{RT} = &\left[\sum n_{B_h} \ln\left(\frac{\Phi_{B_h}}{h}\right) + \sum n_{B_h A} \ln\left(\frac{\Phi_{B_h A}}{h+r}\right) + n_{A_1} \ln \Phi_A \right. \\
&- \left(\sum n_{B_h} (h-1) + \sum n_{B_h A} h + n_A (r-1) \right) (\ln (z-1) -1) \\
&\left. + \left(\sum n_{B_h} + \sum n_{B_h A} \right) \ln \sigma \right] \\
&- \left[\sum n_{B_h}^\circ \ln\left(\frac{\Phi_{B_h}^\circ}{h}\right) - \left(\sum n_{B_h}^\circ (h-1) \right)(\ln (z-1) - 1) \right. \\
&\left. + \sum n_{B_h}^\circ \ln \sigma \right]
\end{aligned} \tag{6.11}
$$

The terms in the first set of square brackets represent the free energy of mixing the distribution of species found in the mixture, while those in the second represent the free energy of mixing with one another the B_h-mers found in pure B. For future convenience we will designate these quantities $\Delta G_H^*/RT$ and $\Delta G_m^0/RT$, respectively, so that:

$$\Delta G_m = \Delta G_m^* - \Delta G_m^\circ \qquad (6.12)$$

Note that for completeness we have now included the condition $V_A/V_B = r \neq 1$ (terms describing the disorientation of the r segments of the A units have been removed by reference to the pure A state). Equation 6.11 is identical in form to the results obtained by Flory (Equation 6' of reference 2), but here $n_{B_h} \neq n_{B_h}^0$ so that terms in z, σ etc., do not cancel from the final expression for the free energy of mixing. The free energy equation is thus not only algebraically long-winded, but of little use, requiring a knowledge of z and the entire distribution of species present in the mixture and in pure B. Fortunately, a solution that can be related to the stoichiometry of hydrogen bonding and infrared spectroscopic measurements can be obtained from the chemical potentials using a result due to Prigogine. Accordingly, we will briefly consider this authors analysis of the thermodynamics of associated solutions (see ref. 4, page 312), and then proceed to modify equation 6.11 into a more useful form.

C. THERMODYNAMIC PROPERTIES OF ASSOCIATED SOLUTIONS

We consider the situation we will most commonly encounter later, where one component, B, self associates to form h-mers, while the second component does not, but has a functional group capable of hydrogen bonding with B units (Prigogine considered the more general case of both A and B being self-associating molecules). If the total number of molecules of A and B in the mixture are n_A and n_B, respectively, then:

$$n_B = \sum_{h=1}^{\infty} n_{B_h} h + \sum_{h=1}^{\infty} n_{B_h A} h \qquad (6.13)$$

$$n_A = n_{A_1} + \sum_{h=1}^{\infty} n_{B_hA}$$ (6.14)

The chemical potentials of the associated chains and non-hydrogen bonded A units are $\mu_{n_{B_h}}$, $\mu_{n_{B_hA}}$ and $\mu_{n_{A_1}}$, respectively, while the macroscopic chemical potentials (the chemical potentials of the stoichiometric components) are μ_A and μ_B. The hydrogen bonded chains are in chemical equilibrium with one another and the "monomers" (i.e., non hydrogen bonded units, A_1 and B_1), which requires:

$$\mu_{B_h} = h\,\mu_{B_1}$$ (6.15)

$$\mu_{B_hA} = h\,\mu_{B_1} + \mu_{A_1}$$ (6.16)

The total differential of the Gibbs free energy of the mixture of these complexes can be written:

$$dG = \sum \mu_{B_h}\,dn_{B_h} + \sum \mu_{B_hA}\,dn_{B_hA} + \mu_{A_1}\,dn_{A_1}$$

$$= \mu_{B_1} \sum h\,dn_{B_h} + \mu_{B_1} \sum h\,dn_{B_hA} + \mu_{A_1} \sum dn_{B_hA} + \mu_{A_1}\,dn_{A_1}$$

(6.17)

so from equations 6.13 and 6.14:

$$dG = \mu_{B_1}\,dn_B + \mu_{A_1}\,dn_{A_1}$$ (6.18)

But, for any binary system at constant T, P:

$$dG = \mu_B\,dn_B + \mu_A\,dn_A$$ (6.19)

and therefore:

$$\mu_A = \mu_{A_1}$$ (6.20)

$$\mu_B = \mu_{B_1}$$ (6.21)

so that the chemical potentials of the stoichiometric components are equal to those of their respective monomers. As Prigogine pointed out, this result is quite general requiring no assumptions concerning the mode of association, the only necessity being that the hydrogen bonded chains or complexes are in chemical equilibrium with one another.

We can now apply Prigogine's result to obtain an expression for the free energy of mixing in a more useful form than equation 6.11 by first differentiating ΔG_H^* with respect to n_{B_h} to obtain $\mu_{B_h} - \mu_{B_h}^*$, the chemical potential of species B_h with respect to Flory's reference state:

$$\frac{\mu_{B_h} - \mu_{B_h}^*}{RT} = \ln\left(\frac{\Phi_{B_h}}{h}\right) + 1 - \frac{h}{n_0}\left\{\sum n_{B_h} + \sum n_{B_h A} + n_{A_1}\right\}$$
$$- (h-1)(\ln(z-1) - 1) + \ln \sigma \qquad (6.22)$$

Similarly, for the species in pure B, with respect to Flory's reference state:

$$\frac{\mu_{B_h}^0 - \mu_{B_h}^*}{RT} = \ln\left(\frac{\Phi_{B_h}^0}{h}\right) + 1 - h\left\{\frac{\sum n_{B_h}^0}{n_B}\right\}$$
$$- (h-1)(\ln(z-1) - 1) + \ln \sigma \qquad (6.23)$$

Subtracting equation 6.23 from 6.22 and putting $h = 1$:

$$\frac{\mu_{B_1} - \mu_{B_1}^0}{RT} = \ln\left(\frac{\Phi_{B_1}}{\Phi_{B_1}^0}\right) - \left\{\frac{\sum n_{B_h} + \sum n_{B_h A} + n_{A_1}}{n_0}\right\}$$
$$+ \left\{\frac{\sum n_{B_h}^0}{n_B}\right\} \qquad (6.24)$$

[Note : there is a typographical error in equation 26 of reference 30.]

The quantities $(\sum n_{B_h} + \sum n_{B_h A}) / n_B$ and $\sum n_{B_h}^0 / n_B$ are the reciprocals of the number average "degrees of association" of the B units in the mixture and pure B, respectively, and can thus be related to the number of end groups, the non-hydrogen bonded units or "fraction free", defined in Chapters 4 and 5. In deriving equation 6.24 the only additional assumption to those inherent in a Flory lattice model is that association is in the form of linear chains. If we now also assume, for example, that in the formation of an h-mer:

$$B_{h-1} + B_1 \overset{K_B}{\rightleftharpoons} B_h \qquad (6.25)$$

the equilibrium constant K_B is independent of h, then using the following equations, obtained from the identities derived in Chapter 4;

$$\sum \frac{n_{B_h}}{n_0} = \frac{\Phi_B \left(1 - K_B \Phi_{B_1}\right)}{\left(1 + \dfrac{K_A \Phi_{A_1}}{r}\right)}$$

$$= \Phi_B \left(1 - K_B \Phi_{B_1}\right) - \frac{\Phi_B \left(1 - K_B \Phi_{B_1}\right)\left(\dfrac{K_A \Phi_{A_1}}{r}\right)}{\left(1 + \dfrac{K_A \Phi_{A_1}}{r}\right)} \qquad (6.26)$$

and:

$$\frac{\sum n_{B_h A} + n_{A_1}}{n_0} = \frac{\Phi_A}{r} \qquad (6.27)$$

Equation 6.24 becomes:

$$\frac{\mu_B - \mu_B^o}{RT} = \frac{\mu_{B_1} - \mu_{B_1}^o}{RT} = \ln\left(\frac{\Phi_{B_1}}{\Phi_{B_1}^o}\right) - \frac{\Phi_A}{r} + \left(1 - K_B \Phi_{B_1}^o\right)$$

$$- \Phi_B \left[\left(1 - K_B \Phi_{B_1}\right)\left\{1 - \frac{\dfrac{K_A \Phi_{A_1}}{r}}{\left(1 + \dfrac{K_A \Phi_{A_1}}{r}\right)}\right\}\right] \qquad (6.28)$$

Similarly:

$$\frac{\mu_A - \mu_A^o}{RT} = \frac{\mu_{A_1} - \mu_{A_1}^o}{RT} = \ln \Phi_{A_1} - \Phi_B$$

$$- r \Phi_B \left[\left(1 - K_B \Phi_{B_1} \right) \left\{ 1 - \frac{\dfrac{K_A \Phi_{A_1}}{r}}{\left(1 + \dfrac{K_A \Phi_{A_1}}{r} \right)} \right\} \right] \tag{6.29}$$

The expressions in square brackets appear to be unnecessarily complicated, but are left in this form because they are directly related to very simple quantities that have a direct physical meaning, as we will show in the following section. Furthermore, combining equation 6.28 and 6.29 we now obtain a relatively simple expression for the free energy of mixing:

$$\frac{\Delta G_m}{RT} = n_B \ln \left(\frac{\Phi_{B_1}}{\Phi_{B_1}^o} \right) + n_A \ln \Phi_{A_1} + n_B \left(K_B \Phi_{B_1} - K_B \Phi_{B_1}^o \right)$$

$$+ n_B \left(1 - K_B \Phi_{B_1} \right) \frac{\dfrac{K_A \Phi_{A_1}}{r}}{\left(1 + \dfrac{K_A \Phi_{A_1}}{r} \right)} \tag{6.30}$$

The quantities K_A, K_B, Φ_B, Φ_{B_1} and Φ_{A_1} can all be determined from the equations describing the stoichiometry of hydrogen bonding and infrared spectroscopic measurements, as described in Chapter 4. An expression for weak interactions (χ terms) can be added in the usual manner. Because of the way it is derived this free energy expression does not explicitly contain a term for the enthalpy change due to hydrogen bonding, but it is a relatively easy task to obtain this quantity from equation 6.30 as we will show later.

It is not necessary to assume that all steps in the equations describing self association are governed by the same equilibrium constant. A particularly important exception is the situation where the formation of dimers differs from that of subsequent h-mers. The principles are the same, but as we showed in Chapter 4 the algebra gets more complicated and we have an additional variable that has to be determined by experiment. The appropriate equations are given in an appendix to this chapter.

Finally, there is an important point that should be made here. Clearly, there are a number of association models that could be used to describe the free energy and we have confined our attention to just two, differing slightly in the mode of self association. As long as the model we choose accurately reproduces the experimentally determined stoichiometry of hydrogen bonding, however, we will obtain a reasonably accurate reflection of the contributions of these interactions to the free energy, given the limitations and assumptions of the Flory lattice model. This follows directly from an examination of equation 6.24, where the chemical potential of the B units is expressed in terms of quantities that are related directly to the degree of association, or the number of hydrogen bonds of each type present, quantities that can be measured spectroscopically (this relationship will be made more explicit in the following section). Accordingly, one condition for the choice of a particular model is that it should reproduce these experimental quantities across the composition range. As we demonstrated in Chapter 4, this model appears to work well for the systems we have examined so far.

D. A SIMPLIFIED EXPRESSION FOR THE FREE ENERGY

In the preceding section we obtained equations for the free energy of mixing and chemical potentials in terms of the equilibrium constants K_A, K_B, and the volume fractions of non hydrogen bonded molecules or "monomers", Φ_{A_1}, Φ_{B_1}. These equations appear complicated, but, as we have repeatedly noted are useful in this form because the parameters can be directly determined from the equations describing the stoichiometry of hydrogen bonding and the relationship to infrared spectroscopic measurements. Here we will demonstrate that these equations have a very simple underlying structure and can be expressed in a much less complicated manner in terms of the probabilities of forming hydrogen bonds. We will illustrate this using the simplest model of self-association, where each step is governed by the same equilibrium constant K_B. The final result is quite general, however, and could be obtained directly from equation 6.24.

We let the probability of self association, or the probability of forming a hydrogen bond between two molecules of type B be p_{BB}. In Chapter 4 we defined the quantity f_F^J, the fraction of free units in a system where one of the end groups can be measured spectroscopically, and determined that this was given by:

$$f_F^J = 1 - K_B \Phi_{B_1}$$ (6.31)

The fraction of hydrogen bonded B groups, equal to the probability of a particular group being hydrogen bonded at some chosen instant of time in a mixture in equilibrium is then simply given by:

$$p_{BB} = (1 - f_F) = K_B \Phi_{B_1}$$ (6.32)

In pure B the probability of a functional group being hydrogen bonded is similarly given by:

$$p_{BB}^o = K_B \Phi_{B_1}^o$$ (6.33)

Accordingly, the first non logarithm term in equation 6.28 can be rewritten:

$$n_B \left(K_B \Phi_{B_1} - K_B \Phi_{B_1}^o \right) = n_B \left(p_{BB} - p_{BB}^o \right)$$ (6.34)

and is simply equal to the change in the number of B--B hydrogen bonds on going from pure B to the mixture. In a similar fashion the second non logarithm term in equation 6.28 can be expressed in terms of the probability of forming a hydrogen bond between a B and an A unit, p_{BA}, by first noting that:

$$p_{BA} = \left(1 - \frac{n_{A_1}}{n_A}\right) \qquad (6.35)$$

where n_{A_1}, is the number of A molecules that are not hydrogen bonded. In terms of volume fractions:

$$p_{BA} = \left(1 - \frac{\Phi_{A_1}}{\Phi_A}\right) \qquad (6.36)$$

Using the stoichiometric relationship for the volume fraction of A units (see Chapter 4).

$$\Phi_A = \Phi_{A_1}\left[1 - \frac{K_A \Phi_{A_1}}{\left(1 - K_B \Phi_{B_1}\right)}\right] \qquad (6.37)$$

we obtain:

$$p_{BA} = \left(\frac{r}{\Phi_A}\right)\left(\frac{K_A \Phi_{A_1}}{r}\right)\left(\frac{\Phi_{B_1}}{1 - K_B \Phi_{B_1}}\right) \qquad (6.38)$$

and using the second stoichiometric relationship:

$$\Phi_B = \frac{\Phi_{B_1}}{\left(1 - K_B \Phi_{B_1}\right)^2}\left(1 + \frac{K_A \Phi_{A_1}}{r}\right) \qquad (6.39)$$

this can be rewritten:

$$p_{BA} = \left(\frac{r}{\Phi_A}\right)\left(\Phi_B\right)\left(1 - K_B \Phi_{B_1}\right)\left[\frac{\dfrac{K_A \Phi_{A_1}}{r}}{\left(1 + \dfrac{K_A \Phi_{A_1}}{r}\right)}\right] \qquad (6.40)$$

or:

$$n_A p_{BA} = n_B \left(1 - K_B \Phi_{B_1} \right) \left[\frac{\dfrac{K_A \Phi_{A_1}}{r}}{\left(1 + \dfrac{K_A \Phi_{A_1}}{r} \right)} \right] \tag{6.41}$$

Equation 6.30 for the free energy can thus be rewritten as:

$$\frac{\Delta G_m}{RT} = n_B \ln \left(\frac{\Phi_{B_1}}{\Phi_{B_!}^o} \right) + n_A \ln \Phi_{A_1} + n_B (p_{BB} - p_{BB}^o) + n_A p_{BA}$$

$$= n_B \ln \left(\frac{p_{BB}}{p_{BB}^o} \right) + n_A \ln \left[\Phi_A (1 - p_{BA}) \right]$$

$$+ n_B (p_{BB} - p_{BB}^o) + n_A p_{BA} \tag{6.42}$$

As we noted above, this result is general and does not depend upon the mode of linear association that we choose to use. Its limitations are those of any Flory lattice model. Clearly, if the model we choose to use accurately reproduces the experimentally determined fraction of hydrogen bonded groups as a function of composition and temperature, then we can be confident (within the limits of experimental error) that we know the quantities p_{BB}, p_{BB}^o and p_{BA}.

E. HYDROGEN BONDING IN POLYMER MIXTURES–LINEAR ASSOCIATION

If we attempt to use a lattice model to describe hydrogen bonding interactions in polymer mixtures we immediately encounter the problem of describing two sets of chains, one covalent the other hydrogen bonded, both of which share the same segments. A formal derivation of an entropy of mixing using, for example, the lattice filling procedure of Flory[2], immediately becomes more complicated. Each segment of a covalent chain belongs (most often) to a *different* hydrogen bonded chain. In our initial work in this area[27-29] we essentially assumed that the formation of hydrogen bonds is unaffected by the linkage of the functional groups involved into covalent chains. An expression for the free energy of mixing was then obtained by adding the result obtained for low molecular weight molecules to a Flory-Huggins expression for polymers. A correction term must be subtracted because the first expression contains an excess combinatorial entropy component (see below).

Lattice Model Depicting
Hydrogen Bonded Polymer Chain

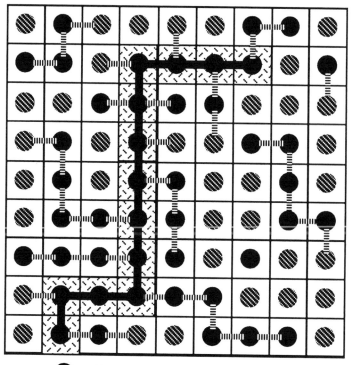

● = B ⅢⅢ = Hydrogen Bond

◍ = A ▬ = Covalent Bond

FIGURE 6.1

In describing the mixing of hydrogen bonded chains of small molecules we established a methodology that depended upon the determination of the probability that a mixture of A and B monomers would spontaneously occur in a configuration equivalent to that of the hydrogen bonded system. We can now apply this to a polymer[30] and in effect determined the probability that the mixture would now exist in a configuration equivalent to that of a hydrogen bonded network. This can be accomplished by taking the equilibrium distribution of *hydrogen bonded* chains that is characteristic of a particular temperature and composition and place it on the lattice. This distribution is the one found in the polymer mixture, *not one that might occur in a mixture of monomer analogues* (more on this later), and in principle, can be determined by appropriate spectroscopic measurements. We then simply take the self-associating (B) units, one from each h-mer, and find the number of ways of linking them to form N_B covalent chains, as illustrated in figure 6.1.

The same procedure is followed for the A units. Obviously, this process depends upon the probability that (M_B - 1) B units are adjacent to a selected B unit, and so on. The corresponding procedure, placing the *covalent* chains on the lattice and linking their segments to form the equilibrium distribution of hydrogen bonded chains, must give the same result and is the method we will actually employ. In effect, we determine the probability that the non-hydrogen bonded covalent chains will spontaneously occur in a configuration equivalent to that of a hydrogen bonded system. We initially considered this to be an original approach, but in order to address the problem of the appropriate choice of reference states we became aware that this is precisely the same concept used by Flory to determine the entropy of network formation in cross-linked rubbers[31,32], although our methodology is somewhat different.

Hydrogen Bonding in Polymers–The Distribution in the Mixture

Following the procedure described in Section 6B, the number of configurations available to a hydrogen bonded mixture of polymers can be written as:

$$\Omega = \Omega_F \Omega_{conn}^P \tag{6.43}$$

where Ω_F is the number of configurations available to the non-hydrogen bonded polymer chains, which in the Flory approximation is given by:

$$\Omega_F = \frac{1}{N_B!} \frac{1}{N_A!} \left\{ \frac{(z-1)^{\left(N_B(M_B-1)+N_A(M_A-1)\right)}}{\sigma^{(N_B+N_A)}} \right\}$$

$$\times \left\{ \frac{n_0!}{n_0^{\left(N_A(M_A-1)\right)} n_0^{\left(N_B(M_B-1)\right)}} \right\} \tag{6.44}$$

As before, Ω_{conn}^P determines the number of these configurations that are available to the hydrogen bonded system. We will require the probability that (h-1) segments are adjacent to a chosen segment, but now assume that each of these units is located on different covalent polymer chains.

Following Guggenheim[33], we can write the frequency of occupation of the (h-1) sites by segments of different chains as:

$$\left(\frac{q_B N_B}{q_A N_A + q_B N_B} \right)^{h-1}$$

where q_B, q_A are related to the number of nearest neighbors of the covalent chains through:

$$q_B z = M_B (z-2) + 2 \tag{6.45}$$

$$q_A z = M_A (z-2) + 2 \tag{6.46}$$

For mixtures of two polymers (M_A, $M_B \gg 1$) we therefore obtain with negligible error:

$$\left(\frac{M_B N_B}{M_A N_A + M_B N_B} \right)^{h-1} = \left(\frac{n_B}{n_0} \right)^{h-1} \tag{6.47}$$

so that as in low molecular weight materials we can assume that the probability of finding adjacent B segments is proportional to their volume fractions (this assumption will clearly be less accurate when considering solutions of polymers in "monomeric" solvents, unless we make the usual assumption that $z \to \infty$).

As before, we first consider the number of configurations available to (or ways of connecting) the jth B_h-mer, after $(j-1)$ hydrogen bonded B_h chains have been formed. Recalling that $n_B = N_B M_B$, the number of unconnected segments is given by:

$$\left(n_B - \sum_{i=1}^{j-1} h_i \right)$$

and this is equal to the number of sites available to the first unit of the jth hydrogen bonded chain. If we assume that the probability that a given segment in a covalent chain has not yet been "connected" (become part of the j-1 B_h-mers) is simply equal to the over-all average fraction of unconnected segments, equivalent to the assumption of the random occupation of sites made by Flory[2,34], then we obtain for υ_j, the number of configurations available to the jth chain:

$$\upsilon_j = \left(\frac{(z-1)^{(h_j-1)}}{\sigma} \right) \frac{\left(n_B - \sum_{i=1}^{j-1} h_i \right)!}{\left(n_B - \sum_{i=1}^{j} h_i \right)!} \left(\frac{1}{n_0} \right)^{(h_j-1)} \qquad (6.48)$$

which is the same as equation 6.5, as it must be given the identity of the assumptions and approximations. The quantity Ω^P_{conn} is then given by:

$$\Omega^P_{conn} = \frac{1}{\prod n_{B_h}! \prod n_{B_h A}! \, n_{A_1}} \left[\frac{(z-1)^{\left(\sum n_{B_h}(h-1) + \sum n_{B_h A} h \right)}}{\sigma^{\left(n_{B_h} + n_{B_h A} \right)}} \right]$$

$$\times \left[\frac{n_B! \, n_A!}{n_0^{\sum n_{B_h}(h-1)} \, n_0^{\sum n_{B_h A}(h-1)} \, n_0^{\sum n_{B_h A}}} \right] \qquad (6.49)$$

The free energy of mixing, relative to Flory's reference state ΔG_m^*, can thus be obtained directly from equations 6.43, 6.44 and 6.49. This equation has an unreasonably large number of terms, but by inspection it can be immediately seen that it breaks down into three parts; the usual Flory Huggins entropy of mixing, which if we for the moment ignore non hydrogen bonding interactions we write as a free energy ΔG_F^* (again relative to Flory's reference state); a contribution from hydrogen bonding ΔG_H^*, that is identical to the quantity defined in equations 6.12 and 6.11 for "small molecules"; a third part arising from the $n_B!$, $n_A!$ terms in equation 6.49. Taken together we obtain:

$$\frac{\Delta G_m^*}{RT} = \frac{\Delta G_F^*}{RT} + \frac{\Delta G_H^*}{RT} - \left(n_B \ln \Phi_B + n_A \ln \Phi_A \right) \qquad (6.50)$$

If the A and B segments are approximately the same size then the final terms in parentheses are equivalent to a regular solution entropy of mixing. Accordingly, subtracting this from the expression for ΔG_H^* gives an excess function; the change in free energy of a hydrogen bonded non-covalently linked system of B and A units relative to the entropy change that would occur if such non-hydrogen bonded units were randomly mixed.

The Free Energy of Mixing–A Choice of Reference States

To obtain our final expression for the free energy of mixing we now need to determine the number of configurations available to the hydrogen bonded chains in pure B. The obvious way of accomplishing this is to simply apply the above methodology to pure B and subtracting the result from equation 6.50. We believe this procedure gives an incorrect result, however, because the reference states used in each derivation are then different. This is a subtle point and is related to Flory's arguments that the probability of a cross-linked state depends upon the volume[31,32] (at any instant of time, of course, most hydrogen bonded systems can be considered to be a densely cross-linked network).

We start our discussion of reference states by asking the reader to turn back the pages of this chapter to re-examine equations 6.2 and 6.9. Both describe the number of configurations available to hydrogen bonded chains on a lattice, but equation 6.9 was obtained by first allowing A and B molecules to randomly mix, then finding the probabilities that these molecules would spontaneously occur in a configuration equivalent to that of the hydrogen bonded system. This equation is identical to equation 6.2 (obtained by the usual lattice filling procedure), but has a term, $(n_0! \, / \, n_B! \, n_A!)$ that has been separated out. Upon taking logarithms and applying Stirling's approximation, this part of the expression would give a term $(n_A \ln \Phi_A + n_B \ln \Phi_B)$ equivalent to an ideal combinatorial entropy of mixing if A and B molecules are the same size. Subtracting this term from the free energy gives an excess function *relative to a mixture of non-hydrogen bonded A and B units. This does not account for the fact that the B units are hydrogen bonded in the pure state.* This is not to say that this is an incorrect reference state, but it is one that does not give the excess function required for our problem.

Moving on to hydrogen bonding in polymers, an examination of equation 6.50 shows that we have obtained an excess free energy term for the hydrogen bonding interactions that is with respect to non-covalently linked, non-hydrogen bonded, randomly mixed A and B segments. If we form the hydrogen bonds characteristic of B segments in pure polymer B and subtract the resulting expression from equation 6.50, this excess function remains the same. It does not measure the change in free energy due to the change in the distribution of hydrogen bonds relative to that found in pure B, but relative to a mixture where there are no hydrogen bonds at all.

The appropriate reference state for our problem can perhaps best be perceived by first considering a situation where the strength of the hydrogen bond, or the equilibrium constant describing association, is very large, so that except in dilute solutions the distribution of species is effectively unaltered. We would then obtain a free energy of mixing (recalling that we are ignoring non-specific interactions) that is given by;

$$\sum_{B_h} n^{\circ}_{B_h} \ln \Phi_B + n_A \ln \Phi_A \qquad (6.51)$$

the usual Flory equation for the combinatorial entropy of mixing covalent polymers that have a distribution of molecular weights. Clearly, the excess function we require should be with respect to this "combinatorial entropy" of mixing expression. In terms of describing the number of configurations available to the system, this excess term is equivalent to forming the hydrogen bonded species, both those found in the mixture and those in pure B *at the same volume*, a process illustrated schematically in figure 6.2.

By forming the hydrogen bonds found in pure B *at the volume of the mixture* we now obtain:

$$\frac{\Delta G^{\circ}_m}{RT} = \frac{\Delta G^{\circ}_H}{RT} - \left[n_B \ln \Phi_B - \sum_{B_h} n^{\circ}_{B_h} \ln \Phi_B \right] \qquad (6.52)$$

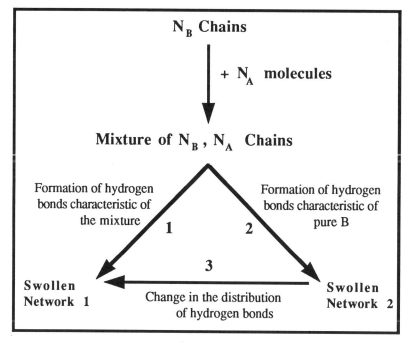

FIGURE 6.2

where ΔG_H^o is the result obtained for non-covalently linked B units (see equations 6.10, 6.11 and 6.12), so that by subtracting equations 6.52 from equation 6.50 we obtain an expression for the free energy of mixing that is given by:

$$\frac{\Delta G_m}{RT} = N_A \ln \Phi_A + N_B \ln \Phi_B + \left[\frac{\Delta G_H^*}{RT} - \frac{\Delta G_H^o}{RT} \right]$$

$$- \left[\frac{n_B}{\overline{h}^o} \ln \Phi_B + n_A \ln \Phi_A \right] \tag{6.53}$$

where $\overline{h}^o$ is the number average degree of association found in pure B and is equal to $n_B / \Sigma n_{B_h}^o$. If self association is described by a single equilibrium constant, K_B, then we can substitute from equation 6.30 and multiply by V_B / V to obtain the free energy per mole of lattice sites:

$$\frac{\Delta G_m'}{RT} = \left[\frac{\Phi_A}{M_A} \ln \Phi_A + \frac{\Phi_B}{M_B} \ln \Phi_B + \Phi_A \Phi_B \chi \right]$$

$$+ \left[\Phi_B \ln \left(\frac{\Phi_{B_1}}{\Phi_{B_1}^o} \right) + \frac{\Phi_A}{r} \ln \Phi_{A_1} + \Phi_B K_B \left(\Phi_{B_1} - \Phi_{B_1}^o \right) \right.$$

$$\left. + \Phi_B \left(1 - K_B \Phi_{B_1} \right) \frac{\left(\dfrac{K_A \Phi_{A_1}}{r} \right)}{\left(1 + \dfrac{K_A \Phi_{A_1}}{r} \right)} \right]$$

$$- \left[\frac{\Phi_B}{\overline{h}^o} \ln \Phi_B - \frac{\Phi_A}{r} \ln \Phi_A \right] \tag{6.54}$$

where a χ term accounting for the dispersion forces between segments has simply been added, following Flory[#2,34]. (The algebraically more complicated case where self-association is described by more than one equilibrium constant is given in an appendix to this chapter.) Left in this form it can be seen that the free energy of mixing consists of three terms. First, the usual Flory expression for the mixing of polymers; second, an expression that is equal to the change in free energy that would occur as a result of the change in hydrogen bonding upon mixing if the segments were not connected to form covalent chains (note that the distribution of hydrogen bonds is defined as that actually found in the polymer, not that which might occur in an equivalent mixture of monomer analogues); finally, the third term "corrects" for the fact that the second term not only gives the free energy change associated with the changing pattern of hydrogen bonding, but also has a "combinatorial" entropy of mixing contribution. This final term in brackets in equation 6.54 is thus the entropy of mixing that would be obtained by mixing non-covalently bonded B segments with A units if there were no change in hydrogen bonding upon mixing (i.e. if the distribution of hydrogen bonds remained the same as that found in pure B).

To reiterate, if we had simply considered the number of configurations available to the hydrogen bonded chains in pure B at the volume of pure B as our reference state, we would have obtained a final term equal to ($n_B \ln \Phi_B + n_A \ln \Phi_A$), approximately a regular solution entropy of mixing. This "correction" term would clearly be excessive, as the initial state of the B units consists of hydrogen bonded chains, not freely mixing independent B units. Accordingly, for hydrogen bonding in polymers a reference state consisting of the initially mixed covalent chains is most appropriate (see figure 6.2) and gives equation 6.54.

[#] Note that for a segment that is hydrogen bonded to two other segments, there are now four neighbors that do not participate in physical interactions and z - 4 that do. If we regard χ as an adjustable parameter this makes no difference, but could lead to errors when we calculate its value from solubility parameters. The effect of such errors will be discussed in Chapter 7.

Finally, in Chapter 2 and in most of our published work, we write equation 6.54 in a simpler form:

$$\frac{\Delta G'_m}{RT} = \frac{\Phi_A}{M_A} \ln \Phi_A + \frac{\Phi_B}{M_B} \ln \Phi_B + \Phi_A \Phi_B \chi + \frac{\Delta G_H}{RT}$$

(6.55)

where $\Delta G_H/RT$ is now the excess function describing the free energy contributions due to the change in hydrogen bonding upon mixing.

Given this simple form, one could well ask why not simply roll the ΔG_H term into an overall χ and thus allow phase behavior to be described by a single adjustable parameter? There are at least three reasons why this is an undesirable approach. In its simplest form the χ term is symmetric with composition. The ΔG_H term has a strong composition dependence, however, that is a consequence of the often great difference in the equilibrium constants describing self-association and inter-association. We will demonstrate this when we consider phase behavior later in this chapter and in Chapter 7. Secondly, we can determine ΔG_H given that equilibrium constants can be determined from infrared spectroscopy. If we now assume that because χ now reflects physical forces only, to a first approximation we can estimate its value using solubility parameters. Given a knowledge of ΔG_H and χ, and assuming free volume effects can be neglected, we should be able to predict phase behavior. We will attempt a preliminary assessment of the value of this approach, again in Chapter 7. Finally, and in our view most compellingly, calculating an overall χ parameter does not provide any fundamental insight into the balance of intermolecular forces. In systems where specific interactions can be unambiguously identified thermodynamic properties can be shown to depend upon the balance between these forces and "physical" or non-specific interactions, as we will seek to demonstrate with the various examples we will present later in this book.

F. CHEMICAL POTENTIALS

For many applications (e.g., depression of melting point in polymer crystals in equilibrium with an amorphous phase that is a miscible blend, calculation of binodals, etc) a knowledge of the chemical potentials is essential. These follow directly in the usual manner and are expressed as a chemical potential per structural unit:

$$\frac{\mu_B - \mu_B^o}{RT} = \left[\frac{\ln \Phi_B}{M_B} - \Phi_A \left(\frac{1}{M_A} - \frac{1}{M_B} \right) + \Phi_A^2 \chi \right] +$$

$$\ln \left(\frac{\Phi_{B_1}}{\Phi_{B_1}^o} \right) - \frac{\ln \Phi_B}{\overline{h}^o} + \Phi_B K_B \left(\Phi_{B_1} - \Phi_{B_1}^o \right) +$$

$$\Phi_B \left(1 - K_B \Phi_{B_1} \right) \frac{\left(\dfrac{K_A \Phi_{A_1}}{r} \right)}{\left(1 + \dfrac{K_A \Phi_{A_1}}{r} \right)} \tag{6.56}$$

and:

$$\frac{\mu_A - \mu_A^o}{RT} = r \left[\frac{\ln \Phi_A}{M_A} - \Phi_B \left(\frac{1}{M_B} - \frac{1}{M_A} \right) + \Phi_B^2 \chi \right] +$$

$$\ln \left(\frac{\Phi_{A_1}}{\Phi_A} \right) + r \Phi_B K_B \left(\Phi_{B_1} - \Phi_{B_1}^o \right) +$$

$$r \Phi_B \left(1 - K_B \Phi_{B_1} \right) \frac{\left(\dfrac{K_A \Phi_{A_1}}{r} \right)}{\left(1 + \dfrac{K_A \Phi_{A_1}}{r} \right)} \tag{6.57}$$

It should be recalled that these expressions seem complicated only because of the presence of the terms:

$$\Phi_B K_B \left(\Phi_{B_1} - \Phi_{B_1}^o \right) \quad \text{and} \quad \Phi_B \left(1 - K_B \Phi_{B_1} \right) \frac{\left(\dfrac{K_A \Phi_{A_1}}{r} \right)}{\left(1 + \dfrac{K_A \Phi_{A_1}}{r} \right)}$$

These are related to the change in the number of B--B hydrogen bonds and B--A hydrogen bonds, respectively (see Section 6D), or the reciprocals of the number average degrees of association, and once seen in that light the correspondence to the more familiar forms of the chemical potentials derived by Flory become apparent. [Of course, these equations would not be valid in dilute solutions.] Chemical potentials corresponding to other forms of association are presented in the appendices.

G. DEPENDENCE OF EQUILIBRIUM CONSTANTS ON MOLECULAR WEIGHT AND CHAIN STIFFNESS

In our discussions so far we have been careful to point out that the distribution of hydrogen bonded species is that found in the polymer, not in a mixture of monomers or low molecular weight analogues that are not covalently linked. For many systems this presents no problem, in that equilibrium constants can be determined directly from spectroscopic studies of the blend components and their mixtures. We will elaborate on this point below. In blends involving poly(vinyl phenol), however, we have used self-association constants determined for phenol, adjusted to account for the difference in molar volume of this molecule and that of the polymer repeat unit. For low molecular weight materials we demonstrated in Chapter 4 that equilibrium constants can be transferred in this manner from one molecule to another by simply accounting for differences in molar volume, providing that there are no significant steric or electronic differences between the molecules. For polymers we have the possibility that the equilibrium constants are also influenced by the presence of more than one interacting functional group per molecule and hence, depend upon the molecular weight of the covalent chain and the chain stiffness.

Stadler[35] has presented an analysis which concludes that the equilibrium constant describing simple pairwise association decreases with increasing chain length for very short chains and then levels off at a value that is significantly less than that found in monomer analogues. In this model, the probability that an unbonded unit can find a partner for complex formation is considered to be restricted by those units that are already associated, so that topological constraints prevent the attainment of the equilibrium distribution found in non-covalently linked units. In our view, this does not account for the dynamic character of hydrogen bonds in the liquid state (lifetimes of the order of 10^{-5} to 10^{-11} secs), which, it seems to us, would allow hydrogen bonds to form according to their intrinsic proclivities, *providing that the chains are perfectly flexible.* Indeed, it can be shown that within the limits of the assumptions inherent in the Flory lattice model there should be no chain length dependence at all. As in Chapter 4, Section B, the standard free energy change, ΔG°, for the process:

$$\text{h-mer} + \text{1-mer} \rightleftharpoons (\text{h+1})\text{-mer} \qquad (6.58)$$

is given by the equilibrium condition:

$$\overline{\Delta G_{h+1}} - \overline{\Delta G_h} - \overline{\Delta G_1} + \Delta G^\circ = 0 \qquad (6.59)$$

Differentiating equation 6.50 to obtain the appropriate partial molar free energies it can be shown that terms involving the covalent degree of polymerization (M_B, M_A) are eliminated, leaving:

$$\ln K_B = \ln \left[\frac{\Phi_{h+1}}{\Phi_h \Phi_1} \cdot \frac{hl}{(h+1)} \right] = \frac{\Delta G^\circ}{RT} + \ln \left[(z-1)\,\sigma \right] \qquad (6.60)$$

which is the equilibrium constant definition obtained earlier for the association of monomeric species (to obtain this result we differentiate with respect to n_{B_h}, n_{B_1}, and $n_{B_{h-1}}$. Writing $\Phi_B = (\Sigma n_{B_h} h/n_o)$ (for self-association only), etc., we find most terms, T, when substituted into equation 6.59, give a result of the form $(h+1)T - hT - 1T$ and thus cancel). Therefore, there is no variation in equilibrium constants with covalent chain length, *providing that Flory's assumptions are valid.* Accordingly, in order to account for any effect due to chain length and stiffness in the context of the association lattice model presented here, we have to consider modifications to the Flory treatment, which will enter through the conditional probabilities of finding appropriate segments of different chains adjacent to one another, such that they could be linked by hydrogen bonds.

The Flory model[34] essentially assumes that such probabilities can be approximated by volume fractions, while the more exact probabilities used by Huggins[36] and Guggenheim[33] employ surface site fractions, defined in terms of the q factors given by equations 6.45 and 6.46. We demonstrated earlier that for mixing two high molecular weight polymers, surface site fractions are approximately equal to volume fractions (see page 331), but for oligomers this is not so, unless we make the assumption that $z \rightarrow \infty$ (where z is the lattice coordination number). Similarly, for a stiff chain we would have to discuss modifications to the conditional probabilities imposed by chain stiffness and presumably follow the treatments given by DiMarzio and Gibbs[37] and DiMarzio[38,39]. We are now pursuing this line of research and so cannot present a formal analysis here. The crucial point that must be grasped, however, is that *it is not necessary to account for chain stiffness (or chain length) if equilibrium constants are obtained by direct spectroscopic studies of the polymers being considered.* This is because the experimentally determined fraction of hydrogen bonded species obviously *includes the influence of such factors*, as will the equilibrium constants then calculated from such measurements. The problem is only significant when we seek to apply equilibrium constants obtained from low molecular weight species (e.g., phenol, cresol) to polymers with similar functional groups (e.g., poly(vinyl phenol)), after adjusting for differences in molar volumes (as discussed in Chapter 4). In our work so far we have only employed monomer analogue equilibrium constants in describing the self-association of poly(vinyl phenol). Inter-association constants between the OH group of this polymer and the ester group of various acrylates and methacrylates were obtained by direct studies of the polymer mixtures. As we demonstrated in Chapter 5, the stoichiometry of hydrogen bonding was then accurately reproduced for a range of systems.

Furthermore, we will show in the following chapter that these equilibrium constants allow a surprisingly accurate prediction of the phase behavior of these blends. Accordingly, for this system factors such as chain stiffness do not appear to be critical. This may not be so for other polymers, but to reiterate, this should not be a problem in blends where equilibrium constants can be measured directly. [If one component is oriented, as in rigid rod/random coil mixtures, the problem again changes and there is the possibility of orientation dependent interactions.]

H. PHASE BEHAVIOR

In the preceding sections we have used an association model to obtain an expression for the free energy of mixing polymer molecules that can hydrogen bond in a specific but very common manner (one polymer self-associates through formation of "chains" of hydrogen bonds; the second does not, but is capable of forming hydrogen bonds with units of the first). The obvious next step is to apply this model to a quantitative description of phase behavior in these systems, so that the model can be experimentally tested. Binodal curves are, of course, calculated by equating the chemical potentials of each component in each phase and solving by numerical methods (equations for the chemical potentials were given in Section F). In contrast, analytical expressions can usually be determined for the spinodal, so curves defining the stability limit are more easily obtained. This remains true when hydrogen-bonding interactions are accounted for by an association model, as we will demonstrate in this section. As we will see, the balance of free energy contributions arising from self-association, inter-association and physical interactions can give rise to a rich variety of phase behavior, as we noted in Chapter 1 and as described in a splendid review by Walker and Vause[40].

Spinodal Equations

For simplicity of presentation we will again assume that self-association is governed by a single equilibrium constant and relegate more complicated association schemes to the appendix. Accordingly, if we neglect free volume and accept the other approximations inherent in the use of a simple Flory lattice model, then the free energy is given by equation 6.54 and an equation describing the spinodal can then be obtained from the condition:

$$\frac{\partial^2 \left(\dfrac{\Delta G_m}{RT} \right)}{\partial \Phi_B^2} = 0 \tag{6.61}$$

The second derivatives of the terms in the first and last set of brackets in equation 6.54 have the familiar form:

$$\left[\frac{1}{M_A \, \Phi_A} + \frac{1}{M_B \, \Phi_B} - 2\,\chi \right] - \left[\frac{1}{\bar{h}^o \, \Phi_B} + \frac{1}{r \, \Phi_A} \right] \tag{6.62}$$

The second derivative of the second set of terms in brackets in equation 6.54 is not so easily obtained, however, and requires some manipulation. The first derivative is given by:

$$\ln \left(\frac{\Phi_{B_1}}{\Phi_{B_1}^o} \right) + \Phi_B \left(\frac{1}{\Phi_{B_1}} \frac{\partial \Phi_{B_1}}{\partial \Phi_B} - \frac{1}{\Phi_{B_1}^o} \frac{\partial \Phi_{B_1}^o}{\partial \Phi_B} \right) +$$

$$\frac{\Phi_A}{r \, \Phi_{A_1}} \frac{\partial \Phi_{A_1}}{\partial \Phi_B} - \frac{1}{r} \ln \Phi_{A_1} + K_B \, \Phi_B \left(\frac{\partial \Phi_{B_1}}{\partial \Phi_B} - \frac{\partial \Phi_{B_1}^o}{\partial \Phi_B} \right) +$$

$$K_B \left(\Phi_{B_1} - \Phi_{B_1}^o \right) - K_B \, \Phi_B \left(\frac{X}{1+X} \right) \left(\frac{\partial \Phi_{B_1}}{\partial \Phi_B} \right) +$$

$$\left(\frac{X}{1+X} \right) \left(1 - K_B \, \Phi_{B_1} \right) + \frac{\Phi_B \left(1 - K_B \, \Phi_{B_1} \right)}{\left(1 + X \right)^2} \frac{\partial X}{\partial \Phi_B} \tag{6.63}$$

Of course, $\Phi_{B_1}^{\circ}$ the concentration of non-hydrogen-bonded species in pure B ($\Phi_B = 1$) is a constant, so that:

$$\left(\frac{\partial \Phi_{B_1}^{\circ}}{\partial \Phi_B}\right) = 0 \tag{6.64}$$

Also, for ease of presentation, we have let:

$$X = \frac{K_A \Phi_{A_1}}{r} \tag{6.65}$$

so, from the stoichiometric relationship:

$$\Phi_{A_1} = (1 - \Phi_B)\left(1 + \frac{K_A \Phi_{B_1}}{1 - K_B \Phi_{B_1}}\right)^{-1} \tag{6.66}$$

the last term in equation 6.63 is given by:

$$\frac{\Phi_B (1 - K_B \Phi_{B_1})^3}{(1 + X)^2 \Phi_{B_1}} - \frac{\Phi_B (1 - \Phi_B)}{(1 + X) \Phi_{B_1}} \frac{\partial \Phi_{B_1}}{\partial \Phi_B} \tag{6.67}$$

The equation can be further simplified by using:

$$\Phi_B = \frac{\Phi_{B_1}}{(1 - K_B \Phi_{B_1})^2} (1 + X) \tag{6.68}$$

From equation 6.66 we can also express $\partial \Phi_{A_1}/\partial \Phi_B$ in terms of $\partial \Phi_{B_1}/\partial \Phi_B$:

$$\frac{\partial \Phi_{A_1}}{\partial \Phi_B} = -\frac{\Phi_{A_1}}{\Phi_A}\left(\frac{K_A \Phi_{A_1}}{(1 - K_B \Phi_{B_1})^2} \frac{\partial \Phi_{B_1}}{\partial \Phi_B} + 1\right) \tag{6.69}$$

By substituting 6.67 and 6.69 into 6.63 it can be shown that terms in $\partial\Phi_{B_1}/\partial\Phi_B$ cancel, so the first derivative of the hydrogen-bonding terms is given by:

$$\ln\left(\frac{\Phi_{B_1}}{\Phi_{B_1}^o}\right) - \frac{1}{r}\ln\Phi_{A_1} - \frac{1}{r} + (1 - K_B\Phi_{B_1}^o) \quad (6.70)$$

The second derivative of this equation is then:

$$\frac{1}{\Phi_{B_1}}\frac{\partial\Phi_{B_1}}{\partial\Phi_B} - \frac{1}{r\Phi_{A_1}}\frac{\partial\Phi_{A_1}}{\partial\Phi_B} \quad (6.71)$$

This is a beautifully simple and intuitively pleasing result. As demonstrated in Section C, the free energy associated with hydrogen-bonding interactions in the blend is derived in terms of the concentrations of "monomers," i.e., those segments that have no hydrogen-bonded partners whatsoever, through substitution of the condition of equality of the chemical potentials of these "monomers" and their stoichiometric counterparts:

$$\mu_{B_1} = \mu_B \; ; \; \mu_{A_1} = \mu_A \quad (6.72)$$

Accordingly, the stability limit should depend upon the balance between the variation of the concentration of these units with composition, a relationship expressed by equation 6.71. Furthermore, on this basis we would also expect the spinodal equation to have the same form if we now characterize the self-association of the B units by two equilibrium constants, one describing the formation of dimers and the second the formation of h-mers (h>2). It is a conceptually straightforward (but algebraically complicated) matter to demonstrate that this is indeed so, and the derivation has been presented in the literature[29]. [The same result is also obtained when self-association is in the form of cyclic complexes - see the final section of this chapter.]

The derivative terms in equation 6.71 can be determined numerically (as can the spinodal directly from the second derivative of equation 6.54), but for simple systems this is unnecessary (and inelegant) in that analytical expressions for $\partial \Phi_{B_1}/\partial \Phi_B$ and $\partial \Phi_{A_1}/\partial \Phi_B$ are readily obtained by taking derivatives of equations 6.66 and 6.68 and solving the two simultaneous equations in $\partial \Phi_{B_1}/\partial \Phi_B$ and $\partial \Phi_{A_1}/\partial \Phi_B$ to give:

$$\frac{\partial \Phi_{B_1}}{\partial \Phi_B} \left[-\frac{X}{1+X}\frac{r}{\Phi_{B_1}}\frac{\Phi_B}{\Phi_A} + \frac{1+X}{X\Phi_{B_1}}\left(\frac{1+K_B\Phi_{B_1}}{1-K_B\Phi_{B_1}}\right)\right] =$$

$$\frac{1}{X\Phi_B} + \frac{1}{\Phi_B\Phi_A} \qquad (6.73)$$

Values for $\partial \Phi_{A_1}/\partial \Phi_B$ are then obtained by substitution in equation 6.69. These equations have a clumsy appearance, however, and equation 6.71 remains the simpler and more illuminating form. We therefore write the equation for the spinodal as:

$$2\chi - \left[\frac{1}{\Phi_{B_1}}\frac{\partial \Phi_{B_1}}{\partial \Phi_B} - \frac{1}{r\Phi_{A_1}}\frac{\Phi_{A_1}}{\partial \Phi_B}\right] + \left[\frac{1}{h^\circ\Phi_B} + \frac{1}{r\Phi_A}\right] =$$

$$\frac{1}{\Phi_B M_B} + \frac{1}{\Phi_A M_A} \qquad (6.74)$$

Temperature Dependence

Phase behavior as a function of temperature and composition is simply obtained for this model. We let χ have the conventional 1/T dependence and as in Chapter 4, note that the equilibrium constants K_B and K_A describing B-B and B-A hydrogen bonds depend upon the enthalpy of these bonds, h_B and h_{BA}, through the usual thermodynamic relationships:

$$\frac{\partial \ln K_B}{\partial(1/T)} = \frac{-h_B}{R} \tag{6.75}$$

$$\frac{\partial \ln K_A}{\partial(1/T)} = \frac{-h_{BA}}{R} \tag{6.76}$$

Hence:

$$K_B = K_B^\circ \exp\left\{-\frac{h_B}{R}\left(\frac{1}{T} - \frac{1}{T^\circ}\right)\right\} \tag{6.77}$$

$$K_A = K_A^\circ \exp\left\{-\frac{h_{BA}}{R}\left(\frac{1}{T} - \frac{1}{T^\circ}\right)\right\} \tag{6.78}$$

where K_A° and K_B° are the values of the equilibrium constants at the absolute temperature T°. Of course, this assumes that the enthalpy of hydrogen-bond formation remains a constant with temperature. Infrared frequency shifts indicate that is a good assumption for the temperature range accessible to most polymers (i.e., between the T_g and the degradation temperature).

Sample Calculations of Spinodals

We will consider the rich variety of phase behaviors calculated for various blends where the components hydrogen bond in Chapter 7. Here, we will simply describe the results of some calculations for hypothetical systems in order to illustrate the balance of forces that determine phase behavior. We will consider a system where the equilibrium constant describing self-association, K_B, has a value of 40, while competing interactions are characterized by a value of $K_A = 4$ (at 300°K). These values are similar to those determined experimentally for an amorphous polyurethane blended with a polyether (see Chapter 7).

The factors in equation 6.74 that vary with temperature are χ and the two terms in brackets, which represent the contribution of the hydrogen-bonding interactions. The simple χ parameter used here has a 1/T dependence, and the variation of the factor 2χ with temperature is plotted in figure 6.3 (a value of $\chi = 0.1$ at 300°K was assumed as a reference point). The hydrogen-bonding terms vary through the dependence of the equilibrium constants upon temperature. We assume the enthalpies of hydrogen-bond formation h_B and h_{BA} are equal, $h_B = h_{BA} = -5$ kcal/mol. The terms in parentheses in equation 6.74 are then evaluated as a function of temperature, using the relationships of equation 6.73 and 6.69 for the derivative terms and the equations defining the stoichiometry of the system, equations 6.66 and 6.68. The resulting curve is also shown in figure 6.3 (for a composition of $\Phi_A = \Phi_B = 0.5$). Points on the spinodal are obtained when the sum of the contributions illustrated in figure 6.3 cross a line defined by $(1/\Phi_A M_A + 1/\Phi_B M_B)$, the right-hand side of equation 6.74.

The curves obtained for two values of r, the ratio of the molar volumes of the chemical repeat units, are illustrated in figure 6.4. [Smaller values of r are equivalent to making the size of the A chemical repeat unit larger, e.g. through the addition of CH_2 groups, thus diluting the system and reducing the number of inter-association (A- - -B) hydrogen bonds that can be formed. We could also have reduced K_A and h_A, thus reducing the strength of this interaction.] For small values of r, the second derivative of the free energy changes sign only once, at low temperature, the curves asymptotically approaching the line defined by the right-hand side of equation 6.74 at high temperature. At larger values of r the curves can cross this line (or the second derivative can change sign) two more times, however (some of the crossover points are off the temperature scale of figure 6.4). This results in the type of closed-loop spinodal curves characteristic of some hydrogen-bonding mixtures (see chapter 1). Note that the change in sign of the curves in figure 6.4 is opposite the change in sign of the second derivative, because of the way equation 6.74 is arranged. Accordingly, the first crossover point at low temperature indicates a transition from an unstable to a miscible mixture.

Variation of the 2χ and Hydrogen-Bonding Terms of Equation 6.74 as a Function of Temperature

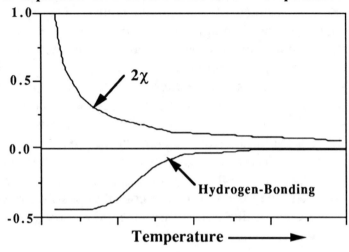

FIGURE 6.3

Sum of the Curves shown in Figure 6.3 as a function of r

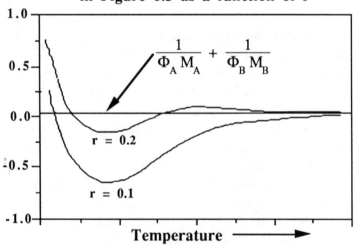

FIGURE 6.4

A variety of phase behaviors are predicted by this model, the specifics obviously depending upon the values of the various parameters. To illustrate this, we will use the equilibrium constants given above, which as we noted, are typical order of magnitude values for urethanes and ethers, and simply consider the effect of varying r, the ratio of the molar volumes of the chemical repeat units. Equation 6.74, together with the appropriate equations describing the stoichiometry at hydrogen bonding, are used to calculate the spinodals.

Figure 6.5 illustrates the effect of changing the relative size of the chemical repeat units through the parameter r. If, for example, we consider mixing a particular amorphous polyurethane with a range of different polyethers, then the values of the equilibrium constants are unaltered, the difference in the size of the species affecting the phase behavior of the blends through the factor K_A/r. Values of $\chi = 0.5$ and $M_A = M_B = 100$ were assumed.

At values of r equal to 0.15 (top right-hand plot) a closed loop characterized by an upper and lower critical solution temperature is obtained. As r is decreased this loop decreases in size and disappears at a double critical point. For r = 0.10 (top left-hand plot), all that is observed is an upper critical solution temperature on the bottom curve. Conversely, as r increases, this curve rises to meet the closed-loop stability curve, which also gets larger with r. For this system, the lower lying upper critical solution temperature (i.e., on the bottom curve) meets the lower critical solution temperature of the closed loop at a value of r between 0.5 and 1.0. At higher values of r the phase diagram then takes on an increasingly hourglass shape.

Clearly, this model predicts the type of reappearing phases considered characteristic of mixtures involving hydrogen-bonding components. In the example presented above, however, the closed-loop region of immiscibility is calculated to be at experimentally unattainable temperatures. But, the precise position of the curves is a sensitive function of, on one hand χ, and on the other, the molecular weight of the blend components and the parameters that define ΔG_H, the equilibrium constants, K, and the ratio of the molar volumes of the chemical repeat

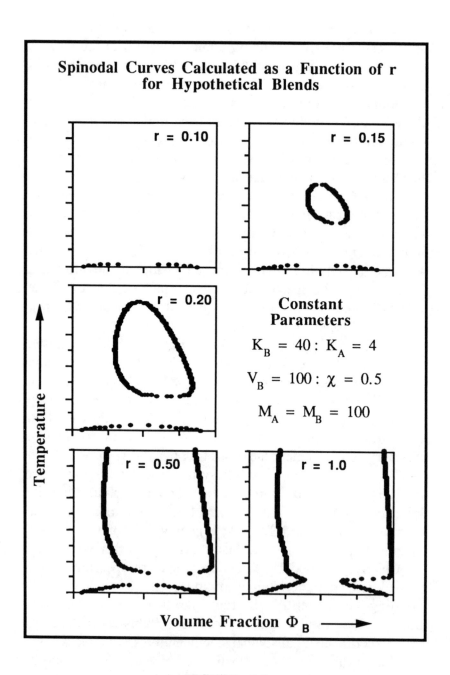

FIGURE 6.5

units, r. Accordingly, on the basis of this model we expect that we should be able to experimentally observe an intriguing variety of phase behaviors through the systematic variation of controllable parameters. Examples will be considered in the following chapter.

The above analysis does not consider in any depth the thermodynamic requirements which must be satisfied in order to observe UCST, LCST and closed loop phase behavior. These have been discussed by Sanchez[41] and for the specific case of hydrogen bonding, Copp and Everett[42]. These latter authors summarize their analysis by stating that UCST's are related to large positive deviations of the *energy* of the system from ideality, while LCST phenomena result from sufficiently large negative deviations of the *entropy* from ideality. Because the enthalpy of individual hydrogen bond formation is relatively large and negative, one might then conclude that in systems that hydrogen bond, UCST's are associated with large positive values of χ describing unfavorable physical interactions.

Surprisingly, there can be, in certain circumstances, large positive endothermic contributions of the enthalpy of mixing from hydrogen bonding. At first this might seem counter-intuitive, but it is a simple consequence of the balance between "breaking" self-association hydrogen bonds and forming unlike (inter-association) bonds. We will consider this in more detail in the following section.

In non-polar polymer mixtures, the negative excess entropy that is the origin of LCST behavior is related to free volume or compressibility effects. In mixtures that hydrogen bond, however, a decrease in the entropy of the mixture also follows from the formation of an A- - -B complex, as rotational degrees of freedom are lost. In most systems at least one component self-associates, so that there is a balance between entropy gained by breaking B--B contacts and entropy lost by the formation of A--B species that has to be taken into account. There are other subtleties, discussed by Copp and Everett, such as the requirement that there should be a small negative heat of mixing (if there is a large negative heat of mixing, the system does not phase separate). Of more concern to us in this book is the relative magnitude of the two factors that can result in a negative excess entropy; free volume and specific interactions. In our treatment so far we have neglected the former. We will examine whether this is a reasonable assumption later in this chapter.

Melting Point Depression in Polymer Blends where One Component Crystallizes–A Demonstration of the Composition Dependence of the ΔG_H Term

An examination of the phase diagrams shown in figure 6.5 reveal a pronounced asymmetry that is a consequence of the composition dependence of the ΔG_H term. This can be more clearly perceived if we examine the effect of hydrogen bonding on the melting point depression of a mixture where one component crystallizes.

Studies of melting point depression have been widely used to study interactions between polymer segments (references 43-56, for example). The data obtained in these studies are usually analyzed in terms of the equation derived by Nishi and Wang[43], which, in turn, is based on using the Flory-Huggins equation[34] to obtain the chemical potential of the crystallizable component in the amorphous phase;

$$\left[\frac{1}{T_m} - \frac{1}{T_m^o}\right] = -\frac{R}{\Delta H_f} \cdot \frac{V_A}{V_B}\left[\frac{\ln \Phi_A}{M_A} + \left(\frac{1}{M_A} - \frac{1}{M_B}\right)\Phi_B + \Phi_B^2 \chi\right] \tag{6.79}$$

where A is the crystallizable polymer, ΔH_f is its heat of fusion and T_m and T_m^o are the equilibrium melting temperatures in the blend and in pure A, respectively. When M_A and M_B are both large, then the left-hand side of equation 6.79 is a linear function of Φ_B^2. The line should (just about) go through the origin and have a slope equal to χ, and therefore, be a useful method for measuring this interaction parameter.

There are problems, of course. Alfonso and Russell[50] determined that, at least in blends of poly(ethylene oxide) (PEO) with poly(methyl methacrylate) (PMMA), the accuracy to which the equilibrium melting points can be determined is comparable to the depression of the melting point, so that an interaction parameter cannot be determined with any degree of certainty. In this system, however, the maximum melting point depression observed was only 2.5°C. In other blends, particularly those involving strong(er) specific interactions, such as hydrogen bonds, the melting point depression can be larger, sometimes 5-10°C or more[43-56], but obviously errors are still a significant factor, casting doubts on the accuracy of any value of χ so determined.

We are not concerned here with these problems, but rather the degree to which the model we are using can predict melting point depression and the degree to which this data illustrates the composition dependence of the ΔG_H term. To this end, we have examined[57] blends of a crystallizable polymer, poly(ethylene oxide), PEO, with three polymers that hydrogen bond; poly(vinyl phenol), (PVPh), an amorphous polyurethane, (APU), and an ethylene-co-methacrylic acid copolymer (containing 55 weight % methacrylic acid, EMMA[55]). The Nishi-Wang equation is easily modified to account for hydrogen bonding by simply including the contribution of ΔG_H to the partial molar free energy:

$$\left[\frac{1}{T_m} - \frac{1}{T_m^o} \right] = - \frac{R\,r}{\Delta H_f^o} \left[\chi\,\Phi_B^2 + \left(\Delta \overline{G_H} \right)_A \right] \qquad (6.80)$$

where we have now assumed that M_A and M_B are large. The Flory-Huggins parameter, χ, can be calculated from:

$$\chi = \frac{V_B}{RT} \left[\delta_A - \delta_B \right]^2 \qquad (6.81)$$

using the group contributions given in Chapter 2, leaving the $(\Delta \overline{G_H})_A$ term to be calculated from spectroscopically determined equilibrium constants.

The hydrogen bonding interactions of APU, PVPh and EMAA [55] in the pure state (self-association) are described in different ways, hence $(\Delta \overline{G_H})_A$ has different forms. The urethane functional groups of APU form chains in such a manner that the equilibrium constant describing self-association, K_B, can be assumed to be independent of h (see Chapter 7).

In contrast, hydrogen bonds between the OH group of PVPh are such that the stoichiometry of hydrogen bonding is more accurately described by using an equilibrium constant for the formation of dimers, K_2, that is different to that of subsequent h-mers K_B. Finally, the carboxylic acid groups of EMAA [55] form strongly hydrogen bonded cyclic dimers (see Chapter 2, page 147, for a schematic representation.) The equations for the partial molar free energies for each of these situations are presented in Section F, the appendix to this chapter and Section E, respectively.

We obtained equilibrium crystallization temperatures using the Hoffman-Weeks extrapolation[58] and the observed melting point depressions are plotted as a function of Φ_B, the volume fraction of APU in the blend, in figure 6.6. Also shown in figure 6.6 is the melting point depression calculated using equation 6.80.

The parameters used in this calculation are those determined on our previous spectroscopic studies of this system[59] and from our recently determined group contributions to cohesive energy densities and molar volumes (see Chapters 7 and 2, respectively). No attempt was made to obtain a better fit between theoretical predictions and experimental observations by adjusting any of these parameters and it can be seen that the agreement between the predictions of the association model and experimental results are very good. Figure 6.7 shows a plot of the left hand side of equation 6.80, multiplied by $(\Delta H_f / R\ r)$, against Φ_B^2. The data points do not fall on a straight line and the association model predicts the composition dependence of the interactions very well.

Similar results were obtained for the PEO - PVPh blends, but over a narrower composition range (crystallization rates were extremely slow at weight fractions of PVPh greater than 0.2). Figure 6.8 again shows a plot of the values of the left hand side of equation 6.80, multiplied by $(\Delta H_f / R\ r)$, plotted as a function of ϕ_B^2 and again compares the results to the theoretical predictions of the association model. A small and systematic

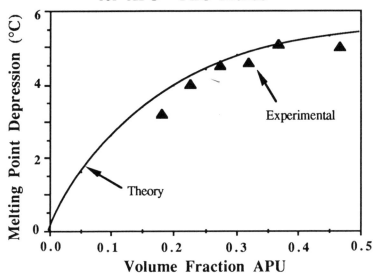

Comparison of Theoretical and Experimental
Depression of Melting Points
for APU - PEO Blends

FIGURE 6.6

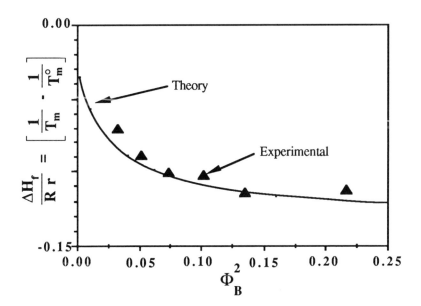

FIGURE 6.7

deviation of theory from experiment is revealed in this plot, possibly due to small inaccuracies in the value chosen for the solubility parameter of PVPh. We will discuss such errors in more detail in Chapter 7. Nevertheless, the model clearly reproduces the non-linear dependence on ϕ_B^2.

Finally, a corresponding comparison of theory and experiment is presented in figure 6.9 for the PEO - EMAA [55] blend. There is excellent agreement, but in this case it is possible to also fit the experimental data points shown in figure 6.9 to a linear plot, albeit one that does not intercept the y-axis at any value close to the origin, as also shown in figure 6.9. This result corresponds closely to those obtained by Jo and Lee[55] for blends of PEO with poly(styrene-co-acrylic acid). The apparent linear relationship that can be obtained is a consequence of the narrow concentration range for which observable PEO crystals can be obtained and the fairly flat nature of the curve over this range.

Unlike PVPh and APU, where self-association results in the formation of chains, only like (acid-acid) and unlike (acid-ether) dimers appear to be formed in this system and it is possible that in the narrow composition range used in this study, a change in the distribution of such paired species does not result in as great an overall composition dependence of the interaction parameters.

To summarize, these results clearly demonstrate two points that are of some significance.

1. In blends where the components interact strongly there is a strong composition dependence of the interaction parameters that is reflected in observed melting point depressions.

2. The association model predicts the observed experimental behavior very well. This, in turn, suggest that in systems where there are strong interactions these dominate the overall composition dependence, as we treat specific interactions separately from those that are non-specific (dispersion forces, etc.) and these latter interactions are assumed to be independent of composition.

We will examine in some detail other predictions of phase behavior in Chapter 7, but this initial success is obviously encouraging.

Comparison of Theoretical and Experimental Depression of Melting Points

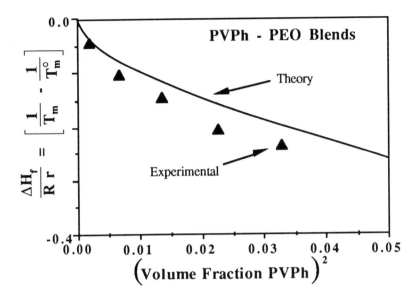

FIGURE 6.8

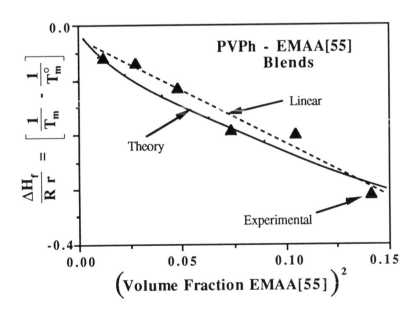

FIGURE 6.9

I. SELF-ASSOCIATION AND THE ENTHALPY AND ENTROPY OF MIXING

If specific interactions were confined to unlike A-B contacts their role in polymer miscibility could be comprehended in a relatively straightforward manner. For situations where the enthalpy of this specific interaction exceeded that of physical interactions, we would obtain miscible systems as long as the entropic changes associated with loss of degrees of freedom (e.g., internal rotational) were not too unfavorable. Negative excess entropy contributions from such interactions, together with free volume effects, could then result in LCST behavior and a contact point theory such as that described by ten Brinke and Karasz[18] or Sanchez and Balazs[19] would no doubt be very successful as in the study of poly (styrene) - poly(vinyl methyl ether) blends[19]. The fact of self-association in many mixtures that hydrogen bond complicates matters and leads to some results that are initially surprising; these we will consider here.

In various studies of polymer blends it has often been stated that because the combinatorial entropy of mixing polymers is small, an exothermic heat of mixing (ΔH_m) is necessary for miscibility and that specific interactions are one source of a negative ΔH_m. In systems that hydrogen bond, however, there can be an endothermic heat of mixing, but the mixture is nevertheless miscible. The origin of this positive ΔH_m is self-association of at least one of the pure components and mixing occurs because of a gain in *entropy* upon mixing. This entropy gain is sometimes assumed to be negligible in treatments of polymer mixtures, but this confuses combinatorial with other contributions to the total entropy of the system.

Although the enthalpy (and entropy) of formation of *individual* hydrogen bonds (or other types of associated complexes) is always negative, the overall enthalpy of mixing will depend upon the balance between three contributions.

1. A *positive* contribution to the enthalpy of mixing that is a result of *"breaking"* hydrogen bonds in the self-associating polymer as it is mixed and diluted with the second component of the mixture.

2. A *negative* contribution to the enthalpy of mixing that is a result of *forming* hydrogen bonds between the self-associating polymer and second component.

3. Contributions from other interactions (van der Waals, dipole forces, etc.)

Our assertion cannot be tested directly because the heats of mixing polymers cannot be measured. Paul and co-workers have used analog calorimetry to determine the heats of mixing of a number of model compounds, however, as in their recent study of mixtures of molecules containing carboxylic acids with those containing esters, where a positive heat of mixing was determined over most of the composition range[60]. In our opinion, this positive heat of mixing has a significant contribution from the large endothermic change associated with breaking acid pair hydrogen bonds relative to the smaller exothermic contribution from forming acid-ester bonds. Because the enthalpy of formation of specific types of hydrogen bonds can be fairly accurately determined by spectroscopic methods, as can the number of hydrogen bonds of each type that are present, we should be able to calculate the contributions of hydrogen bonds to the heat of mixing with a reasonable degree of confidence. The enthalpy of mixing can be directly written in terms of the fractions of hydrogen bonded species (c.f. section D)[#]:

[#] This method of directly counting the change in the number of hydrogen bonds of each type is described by Acree[26], as is an alternative derivation based on determining:

$$R \left[\frac{\partial (\Delta G^{ex}/RT)}{\partial (1/T)} \right] = R \left[\frac{\partial (\Delta G^{ex}/RT)}{\partial K} \right] \left[\frac{\partial K}{\partial (1/T)} \right]$$

$$= R \left[\frac{\partial (\Delta G^{ex}/RT)}{\partial K} \right] [-h K]$$

This can be applied to polymers (e.g., equation 6.54) providing that we keep in mind the necessity of considering the *excess* function, i.e., we remove the last two terms in this equation or treat $\bar{h}^o$ as a constant in the differentiation.

$$\Delta H_m = n_B \left(p_{BB} - p_{BB}^\circ \right) h_B + n_A \, p_{BA} \, h_{BA} + RT \left(\frac{V}{V_B} \right) \Phi_A \, \Phi_B \, \chi$$

$$(6.82)$$

where, as before, n_B and n_A are the number of moles of segments of the self-associating and non-self-associating species, respectively, where a segment is defined so as to contain just one functional group capable of forming a hydrogen bond; p_{BB} is the fraction of functional groups of type B that are hydrogen bonded to other B segments in the mixture (equal to the probability that a B group, taken at random, is hydrogen bonded in this fashion); p_{BB}° is the fraction of hydrogen bonded B segments in pure B; p_{BA} is the fraction of A segments that are hydrogen bonded to a B segment; h_B and h_{BA} are the enthalpies of individual B---B and B---A hydrogen bond formation, respectively; V is the molar volume of the mixture, while V_B is the molar volume of the B segments, which we use as a reference volume to define χ. The quantities p_{BB}, p_{BB}° and p_{BA} can be calculated from the equilibrium constants describing the stoichiometry of hydrogen bonding and will depend upon the form of association that is assumed (e.g., linear association with the equilibrium constant independent of chain length, formation of cyclic pairs, etc.).

Unfortunately, we cannot use this equation to calculate the heats of mixing for the mixtures studied by Brannock et al[60], because at this time we have yet to spectroscopically characterize this system. We can consider data from the literature for the phenol - cyclohexane system[61], however, as mixtures of these molecules have been studied in detail by Whetsel and Lady, as we discussed in Chapter 5.

Applying the equilibrium constants and enthalpies of hydrogen bond formation determined by these authors and approximating the contribution of physical forces to the heat of mixing using solubility parameters, the continuous line shown in figure 6.10 was calculated. Also shown in this figure are the separate contributions from hydrogen bonding and physical interactions (see equation 6.82). Considering the errors involved, particularly those inherent in using solubility parameters, there is good agreement.

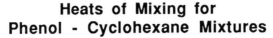

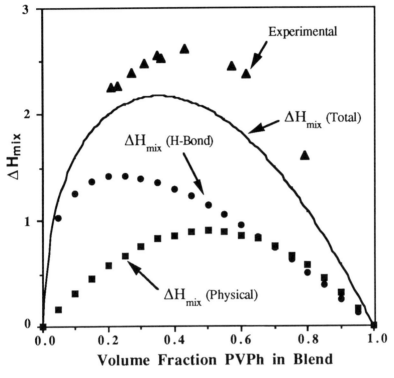

FIGURE 6.10

The crucial point, however, is the significant contribution to the heat of mixing that is a consequence of breaking OH---OH hydrogen bonds as phenol is diluted with cyclohexane. Although this latter molecule does not hydrogen bond, it is clear that an exothermic heat of mixing will only be obtained upon mixing with a molecule that can form a relatively strong hydrogen bond with phenolic OH groups.

We can perform the same calculations for various polymer mixtures and as examples we will consider blends of poly(vinyl phenol), (PVPh), with poly(n-butyl methacrylate), (PBMA), and poly(vinyl pyridine) (PVPy), as we have studied systems of this type in some depth and obtained appropriate equilibrium constants that accurately reproduce the measured fraction of hydrogen bonded species, i.e., p_{BB} etc. (see Chapters 5 and 7).

Figure 6.11 shows the calculated heats of mixing (at 25°C) for blends of PVPh with PBMA. They are positive throughout the composition range. Also shown in figure 6.11 are the separate contributions of the χ term, calculated from the values of the solubility parameters listed in Chapter 2, and the contribution of the hydrogen bonding interactions. This latter term is positive over *most* of the composition range, because of the balance of the terms in equation 6.82. The fraction of B--B (phenolic O-H--O-H) hydrogen bonds in pure B is always greater than that in the mixture, $p_{BB}^0 > p_{BB}$. This would remain true even if the second component did not hydrogen bond, as p_{BB} is reduced by simple dilution. The first term, $n_B(p_{BB}-p_{BB}^0)h_B$ is therefore, always positive (as h_B is negative).

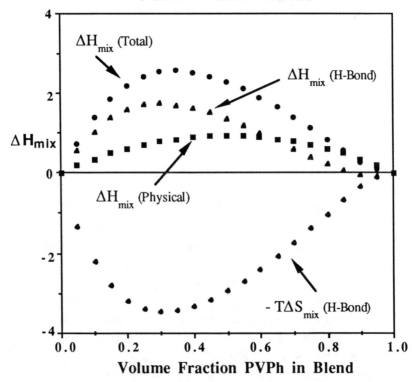

Calculated Heats of Mixing for PVPh - PBMA Blends

FIGURE 6.11

For this system we calculate that the negative contribution to the free energy from the formation of OH--ester (B- - -A) hydrogen bonds is not large enough to overcome this effect of self-association, except at low concentrations of the ester containing polymer. This is because both K_A and h_A, the equilibrium constant describing B---A interactions and the enthalpy of such interactions, are both smaller than the equivalent quantities that describe self-association. [Note that the equilibrium constant K_A and enthalpy of hydrogen bond formation h_A were determined *experimentally* for the polymer repeat units, as described in Chapter 5. For low molecular weight analogues of equivalent molar volume, there could be different values of K_A, h_A, etc., as these quantities may depend upon chain stiffness, as discussed in Section G of this chapter.]

Given this result, one might at first conclude that there is no reason on earth for these two polymers to mix; but they are theoretically predicted to be miscible over the entire composition range, using the same parameters for the calculations of the free energy and spinodal, and experiments confirm these predictions, as we will show in Chapter 7 (up to temperatures of about 160°C, where an LCST occurs). This is because there are significant entropic in addition to enthalpic changes as a result of hydrogen bond formation, or for that matter, the formation of any associated complex. This remains true in polymers, even though the *combinatorial* entropy of mixing is negligible, because the number of configurations available to the polymer changes with concentration (i.e., the number and type of hydrogen bonds that are present). This can be perceived by considering the entropy of mixing two polymers, relative to a reference state of the separate and oriented species, as the sum of two parts:

$$\Delta S_m = \Delta S_{comb} + \Delta S_{dis} \qquad (6.83)$$

the first involving the usual combinatorial entropy, negligible for mixing two polymers, while the second describes an entropy of disorienting these molecules. For weak non-polar forces ΔS_{dis} is assumed independent of composition and is eliminated when converting to a reference state of the pure polymers. The formation of hydrogen bonding complexes limits the number of configurations available to the chain, however, and in the association model this is essentially accounted for in the ΔS_{dis} term. The distribution of hydrogen bonded species in a mixture is usually very different to that in the pure components and this results in a significant change in the entropy of the system (for small molecules, the formation of an A---B hydrogen bond leads to a loss of rotational freedom of the molecule as a whole; for a polymer, entropy changes will be related to loss of degrees of internal rotational freedom of segments).

As might be expected, the entropy changes associated with specific interactions are usually opposite in sign and effect to the heat of mixing (although these also depend upon the balance between the changes corresponding to breaking "like" hydrogen bonds and forming "unlike" ones, c.f. equation 6.82 for the enthalpy of mixing). From the expression for the ΔG_H term (see equation 6.55), the $T\Delta S$ component of the hydrogen bonding interactions was calculated (by subtracting the enthalpy of mixing) and is also presented in figure 6.11. It can be seen that the increase in entropy corresponding to the "breaking" of B---B hydrogen bonds, less the entropy loss from forming A---B bonds, gives a larger negative contribution to the free energy than the positive heat of mixing (otherwise these polymers would not have mixed in the first place). Accordingly, it would appear that it is not necessary to have an exothermic heat of mixing to obtain polymer miscibility, the entropy changes corresponding to the change in the number of self-association interactions is also crucial. It follows that the assumption that self-association can be ignored in theories of specific interactions in polymers neglects an important factor in determining the miscibility of many systems. Furthermore, heat of mixing measurements on model systems may not be good guides to the miscibility of certain types of blends.

Finally, for completeness, we present in figure 6.12 an example of a system where the enthalpy of mixing is negative and the entropic changes corresponding to the change in the distribution of hydrogen bonded species are now unfavorable to mixing over most of the composition range. A large exothermic heat of mixing is calculated for blends of PVPh with poly(vinyl pyridine), PVPy. The "strength" of the hydrogen bond formed between the phenolic OH groups and the basic nitrogen of the pyridine ring is much stronger than the forces of self-association between the phenolic OH groups, and this is reflected in both the experimentally determined enthalpy of hydrogen bond formation and equilibrium constant. In this case, the favorable enthalpic terms are larger than the unfavorable entropic changes and this system is thus also miscible.

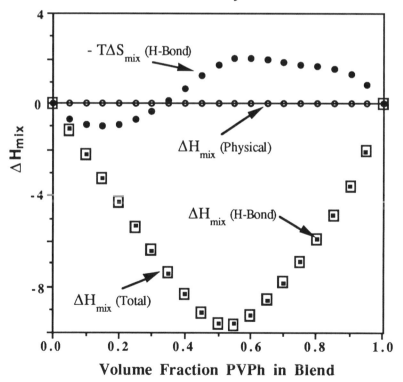

FIGURE 6.12

J. FREE VOLUME EFFECTS IN THE MIXING OF POLYMERS THAT HYDROGEN BOND

In Chapter 1 we observed that both specific interactions and free volume effects can result in the negative deviations of the entropy from ideality that lead to LCST behavior. In our treatment of systems with *strong* specific interactions, hydrogen bonds, we have, in effect, assumed that the former dominate and the latter can be neglected. On the basis of some recent work we believe that this is a good assumption[62] and we will present a brief summary of our results here. The original paper has an mind numbing number of equations, some of which have an appearance of alarming complexity, but there is a relatively simple underlying structure and we have introduced no new principles, merely extended a treatment originally given by Sanchez and Lacombe[63-66].

The Flory treatment[67-70] essentially assumes that the systems free volume is distributed amongst each of the systems lattice sites or cells, whereas the Sanchez-Lacombe lattice fluid model allows for empty lattice sites or holes. On more-or-less intuitive grounds we believed that it would be easier to combine an association model to account for specific interactions with a free volume model of this latter type.

Sanchez and Lacombe's treatment of a system where there are no specific interactions starts from Guggenheim's[33] derivation of the number of configurations available to a chain of N_A polymers of degree of polymerization M_A, N_B polymers of length M_B and n_e "holes" or empty lattice sites. Assuming $z \rightarrow \infty$, the configuration partition function reduces to Flory's result:

$$\Omega = \left[\frac{1}{f_e}\right]^{n_e} \left[\frac{w_A}{f_A}\right]^{N_A} \left[\frac{w_B}{f_B}\right]^{N_B} \qquad (6.84)$$

where f_i is now the volume fraction of the ith components and w_i is its chain disorientation parameter.

With the configurational partition function known, the free energy of mixing can be determined once the energy of the lattice is derived. It is assumed that the empty sites mix athermally, so the interactions, hole-hole and hole-mer have zero energy. The other possible interactions, A mer - A mer, B mer - B mer, and A mer - B mer, have the respective energies ε_{AA}^*, ε_{BB}^* and ε_{AB}^*. The quantity ε_{ij}^* represents the energy of a mer belonging to component i when it is surrounded by mers belonging to component j. Summing up the total number of interactions gives the total lattice energy:

$$E = -\left(n_e + N_A M_A + N_B M_B\right) \sum_i \sum_j f_i f_j \varepsilon_{ij}^* \quad (6.85)$$

After restating the definition of the Gibbs free energy:

$$G = E + PV - kT \ln \Omega \quad (6.86)$$

it becomes obvious[#] that the number of configurations and the total lattice energy can be substituted into the Gibbs free energy relationship to give the reduced free energy of mixing:

$$\tilde{G}_{mix} = \left[-\frac{\tilde{\rho}\,\varepsilon^*}{RT} + \frac{P\upsilon^*}{\tilde{\rho}\,\varepsilon^*} \right] + \left[\left(\frac{1}{\tilde{\rho}} - 1\right) \ln(1 - \tilde{\rho}) + \frac{1}{M} \ln \tilde{\rho} \right]$$
$$+ \left[\frac{\Phi_A'}{M_A} \ln \Phi_A + \frac{\Phi_B'}{M_B} \ln \Phi_B \right] \quad (6.87)$$

For illustrative purposes, the free energy of mixing has been broken up into three terms, the last one representing the combinatorial entropy of mixing solvent with polymer in the hard core state (i.e., no free volume). The volume fractions Φ' are now hard core volume fractions. The middle term is the combinatorial entropy of mixing the polymer-solvent system with free volume which is characterized by the reduced density term, $\tilde{\rho}$. The reduced density has a value of one in the hard core state and a value that approaches zero as the free volume of the

---

[#] We have observed that obvious is a word used extensively in formidable theoretical papers.

system increases. The reduced density is chosen so that it minimizes the free energy according to the equation of state:

$$\tilde{\rho}^2 + \tilde{P} + \tilde{T}\left[\ln(1-\tilde{\rho}) + \left(1 - \frac{1}{M}\right)\tilde{\rho}\right] = 0 \tag{6.88}$$

The molecular size parameter, M, is the mean M-mer length of the polymer-solvent system and is defined as:

$$\frac{1}{r} = \frac{\Phi_A'}{M_A} + \frac{\Phi_B'}{M_B} \tag{6.89}$$

The first term in the free energy expression accommodates the system's enthalpic contribution, which is subdivided into the internal energy and the pressure-volume work. The pressure-volume work contains the hard core volume term, $\upsilon*$, which characterizes the volume of a segment when no free volume is present. This work term is usually insignificant at normal pressures. The term $\varepsilon*$ is defined as:

$$\varepsilon^* = \sum_i\sum_j \Phi_i' \Phi_j' \varepsilon_{ij}^* \tag{6.90}$$

As might be expected, when accounting for hydrogen bonding using an association model, we obtained equation 6.87 with an additional term. This term accounts for hydrogen bonding through the modification to the number of configurations available to the hard core chains in exactly the same manner as described in Section E of this chapter. There are some subtleties involving equilibrium constant definitions and so-called combining rules and the interested reader is referred to the original paper[62]. Essentially, as before, the free energy contributions due to hydrogen bonding interactions were calculated using the equilibrium constants determined from infrared spectroscopy. Then, in order to directly compare the phase behavior predicted by the lattice fluid model to that calculated from the simpler approach that ignores free volume, the energy term of equation 6.87 was calculated from solubility parameters by making the geometric mean assumption and using:

$$\delta_i = (CED_i)^{0.5} = \left(\frac{\Delta E_{vap,i}}{V_i}\right)^{0.5} = \left(\frac{\varepsilon_{ii}^*}{\upsilon_i^*} \cdot \tilde{\rho}_i^2\right)^{0.5} \tag{6.91}$$

Pure component lattice volumes, υ^*, were obtained by fitting the equation of state to density data.

The phase diagram that is calculated using the Sanchez-Lacombe model to account for free volume is shown in figure 6.13 for blends of PVPh with PBMA, where it is compared to the cloud point curve reported by Serman et al.[71]. The number of repeat units for each polymer was fixed at 500.

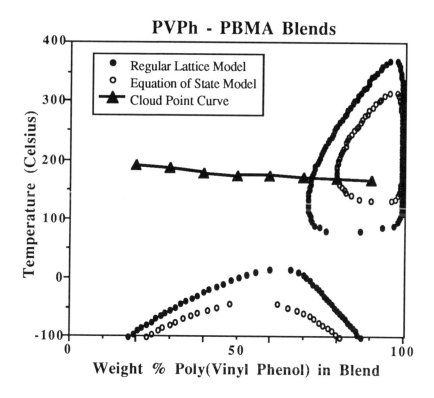

FIGURE 6.13

It appears from figure 6.13 that both models produce similar results in predicting a closed loop and UCST phase behavior. These characteristics arise from the balance of the "physical" interaction term and the hydrogen bonding terms found in both the "regular" lattice model (i.e., a simple lattice model that neglects free volume) and the equation of state model. However, the equation of state model allows the density of the mixture to decrease as the temperature increases, the result being a reduction in both the number of "physical" and hydrogen bond interactions in the mixture. Because the rates of decrease in "physical" and hydrogen bond interactions with density differ, there is an overall *increase* in the predicted miscibility relative to the "regular" lattice model for this system. In other words, the amount of unfavorable "physical" forces are decreasing faster with temperature than the number of favorable hydrogen bond interactions, for this example.

There can be significant errors in the calculation of solubility parameters (see Chapter 2) which can lead to changes in the predicted phase behavior (more on this later in Chapter 7). Figure 6.14 presents the effect of the cohesive energy density error in predicting the spinodal curves for PBMA and PVPh using the equation of state model (top plot) and the "regular" lattice model (bottom plot). The 25°C solubility parameter for poly(vinyl phenol) is changed from 10.6 up to 11.0 and down to 10.4. As figure 6.14 illustrates, the errors in determining the cohesive energy densities are much more significant than the errors introduced when ignoring compressibility effects when predicting polymer blend phase behavior in systems that have strong specific interactions.

These results are perhaps not that surprising. As we noted in Chapter 1, Patterson and Robards[72] pointed out a number of years ago that free volume differences between polymers are small, while differences between polymers and most solvents are large. Accordingly, free volume factors are important in determining the phase behavior of non-polar polymer solutions and in blends characterized by small negative values of χ. For example, in the poly(styrene)-poly(vinyl methyl ether) system, free volume effects would be crucial, as we are dealing with the

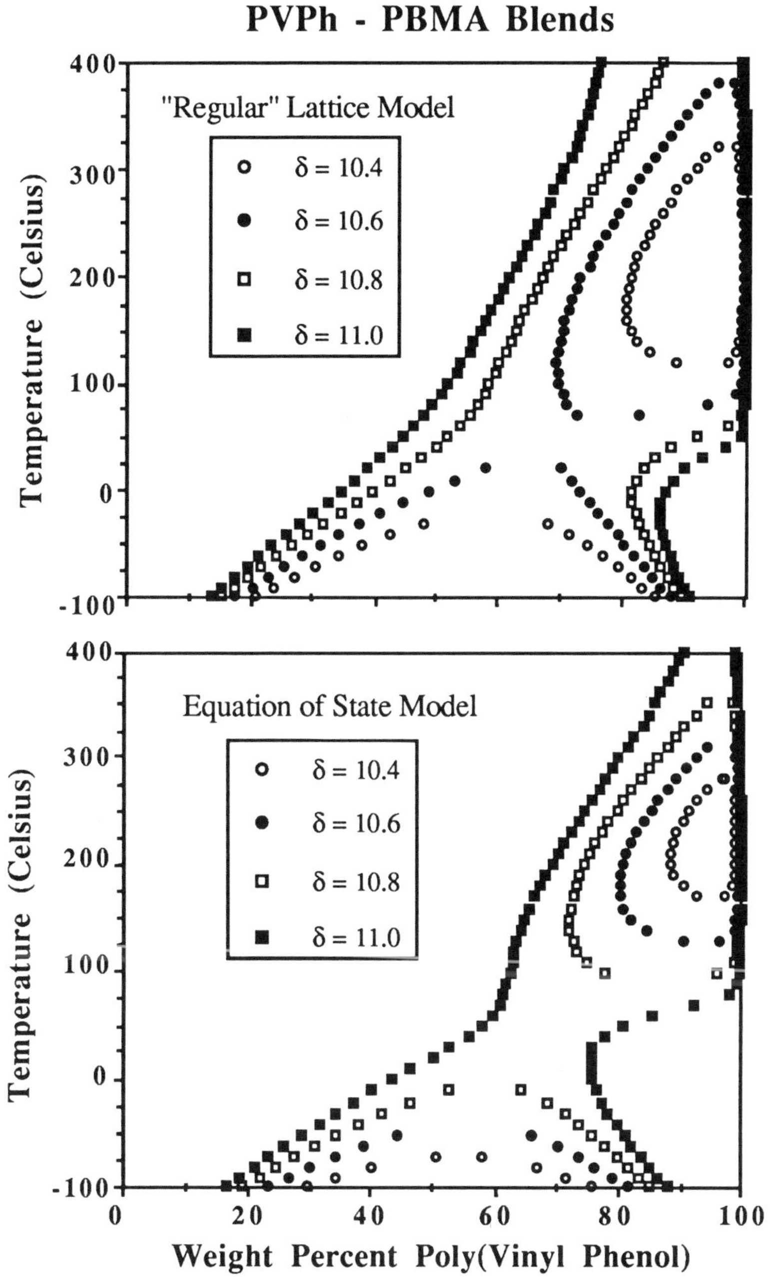
PVPh - PBMA Blends
"Regular" Lattice Model
δ = 10.4
δ = 10.6
δ = 10.8
δ = 11.0
Temperature (Celsius)
400
300
200
100
0
-100
Equation of State Model
δ = 10.4
δ = 10.6
δ = 10.8
δ = 11.0
Temperature (Celsius)
400
300
200
100
0
-100
0
20
40
60
80
100
Weight Percent Poly(Vinyl Phenol)

FIGURE 6.14

balance between relatively small forces. In such systems we would also expect to see significant shifts in LCST's as a result of fairly minor changes in other parameters (e.g., molecular weight). In contrast, in systems that hydrogen bond, we usually have a large negative contribution to the free energy of mixing from hydrogen bonding and a sizeable unfavorable (to mixing) contribution from physical forces. Accordingly, at certain temperatures and compositions the balance between these competing forces can result in phase separation, but it is a balance between large terms that vary sharply with temperature. Small free volume differences in polymers contribute in only a relatively minor fashion and are therefore negligible in these systems.

K. REFERENCES

1. Huggins, M. L., *J. Org. Chem.*, 1936, **1**, 407.
2. Flory, P. J., *J. Chem. Phys*, 1944, **12**, 425.
3. Flory, P. J., *J. Chem. Phys.*, 1946, **14**, 49.
4. Prigogine, I., with Bellemans, A. and Mathot, V., *"The Molecular Theory of Solutions,"* North Holland Publishing Co., 1957.
5. Prigogine, I. and Defay, R., *"Chemical Thermodynamics,"* translated by Everett, D., Longmans, 1954.
6. Dolezalek, F., *Z. Physik Chem.*, 1908, **64**, 727.
7. Van Laar, J. J., *Z. Physik Chem.*, 1910, **72**, 723 and 1913, **83**, 599.
8. Timmermans, J., *J. Chim. Phys.*, 1921, **19**, 169.
9. Scatchard, G., *Chem. Rev.*, 1931, **8**, 321.
10. Hildebrand, J. H., Scott, R. L., *"The Solubility of Nonelectrolytes,"* 3rd ed.; Reinhold: New York, 1950.
11. Guggenheim, E. A., *"Mixtures,"* Clarendon Press, Oxford, 1952.
12. Kohler, F., *"The Liquid State,"* Verlag Chemie, 1972.
13. Marcus, Y., *"Introduction to Liquid State Chemistry,"* John Wiley and Sons, New York, 1977.
14. Marsh, K. and Kohler, F., *J. of Mol. Liq.*, 1985, **30**, 13.
15. Barker, J., A., *J. Chem. Phys.*, 1952, **20**, 794.
16. Barker, J., A., *J. Chem. Phys.*, 1952, **20**, 1526.
17. Tompa, H., *J. Chem Phys.*, 1953, **21**, 1526.
18. ten Brinke, G. and Karasz, F. E., *Macromolecules*, 1984, **17**, 815.
19. Sanchez, I. C. and Balazs, A. C., *Macromolecules*, 1989, **22**, 2325.
20. Han, C. C., Bauer, B. J., Clark, J. C., Muroga, Y., Matsushita, Y., Okada, M., Tran-cong, Q., Chang, T. and Sanchez, I. C., *Polymer*, 1988, **29**, 2002.
21. Kretschmer, C. B. and Wiebe, R., *J. Chem. Phys.*, 1954, **22**, 1697.

22. Renon, H. and Prausnitz, J. M., *Chem. Eng. Sci.*, 1967, **22**, 299; *Erratum*, 1967, **22**, 1891.
23. Nagata, I., *Z. Phys. Chem.*, (Leipzig), 1973, **252**, 305.
24. Hwa, S.C.P.and Ziegler, W. T., *J. Phys. Chem.*, 1966, **70**, 2572.
25. Kehiaian, H. and Treszczanowicz, A., *Bull. Soc. Chim. Fr.*, 1969, **18**, 1561.
26. Acree, W.E., *"Thermodynamic Properties of Non-electrolyte Solutions,"* Academic Press, New York, 1984.
27. Painter, P. C., Park, Y., Coleman, M. M., *Macromolecules*, 1988, **21**, 66.
28. Painter, P. C., Park, Y., Coleman, M. M., *Macromolecules*, 1989, **22**, 570.
29. Painter, P. C., Park, Y., Coleman, M. M., *Macromolecules*, 1989, **22**, 580.
30. Painter, P. C., Graf, J. and Coleman, M. M., *J. Chem. Phys.*, 1990, **92**, 6166.
31. Flory, P. J., *J. Chem. Phys.*, 1950, **18**, 108.
32. Wall, F. T., Flory, P. J., *J. Chem. Phys.*, 1951, **19**, 1435.
33. Guggenheim, E. A., *"Mixtures,"* Claverdon Press, Oxford, 1952, 194.
34. Flory, P. J., *"Principles of Polymer Chemistry,"* Cornell University Press, 1953.
35. Stadler, R., *Macromolecules*, 1988, **21**, 121.
36. Huggins, M. L., *J. Phys. Chem.*, 1942, **46**, 151; *Ann New York Acad. Sci.*, 1942, **41**, 1; *J. Am. Chem. Soc.*, 1942, **64**, 1712.
37. DiMarzio, E. A. and Gibbs, J. H., *J. Chem. Phys.*, 1958, **28**, 807.
38. DiMarzio, E. A., *J. Chem. Phys.*, 1961, **35**, 658.
39. DiMarzio, E. A., *J. Chem. Phys.*, 1977, **66**, 1160.
40. Walker, J. S. and Vause, C. A., *Sci. Am.*, 1987, May issue, p. 98.
41. Sanchez, I. C., *"Polymer Compatibility and Incompatibility: Principles and Practices,"* Solc, K., Ed., Harwood, New York, 1982. MMI Symp. Series, Vol. 3.
42. Copp, J. L. and Everett, D. H., *Discussions of the Farad. Soc.*, 1953, **15**, 174.
43. Nishi, T. and Wang, T. T., *Macromolecules*, 1975, **8**, 909.
44. Paul, D. *"Polymer Blends and Mixtures,"* D. J. Walsh, J. S. Higgins and A. Maconnachie, Eds., Martinus Nijhoff Publishers, Dordrecht, 1985, 1.
45. Morra, B. and Stein, R. S., *J. Polym. Sci.-Polym. Phys.*, 1982, **20**, 2243.
46. Plans, J., MacKnight, W., Karasz, F., *Macromolecules*, 1984, **17**, 810.
47. Avella, M. and Martuscelli, E., *Polymer*, 1988, **29**, 1731.
48. Martuscelli, E., Silvester, C. and Gismondi, C., *Makromol. Chem.*, 1985, **186**, 2161.
49. Greco, P. and Martuscelli, E., *Polymer*, 1989, **30**, 1475.
50. Alfonso, G. C. and Russell, T. P., *Macromolecules*, 1986, **19**, 1143.

51. Cimmino, S., Martuscelli, E., Silvester, C., Canetti, M., DeLalla, C. and Seves, A., *J. Polym. Sci.-Polym. Phys.*, 1989, **27**, 1781.
52. Rim, P. B. and Runt, J. P., *Macromolecules*, 1984, **17**, 1520.
53. Iriarte, M., Iribarren, J. I., Etxeberria, A., Irwin, J. J., *Polymer*, 1989, **30**, 1160.
54. Nishio, Y., Haratani, T., and Takahashi, T., St. John Manley, R., *Macromolecules*, 1989, **22**, 2547.
55. Jo, W. H. and Lee, S. C., *Macromolecules*, 1990, **23**, 2261.
56. Fernandes, A. C., Barlow, J. W. and Paul, D. R., *Polymer*, 1986, **27**, 1799.
57. Painter, P. C., Shenoy, S. L. and Coleman, M. M., *Macromolecules* (submitted for publication).
58. Hoffman, J. D., Weeks, J. J., *Polymer, J. Chem. Phys.*, 1962, **37**, 1723.
59. Coleman, M. M., Hu, J., Park, Y. and Painter, P. C., *Polymer*, 1988, **29**, 1659.
60. Brannock, G. R., Barlow, J. W. and Paul, D.R., *J. Pol. Sci.-Poly. Phys.*, 1990, **28**, 871.
61. Christensen, C., Gonehling, J., Rasmussen, P. and Weidlich, V., *"Heats of Mixing Data Collection,"* Chemistry Data Series, Vol. III, part 2, Diechema.
62. Graf, J. F., Coleman, M. M. and Painter, P. C., *J. Phys. Chem.* (submitted for publication).
63. Sanchez, I. C. and Lacombe, R. H., *J. Phys. Chem.*, 1976, **80**, 2352.
64. Lacombe, R. H. and Sanchez, I. C., *J. Phys. Chem.*, 1976, **80**, 2568.
65. Sanchez, I. C. and Lacombe, R. H., *Polym. Letters Ed.*, 1977, **15**, 71.
66. Sanchez, I. C. and Lacombe, R. H., *Macromolecules*, 1978, **11**, 1145.
67. Flory, P. J., Orwoll, R. A. and Vrij, A., *J. Am. Chem. Soc.*, 1964, **86**, 3515.
68. Flory, P. J., *J. Am. Chem. Soc.*, 1965, **87**, 1833.
69. Eichinger, B. E. and Flory, P. J., Trans. Faraday Soc., 1968, **64**, 2035.
70. Flory, P. J., *Discuss. Faraday Soc.*, 1970, **49**, 7.
71. Serman, C. J., Xu, Y., Painter, P. C. and Coleman, M. M., *Polymer* (in press).
72. Patterson, D. and Robards, A., *Macromolecules*, 1978, **11**, 690.

APPENDIX 1: SELF-ASSOCIATION THROUGH THE FORMATION OF CYCLIC DIMERS (E.G., CARBOXYLIC ACIDS)

The stoichiometry of hydrogen bonding when self-association is in the form of cyclic pairs was discussed in Chapter 4. Here, we simply present the equations for ΔG_H (per mole of lattice site) and the contribution of hydrogen bonds to the partial molar free energies:

$$\frac{\Delta G_H}{RT} = \Phi_B \ln\left(\frac{\Phi_{B_1}}{\Phi_{B_1}^\circ}\right) + \frac{\Phi_A}{r} \ln \Phi_{A_1} +$$

$$\Phi_B \left[\Phi_{B_1}^\circ \left(1 + K_B \Phi_{B_1}^\circ\right)\right] - \Phi_{B_1}\left(1 + K_B \Phi_{B_1}\right)$$

$$- \left[\frac{\Phi_B}{\overline{h^\circ}} \ln \Phi_B + \frac{\Phi_A}{r} \ln \Phi_A\right] \tag{A-1}$$

where:

$$\overline{h^\circ} = \frac{1 + 2 K_B \Phi_{B_1}^\circ}{1 + K_B \Phi_{B_1}^\circ} = \frac{2}{(1 + f_F)} \tag{A-2}$$

[Note: these equations were presented in Polymer, 1989, **30**, 1298, but there are typographical errors in equations 11 and 14 of this paper (see also Chapter 4)].

The contribution of hydrogen bonds to the partial molar free energies are given by:

$$\frac{(\overline{\Delta G_H})_A}{RT} = \left[\ln \Phi_{A_1} + 1 - r \Phi_{B_1}\left(1 + K_B \Phi_{B_1}\right) - \Phi_A\right]$$

$$- \left[\ln \Phi_A + \Phi_B\left(1 - \frac{r}{\overline{h^\circ}}\right)\right] \tag{A-3}$$

and:

$$\frac{(\Delta \overline{G}_H)_B}{RT} =$$

$$\left[\ln \left(\frac{\Phi_{B_1}}{\Phi_{B_1}^o} \right) + \Phi_{B_1}^o \left(1 + K_B \Phi_{B_1}^o \right) - \Phi_{B_1} \left(1 + K_B \Phi_{B_1} \right) - \frac{\Phi_A}{r} \right]$$

$$- \left[\frac{\ln \Phi_B}{h^o} + \Phi_A \left(\frac{1}{h^o} - \frac{1}{r} \right) \right] \tag{A-4}$$

where in each expression the terms in the first set of parentheses are the contributions to the chemical potentials from hydrogen bonding of the segments as if they were free molecules. The terms in the second set of parentheses turns this into an excess function relative to a reference state where the B units have the distribution of hydrogen bonds found in pure B.

APPENDIX 2: LINEAR SELF-ASSOCIATION DESCRIBED BY TWO EQUILIBRIUM CONSTANTS

The stoichiometry of open-chain association when dimer formation is described by a different equilibrium constant to subsequent h-mer formation was presented in Chapter 4. The free energy (per mole of lattice sites) of mixing molecules that hydrogen bond in this fashion is given by:

$$\frac{\Delta G}{RT} = \Phi_B \ln \left(\frac{\Phi_{B_1}}{\Phi_{B_1}^o} \right) + \frac{\Phi_A}{r} \ln \Phi_{A_1}$$

$$+ \Phi_B \left[\left(\frac{\Gamma_1^o}{\Gamma_2^o} \right) - \left(\frac{\Gamma_1}{\Gamma_2} \right) \right] + \Phi_B \left(\frac{\Gamma_1}{\Gamma_2} \right) \left(\frac{X}{1+X} \right) \tag{A-5}$$

where Γ_1, Γ_2, etc., are given by equations 4.94 and 4.95 and:

$$X = \frac{K_A \Phi_{A_1}}{r} \qquad (A\text{-}6)$$

The ΔG_H excess function is obtained by subtraction of:

$$\frac{\Phi_B}{\overline{h}^\circ} \ln \Phi_B + \frac{\Phi_A}{r} \ln \Phi_A$$

from equation A-5, where:

$$\overline{h}^\circ = \frac{\Gamma_2^\circ}{\Gamma_1^\circ} \qquad (A\text{-}7)$$

The contribution of the excess function (ΔG_H) to the partial molar free energy of mixing two polymers that hydrogen bond is then:

$$\frac{(\Delta \overline{G_H})_A}{RT} = \left[\ln \Phi_{A_1} + 1 - r \left(\Phi_{B_1} \Gamma_1 + \frac{\Phi_A}{r} \right) \right]$$
$$- \left[\ln \Phi_A + \Phi_B \left(1 - \frac{r}{\overline{h}^\circ} \right) \right] \qquad (A\text{-}8)$$

and:

$$\frac{(\Delta \overline{G_H})_B}{RT} = \left[\ln \left(\frac{\Phi_{B_1}}{\Phi_B^\circ} \right) + \left(\Phi_{B_1}^\circ \Gamma_1^\circ - \Phi_{B_1} \Gamma_1 \right) + \frac{\Phi_A}{r} \right]$$
$$- \left[\ln \frac{\Phi_B}{\overline{h}^\circ} + \Phi_A \left(\frac{1}{\overline{h}^\circ} - \frac{1}{r} \right) \right] \qquad (A\text{-}9)$$

## CHAPTER 7

The Calculation of Phase Diagrams for Strongly Interacting Polymers

A. INTRODUCTION

In this chapter we will consider how well we can predict the phase behavior of polymer blends using the association model described in the preceding chapters. We will also consider as yet untested predictions and present our results in the context of the second computer program accompaning with this book. Unlike the qualitative **Miscibility Guide** presented in chapter 2, however, this computer program, called **Phase Calculator**, is much more comprehensive, but still based on an expression for the free energy of mixing that can be written in the following form:

$$\frac{\Delta G_m}{RT} = \frac{\Phi_A}{M_A} \ln \Phi_A + \frac{\Phi_B}{M_B} \ln \Phi_B + \Phi_A \Phi_B \chi + \frac{\Delta G_H}{RT} \tag{7.1}$$

By now we anticipate that readers of this book will fall into one of two catagories. The first will have digested (perhaps with a large dose of antacids) the theoretical concepts and assumptions made in deriving equation 7.1 from a lattice model (chapters 4 and 6) and is now interested in how well the theory fits experiment. He or she will have understood the basis for separating the "physical" interaction parameter χ from the "chemical" ΔG_H term; why we can to a first approximation neglect free volume effects; how the ΔG_H term is expressed in terms of various concentration terms and equilibrium constants that describe the distribution of hydrogen bonded species present at a particular concentration and temperature; how we can determine these equilibrium constants by independent experimental infared spectroscopic measurements; and how we can obtain initial estimates of χ from non-hydrogen bonded solubility parameters.

The second type of reader perhaps does not have the time for, nor is particularly interested in, theoretical details and accepts for now that the authors know what they are doing (a dubious assumption some might say). He or she may have used the qualitative **Miscibility Guide**, hopefully with some success, and is now interested in quickly determining whether or not it is worthwhile to calculate the more rigorously derived phase diagrams, miscibility windows and maps. Thus we are dealing with individuals who may have very different goals, and at the risk of being called simple-minded, we decided to err on the side of "overkill" and lead the reader through a description of the computer program **Phase Calculator** that calculates the free energy of mixing, phase diagrams, miscibility windows and maps of hydrogen bonded polymer blend systems. The more sophisticated reader should be able to skip that which is obvious to him or her without loss of continuity. We will commence with a description of the phase diagrams of polymer blend systems that involve only homopolymers and then get progressively more complicated as we describe miscibility windows of homopolymer - copolymer systems and then copolymer - copolymer blend systems.

B.   HOMOPOLYMER (Self-Associated)– HOMOPOLYMER (Non Self-Associated) SYSTEMS

Polyurethane and Polyamide Blends

Amorphous Polyurethane (APU) with Polyethers

The first system that we will consider is the amorphous polyurethane (APU) - polyether blends[1,2] considered in previous chapters (A schematic representation of the chemical structure of APU is shown on page 134). The reader may recall that in Chapter 2 we demonstrated that results obtained using the

Miscibility Guide accurately predicted that APU should be miscible with PEO, but immiscible with PVME. These particular polymer blends were some of the first systems we used to compare theory with experiment and represent a convenient starting point for our calculations of the free energy of mixing and phase diagrams. In this first section we will emphasize the mechanics of using the program, rather than a detailed examination of our results (which were quite simple, APU - PEO blends are miscible at 110°C, APU - PVME blends are not. More detailed and convincing studies appear in later sections.

The **Phase Calculator** computer program that calculates such information is available for both Apple and IBM compatible computers. However, we will again restrict ourselves here to examples derived from the program written for the Macintosh II computer. "Double clicking" on the icon **Phase Calculator 020** produces the **Calculation Type** window shown in figure 7.1. There are four choices, and for this problem we select here the usual temperature/composition phase diagram calculation represented by the icon shown highlighted in the top left hand corner of the figure. The remaining three icons represent calculations for different miscibility windows and maps that will be used later. Selecting the box with the arrow --> produces the **Association Equilibrium Model** window shown in figure 7.2.

It is here that we can select the type of association model that is applicable and there are three choices. For the APU - polyether blends we select a model described by two equilibrium constants depicted by the highlighted icon on the left hand side of figure 7.2. The two equilibrium constants, K_B and K_A, describe, respectively, the self-association of APU urethane segments into "chain-like" hydrogen bonded structures (see figure 3.1 - page 165) and the inter-association of the APU urethane with the polyether oxygen segments.

The remaining two icons in figure 7.2 represent models that we will consider later. Once again, selecting the box with the arrow ---> produces the window shown in figure 7.3. This is the **Segment Information** window where we input data pertaining to the molar volume, molecular weight and non-hydrogen bonded solubility parameters of the polymers. **Segment A** (the *specific* repeat, defined in Chapter 2, of the

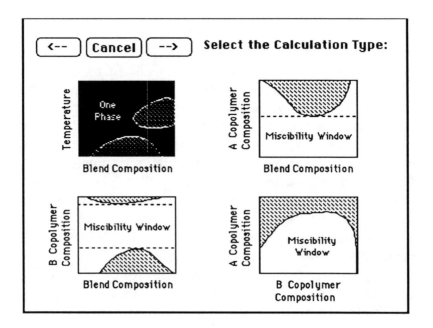

FIGURE 7.1

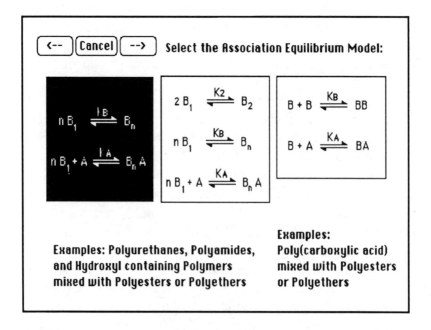

FIGURE 7.2

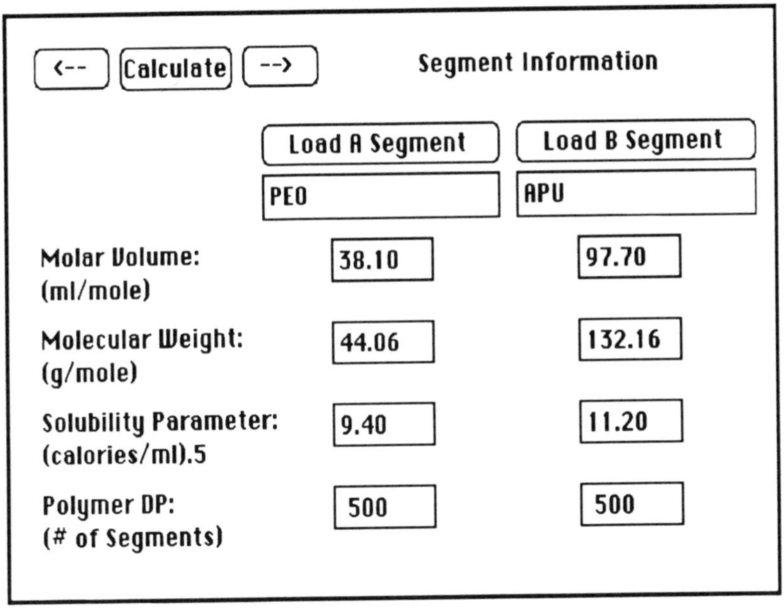

```
┌─────────────────────────────────────────────────────────────────────┐
│  ┌─────┐ ┌─────────┐ ┌─────┐          Segment Information            │
│  │ <-- │ │Calculate│ │ --> │                                         │
│  └─────┘ └─────────┘ └─────┘                                         │
│                                                                       │
│                  ┌──────────────────┐   ┌──────────────────┐          │
│                  │  Load A Segment  │   │  Load B Segment  │          │
│                  └──────────────────┘   └──────────────────┘          │
│                  ┌──────────────────┐   ┌──────────────────┐          │
│                  │ PEO              │   │ APU              │          │
│                  └──────────────────┘   └──────────────────┘          │
│  Molar Volume:        ┌─────────┐         ┌─────────┐                 │
│  (ml/mole)            │ 38.10   │         │ 97.70   │                 │
│                       └─────────┘         └─────────┘                 │
│  Molecular Weight:    ┌─────────┐         ┌─────────┐                 │
│  (g/mole)             │ 44.06   │         │ 132.16  │                 │
│                       └─────────┘         └─────────┘                 │
│  Solubility Parameter:┌─────────┐         ┌─────────┐                 │
│  (calories/ml).5      │ 9.40    │         │ 11.20   │                 │
│                       └─────────┘         └─────────┘                 │
│  Polymer DP:          ┌─────────┐         ┌─────────┐                 │
│  (# of Segments)      │  500    │         │  500    │                 │
│                       └─────────┘         └─────────┘                 │
└─────────────────────────────────────────────────────────────────────┘
```

FIGURE 7.3

essentially non self-associating polymer; e.g. PEO) and **Segment B** (the *specific* repeat of the self-associated polymer; e.g. APU) may be introduced in one of two ways.

First, by simply selecting the appropriate box and entering the value. Alternatively, if the segment data was calculated from the **Miscibility Guide** program and saved in a file, simply selecting the box labelled **Segment A** (or **Segment B**) will access these files and the data can be transferred directly. The reader may recall that the *specific* repeat of APU contains only one urethane group (see page 134).

Accordingly, it has a calculated molar volume of 97.7 cm^{-3} mole^{-1} and a molecular weight of 132.3 g mole^{-1} with an estimated non-hydrogen bonded solubility parameter of 11.2 (cal. cm^{-3})$^{0.5}$. The corresponding values for PEO are 38.1 cm^{-3} mole^{-1}, 44.1 g mole^{-1} and 9.4 (cal. cm^{-3})$^{0.5}$. Also in this window are boxes to input the number of segments in the polymers (degree of polymerization) - a default value of 500 is commonly used in preliminary investigations. Continuing; selecting the box with the arrow --> produces the window shown in figure 7.4.

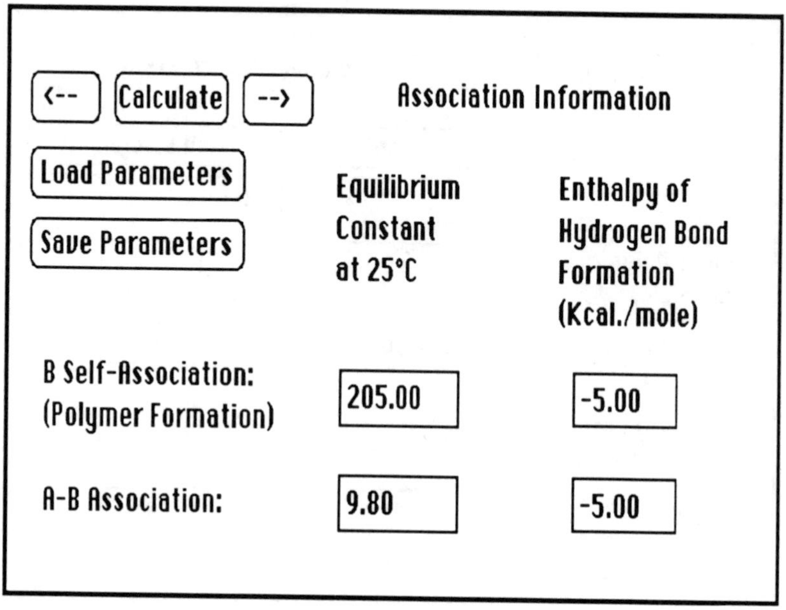

	Association Information
<-- Calculate -->	
Load Parameters	Equilibrium Constant at 25°C · Enthalpy of Hydrogen Bond Formation (Kcal./mole)
Save Parameters	

	Equilibrium Constant at 25°C	Enthalpy of Hydrogen Bond Formation (Kcal./mole)
B Self-Association: (Polymer Formation)	205.00	-5.00
A-B Association:	9.80	-5.00

FIGURE 7.4

This is the **Association Information** window where we imput the values of the two equilibrium constants, K_B (self-association of APU) and K_A (inter-association of the APU with PEO) and the corresponding enthalpies of hydrogen bond formation, h_B and h_A. In chapter 5, page 261, we have described how we can obtain a quantitative infrared analysis of the fraction of "free" C=O groups present in pure APU ($f_F^o = 0.163$) at 110°C (above the glass transition temperature)[1,2]. From this data we can easily calculate the equilibrium constant describing the self-association of APU ($K_B = 31.5$ at 110°C, equation 5.31, page 275). For now we will assume a value of $h_B = -5.0$ kcal. mole[-1], (later we will see how we can determine h_B for certain systems directly from infrared temperature measurements), and using the van't Hoff relationship (equation 4.70, page 190) we determine that $K_B = 205$ at our reference temperature of 25°C. It is this value that is introduced into the window shown in figure 7.4.

In a similar manner, the value of the equilibrium constant describing the inter-association of urethane and ether segments, K_A, can also determined directly from infrared measurements[1,2] at 110°C (chapter 5). We have described the stoichiometry of such systems (see equations 5.44 and 5.45, page 286) and the graph in figure 5.28, page 288, shows the experimental values of the fraction of "free" urethane carbonyl groups (f_F) as a function of the volume fraction of APU in miscible blends with an ethylene oxide-co-propylene oxide copolymer (EPO).

Using a least-squares procedure in conjunction with the previously determined K_B value at 110°C, a value of $K_A = 1.47$ at 110°C was found from the best fit of the experimental data (chapter 5, page 288). Assuming, for now, that $h_A = h_B$, (infrared measurements indicate that they are indeed very similar), we arrive at the conclusion that $K_A = 9.8$ at the reference temperature of 25°C. We should perhaps reiterate that all things being equal, (i.e. no major changes in chemistry, steric hindrance etc.), we theoretically expect to find identical values of K_A for similar APU - polyether blends (see chapter 4, page 214). The difference in the size of the chemical repeat units is accounted for by the factor K_A / r, where r is the ratio of the molar volumes of the segments V_A / V_B. Having introduced the values of the equilibrium constants and enthalpies of hydrogen bond formation we can now move on by selecting the box with the arrow ---> which produces the window shown in figure 7.5.

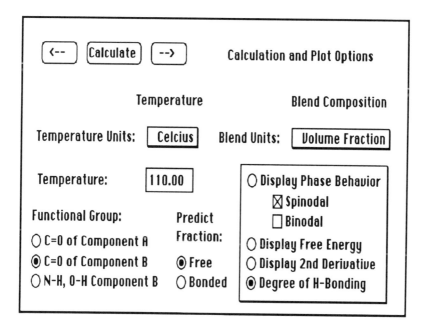

FIGURE 7.5

This is the **Calculation and Plot Options** window which will be featured prominantly in the forthcoming discussion since there are many options. In the bottom right hand side of the figure 7.5 is a large box containing four radio buttons related to the display of the phase diagrams (in conjunction with additional options for spinodal and binodal calculations), free energy, the second derivative of the free energy and the degree of hydrogen bonding. We will commence by selecting the button **Degree of H-Bonding**. This option calculates the fraction of free or hydrogen bonded functional groups in the blend as a function of blend composition from the equations describing the stoichiometry of the system (chapter 4, section C), *assuming that it is single phase (miscible) and in equilibrium.*

Selecting the functional group of interest depends upon the polymer blend considered. In the case of APU - polyether blends the most relevent information is the fraction of free APU carbonyl groups[1,2] (see chapter 4, page 185) and so we select the buttons labelled **Free** and **C=O of Component B** (APU). One may also select the temperature at which the calculation is performed - we will employ 110°C here which is above the T_g of APU and the T_m of PEO - and the units of temperature and blend composition by selecting the appropriate buttons. Now we select **Calculate** which should briefly produce the window shown in figure 7.6 followed by the plot depicted in figure 7.7. It can be seen from the theoretical plot in figure 7.7 that as we increase the concentration of APU in the mixture, the fraction of free urethane carbonyl groups decreases, finally approaching the value for pure APU at 110°C.

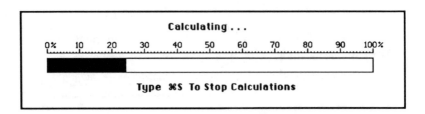

FIGURE 7.6

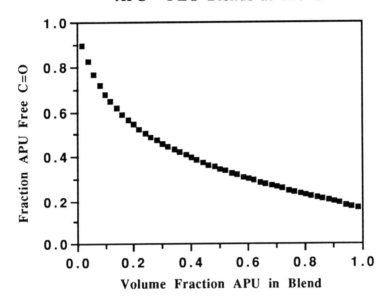

FIGURE 7.7

Theoretically, there should be, for example, some 55 % free urethane carbonyl groups in an amorphous *miscible* 20:80 volume % blend of APU and PEO in equilibrium at 110°C which can immediately be seen to be roughly correct from an examination of the infrared spectrum of such a blend (figure 5.15, page 252). The important point here is that if we know the magnitude of the equilibrium constants K_B and K_A, the enthalpies of hydrogen bond formation h_B and h_A, and the ratio of the molar volumes of the segments V_A / V_B, we can readily calculate the fraction of free urethane carbonyl groups at any given temperature. This is very useful information, (especially when we consider the more complex copolymer - copolymer miscibility windows), since we can compare these calculated results to those obtained experimentally by infrared spectroscopy and determine whether or not a particular blend of a given composition is a single or multiphased material. Of course, for the APU - PEO system we used the infrared spectra to obtain the equilibrium constants, so the agreement was foreordained, but we will consider a number of examples where we transfered equilibrium constants from other systems and in such cases the match of calculation and experimental observation can be used as a probe of phase behavior (see Chapter 5 also).

An example of an immiscible blend, that is grossly phase separated, with the two individual phases composed of essentially the pure components, is the APU - PVME system. To calculate the corresponding fraction of free urethane carbonyl groups for a *theoretically miscible* APU - PVME blends, one simply returns to the **Segment Information** window (figure 7.3) under the **Calculate** menu and repeats the process, replacing the PEO segment data with that of PVME (this is most effeciently performed if the data is already stored in a segment file - selecting the box labelled **Segment A** will access these files and the data is transferred directly) and then selecting **Calculate**.

The results, displayed in figure 7.8, are almost identical to those obtained for PEO (as they should be since the only variable changed in the calculation is the molar volume of the polyether which does not change very much).

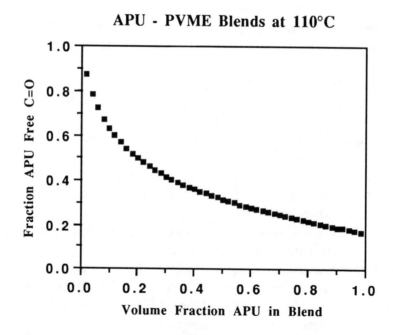

FIGURE 7.8

We must keep in mind, however, that these calculations are telling us what the fraction of free carbonyl groups would be *if the blend is in equilibrium and exists in a single phase.* Even a cursory glance at the infrared spectra of APU - PVME blends of varying composition[1] at 110°C (see, for example, figure 5.15) immediately shows that the system is immiscible, as the fraction of free carbonyl groups corresponds essentially to that found in the pure APU polymer at 110°C.

Returning to the **Calculation and Plot Options** window via the **Calculate** menu we will now calculate the free energy of mixing by selecting the **Display Free Energy** button which should produce figure 7.9. Again, the temperature at which the calculation is performed and the units of temperature and blend composition may be changed by selecting the appropriate buttons. The free energy of mixing calculated at 110°C as a function of the volume fraction of APU in blends with PEO is displayed in figure 7.10 (for clarity the x-axis has been scale expanded - see operators manual). Returning to the **Calculation and Plot Options** window (figure 7.9) via the **Calculate** menu it is easy to now calculate the second derivative of the free energy with respect to APU volume fraction.

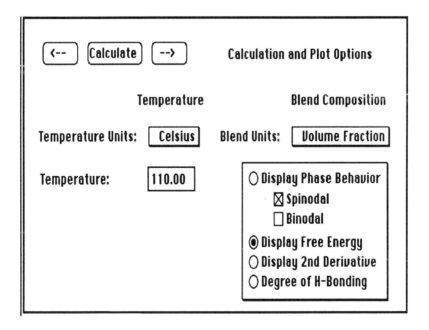

FIGURE 7.9

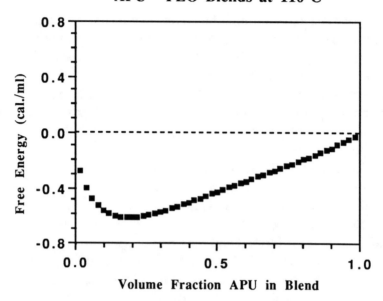

FIGURE 7.10

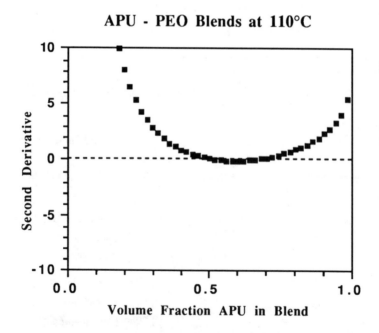

FIGURE 7.11

This is displayed in figure 7.11, by selecting the **Display 2nd Derivative** button. The calculated results predict that at 110°C amorphous bends of APU with PEO are barely miscible. The free energy of mixing is negative and the second derivative is essentially positive throughout the whole composition range. In marked contrast, the amorphous APU - PVME blend system at 110°C is predicted to be definitely immiscible. To calculate the corresponding free energy and second derivative plots for APU - PVME blends, one simply returns to the **Segment Information** window under the **Calculate** menu and repeats the process, replacing the PEO segment data with that of PVME. The results are displayed in figures 7.12 and 7.13, respectively. The free energy is not negative throughout the entire composition range while the second derivative assumes negative values over a wide composition range.

If we wish to see what happens to the phase behavior of polymer blends as a function of temperature we need to calculate a phase diagram.

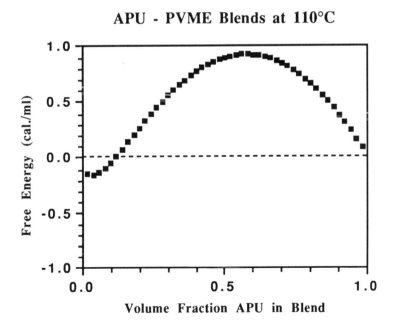

APU - PVME Blends at 110°C

Volume Fraction APU in Blend

FIGURE 7.12

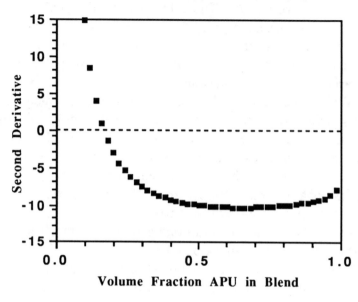

FIGURE 7.13

As we have described elsewhere in this book (chapter 1, page 19), the spinodal may be readily determined by equating the second derivative of the free energy to zero while the binodal may be calculated by searching for equivalent chemical potentials outside the spinodal consistent with a common tangent to the free energy curve.

Once again let us return to the **Calculation and Plot Options** window via the **Calculate** menu and now calculate the phase diagrams by selecting the **Display Phase Behavior** button and choosing either or both of the options for calculating the **Spinodal** and **Binodal** (figure 7.14). The temperature range over which the phase diagram is to be calculated is defined by the values introduced into the boxes labelled **Starting Value** and **Ending Value**. Additionally, the number of different temperatures that will be considered is set by the value introduced in the box labelled **Stepping Value**. In the example shown in figure 7.14, spinodal and binodal calculations were performed every 10°C, starting at -100 and ending at 250°C.

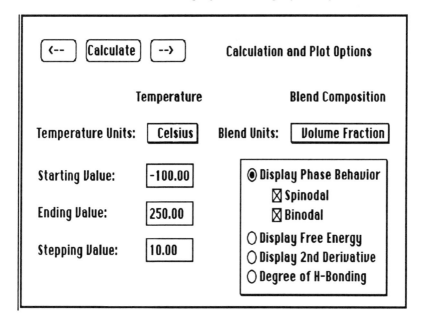

FIGURE 7.14

Following the selection of **Calculate** there is a pause while the calculations are performed, (the window shown in figure 7.6 will be displayed, and at the risk of belaboring the obvious, the time required for computation increases if the binodal or both the spinodal and binodal are required and also increases the greater the temperature range and/or the greater the number of different temperatures considered). *Theoretical* phase diagrams (both spinodal and binodal - the latter naturally lying outside that of the former) of the amorphous APU - PEO system are shown in figure 7.15. A "window of miscibility" is predicted to exist across the entire composition range from about -30°C (the theoretical upper critical solution temperature (UCST)) to 110°C (the theoretical lower critical solution temperature (LCST)). To calculate the corresponding phase diagrams for APU - PVME blends, one simply returns to the **Segment Information** window under the **Calculate** menu and repeats the process, replacing the PEO segment data with that of PVME. The results of such an exercise are shown in figure 7.16. Here we observe the extreme case of a classic "hour glass" shaped phase diagram typical of a grossly phase separated (immiscible) polymer blend system.

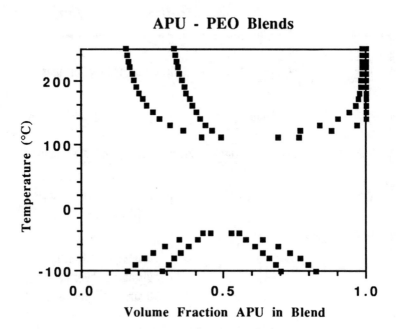

APU - PEO Blends

FIGURE 7.15

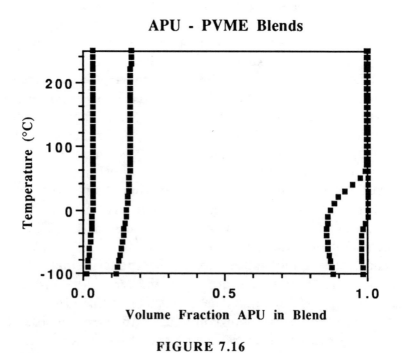

APU - PVME Blends

FIGURE 7.16

It should perhaps be emphasized that modest changes to the values of the parameters employed in the calculations will not significantly affect the general features of the phase diagram shown in figure 7.16. This is readily understood if one considers the overall balance of equation 7.1. In a case where both polymers have high molecular weights, (the combinatorial entropy terms being therefore negligible), the free energy of mixing is determined by the balance between the $\Phi_A\Phi_B\chi$ and hydrogen bonding terms. If the former should *dominate,* then a phase diagram of type shown in figure 7.16 will result and moderate variations of the parameters do not make a lot of difference. However, when these two principal terms are roughly equivalent in magnitude, as is the case for the phase diagram calculated for the APU - PEO blends (figure 7.15), modest changes in the equilibrium constants, solubility parameters etc. can significantly alter the appearance of the phase diagram and affect the "windows of miscibility".

APU-PEO BLENDS

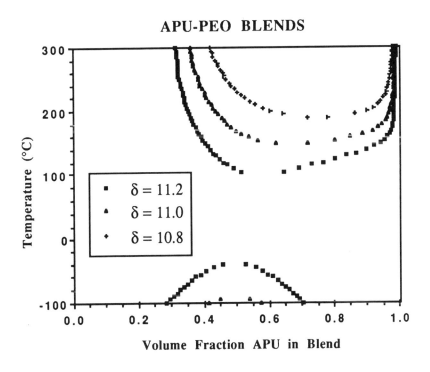

Volume Fraction APU in Blend

FIGURE 7.17

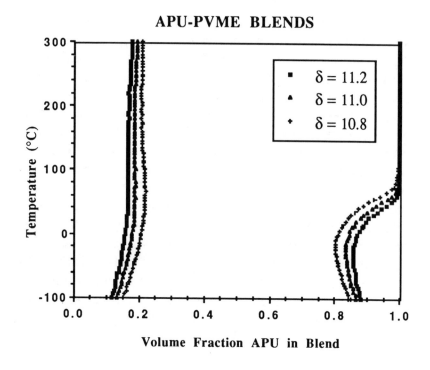

FIGURE 7.18

To illustrate, the non-hydrogen bonded solubility parameter of APU was *estimated* at 11.2 (cal. cm^{-3})$^{0.5}$ (chapter 2, page 134). Let us consider the effect of lowering this value, but still within the range of the estimated error ($\pm$ 0.4 (cal. cm^{-3})$^{0.5}$.

This is readily accomplished by returning to the **Segment Information** window under the **Calculate** menu and replacing the APU solubility parameter value with an alternative value and recalculating. One can store the data from each calculation under the **File** menu by selecting **Save Data (TEXT)** which can then be imported into a commercial software program such as Cricket Graph® to produce multiple graphs like those displayed in figures 7.17 and 7.18. These figures show the theoretical spinodal phase diagrams of APU - PEO and APU - PVME blends, respectively, using values of 11.2, 11.0 and 10.8 (cal. cm^{-3})$^{0.5}$ for the non-hydrogen bonded solubility parameter of APU. Reducing the solubility parameter of APU by 0.2 and 0.4 (cal. cm^{-3})$^{0.5}$ significantly increases the theoretical temperature range over which the system the APU - PEO blend is miscible. The LCST, for example, increases by about 100°C while the UCST decreases in a like manner. Conversely, the calculated spinodal phase diagrams of the APU - PVME system are hardly affected. The main point here is to emphasize the sensitivity of systems that are finely balanced to values of χ. It is perhaps best to regard the value of χ that is determined from the corresponding non-hydrogen bonded solubility parameters as a first *estimate*, albeit a surprizingly good one, that can be modified if we gather more experimental data (more on this later when we discuss PVPh blends).

Aromatic / Aliphatic Polyamide–PEO Blends

In Chapter 2, we demonstrated how the **Miscibility Guide** could be used to ascertain major trends in the phase behavior of a series of polyisophthalamide blends with the polyethers PEO and PVME (figures 2.55 - 2.58, page 131). In the following examples we switch to an analogous series of blends of aromatic / aliphatic polyamides[3,4] where the amide group has been reversed, the so called nylon MPDn series (where n is the number of carbon atoms between nitrogens in the chemical repeat unit and MPD denotes the precursor, *meta*-phenylene diamine).

Nylon MPD5

We do this to make an important point. In our theoretical calculations of the phase diagrams of polymer blends in which hydrogen bonds play a significant role, χ, the polymer / polymer *physical* interaction parameter is estimated from *non-hydrogen bonded* (unassociated) solubility parameters, which in turn are determined from molar attraction and molar volume group contributions. Thus isomorphous isophthalamide and MPDn polymers have identical solubility parameters in our scheme, since we make no distinction between the direction of substitution of the amide group. The isomorphic equivalent to the polyisophthalamide 3-I is MPD5, which is shown schematically overleaf.

The reader will observe that the chemical repeat of MPD5 contains one disubstituted benzene ring, two amide and three methylene groups. However, the *specific repeat unit* which we employ in our calculations is defined in terms of one interacting group (the amide group) and thus contains one half of a disubstituted benzene ring, one amide and 1.5 methylene groups. For subsequent phase behavior calculations of a homologous series of MPD polymers it is convenient to prepare the segment information in advance by using the **Miscibility Guide** and saving the information in individual files labelled MPD5, MPD6 etc ,as shown in the example in figure 7.19.

Before we run the **Phase Calculator** program and calculate the phase diagrams of MPD - PEO we require values of the equilibrium constants K_B and K_A, and the enthalpies of hydrogen bond formation, h_B and h_A.

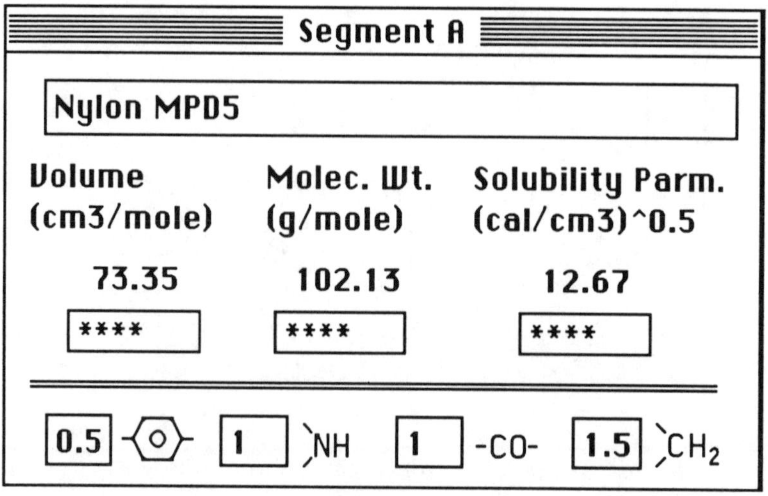

FIGURE 7.19

This needs more discussion. In a manner similar to that described for the APU polymer above, we have independently determined[4] the equilibrium constant describing the self-association of pure MPD6 from the Amide 1 region (1625-1725 cm^{-1}) of the infrared spectrum of pure MPD6, recorded as a function of temperature from 70 to 230°C.

The fraction of free carbonyl groups, f_F^o, were determined from the relative areas of the "free" and hydrogen bonded carbonyl bands. Assuming that equilibrium conditions can be attained, the equilibrium constant (at a given temperature) describing self-association of MPD6, K_B, is readily calculated (chapter 5, page 275) from f_F^o using the relationship:

$$K_B = \frac{\left[1 - f_F^o\right]}{\left[f_F^o\right]^2} \qquad (7.2)$$

The T_g of MPD6 is approximately 150°C and only those data obtained in excess of this temperature can be expected to approach true equilibrium conditions. From a least squares fit of the data obtained above 150°C a van't Hoff plot of ln K_B versus T^{-1}, yielded the relationship:

$$\ln K_B = -0.75 + \frac{1600}{T} \qquad (7.3)$$

This is equivalent at our reference temperature of 25°C to a theoretical value of approximately 100 for K_B. Furthermore, from equation 7.3, the value of the enthalpy of hydrogen bond formation, h_B, is approximately 3.2 kcal. mole^{-1}. The equilibrium constants for the other MPDn polymers containing different numbers (n) of methylene units may be readily obtained by scaling according to the ratio of the molar volumes of the different average repeat units (see Chapter 5, page 219). Thus K_B for nylon MPDn series is simply given by:

$$K_B^{MPDn} = K_B^{MPD6} \cdot \frac{V_B^{MPD6}}{V_B^{MPDn}} \qquad (7.4)$$

The magnitude of K_B varies with the number of methylene units in repeat unit of the MPDn polymer, but the ratio of the two equilibrium constants $Z = K_A / K_B$ is theoretically predicted to be constant (see Chapter 5, page 285). The adjustment for the different size of the A and B chemical repeat units enters separately through the factor $r = V_A / V_B$. Accordingly, having determined the values of K_B and K_A for a *particular* nylon MPDn - polyether blend, it is a straightforward matter to calculate the new values of the equilibrium constants for a blend involving the same polyether with a different n-I nylon. We have established a value of $Z = K_A / K_B = 0.1$ for the PIPA - polyether blends - similar in magnitude to that determined experimentally for the APU - polyether blend system. Values of V_B, K_B and K_A for the MPD polymers that we will use in our calculations are listed in Table 7.1.

Now we are ready to calculate the phase diagrams of selected MPDn - PEO polymers. Let us commence with the MPD5 - PEO blend system. Assuming that the computer program **Phase Calculator** is still active, we select **New Segments** (figure 7.3) from the **Calculate** menu and introduce the segment information of PEO and MPD5 into **Segment A** and **Segment B**, respectively, from the segment files. Now we select the --> arrow to produce the **Association Information** window (figure 7.4) and input the values of K_B and K_A for MPD5 given in Table 7.1, together with values of 3.2 kcal. mole^{-1} for both h_B and h_A.

Table 7.1
Parameters for MPD Nylons

Nylon	V_B (cm^3/mol)	K_B at 25°C	K_A at 25°C
MPD2	48.6	168	16.8
MPD5	73.4	111	11.1
MPD6	81.6	100	10.0
MPD7	89.9	91	9.1
MPD8	98.1	83	8.3

Again we select the --> arrow to now give the **Calculation and Plot Options** window (figure 7.14) and calculate the binodal phase diagrams by selecting the button **Calculate** after choosing the **Display Phase Behavior** button with the option **Binodal**. The result is displayed in figure 7.20. We now repeat the process, replacing, in succession, the segment information and equilibrium constants of MPD6, MPD7 and MPD8 into the **New Segments** and **Association Information** windows and recalculating. The results for the PEO blends with MPD6, MPD7 and MPD8 are also displayed in figure 7.20.

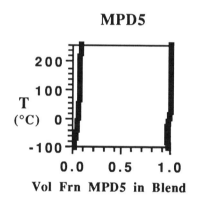

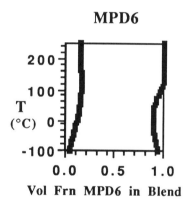

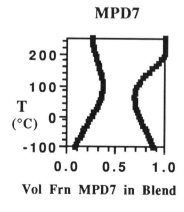

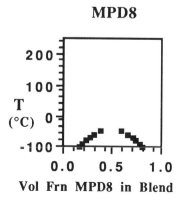

FIGURE 7.20

Not only do such calculations lead to correct predictions of major trends in the phase behavior of aromatic / aliphatic polyamide - polyether blends, but they also show how very sensitive the predicted phase behavior is to the addition or subtraction just a single methylene group in the transitional repeat (or, in our terms, the equivalent of 1/2 a CH_2 group per chemical repeat). As indicated in figure 7.20, blends of PEO with MPD8 are predicted to be thermodynamically stable throughout the entire temperature range (-100 to 250°C) except for a small region of immiscibility below –50°C. Calculations for the PEO - MPD7, MPD6 and MPD5 systems predict a two-phase region resulting from the merger of the upper and lower phase boundaries, with the breadth of the two-phase region becoming progressively larger. This is in good agreement with experimental observations. The trend observed here is identical to that reported previously for isomorphous PIPA blends with PEO [3]. What is so remarkable, and bears repetition, is the fact that only a marginal change in the number of CH_2 groups in the chemical repeat of the MPD polymer radically changes the phase diagram from one reflecting gross immiscibility to one that reflects an essentially miscible blend over the accessible temperature range.

Poly(Vinyl Phenol) (PVPh) Blends

Preamble

Some of our finest examples of experimental and theoretical studies of hydrogen bonded polymer blends involve the strongly self-associated polymer, PVPh, and we have conducted numerous experimental studies concerned with assessing the predictive capabilities of our association model using this polymer. In selecting blend candidates for these studies we have found it beneficial to concentrate on those systems in which one component's chemical structure, and thus the overall free energy of mixing, may be systematically varied. This can be accomplished simply through the use of a homologous series of polymers. In certain systems, this ability to progressively alter the free energy of mixing results in the detection of miscibility limits which then serve as the basis for comparison with model predictions. It also present us with an opportunity to "tinker" with some of the values of the parameters, within the bounds of known error, and assess their relative importance. Using this approach, we have been able to assess the model's ability to predict trends in the miscibility behavior of a number of systems, including blends of PVPh with a series of polyacetates, polylactones, polyacrylates, polymethacrylates and polyethers. We will start by considering blends of PVPh with an homologous series of poly(n-alkyl methacrylates)[5-8].

PVPh - Poly(alkyl methacrylate) Blends

In keeping with the general philosophy of this Chapter, i.e. to describe the calculation of phase diagrams and lead the reader through the **Phase Calculator** program, let us return to the **Association Equilibrium Model** window via the **Calculate** menu. For PVPh blends we need to select the icon for the "3K" model in the middle of the window as highlighted in figure 7.21. As discussed in Chapter 4, page 191, two self-association equilibrium constants, K_B and K_2, which describe the formation of "chain-like" and dimer structures, respectively, are appropriate for PVPh.

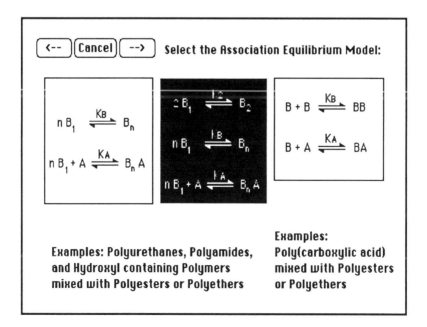

FIGURE 7.21

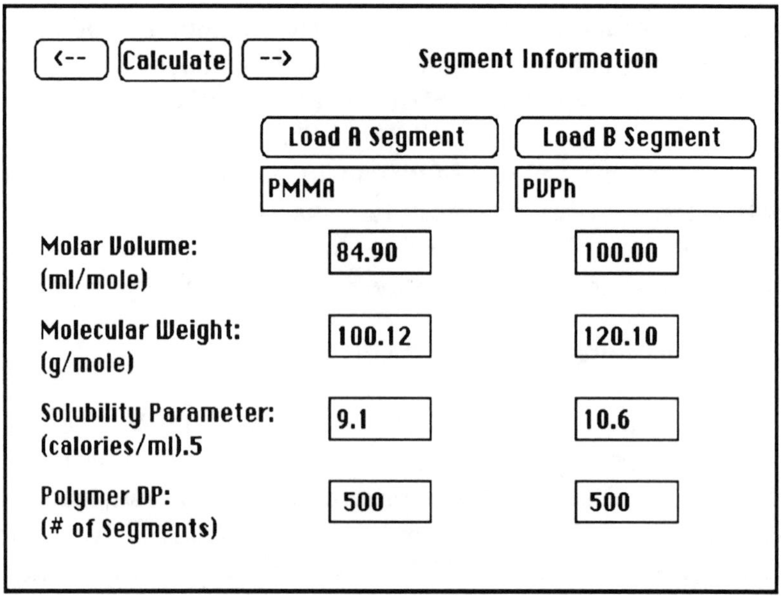

<--	Calculate	-->	Segment Information	

Load A Segment	Load B Segment
PMMA	PVPh

	A	B
Molar Volume: (ml/mole)	84.90	100.00
Molecular Weight: (g/mole)	100.12	120.10
Solubility Parameter: (calories/ml).5	9.1	10.6
Polymer DP: (# of Segments)	500	500

FIGURE 7.22

Selecting the --> arrow produces the (hopefully by now) familar **Segment Information** window shown in figure 7.22, and we now introduce the segment information of PMMA and PVPh into **Segment A** and **Segment B**, respectively, from the segment files. The segment information for PVPh is the same as that used in the **Miscibility Guide** and this requires some elaboration and justification.

It may be recalled that the segment information of the self-associated amorphous polyurethane and the aromatic / aliphatic polyamides considered at the beginning of this chapter was estimated using a combination of the individual -OCO- (or -CO-) and >N-H group contributions (F* and V* - Chapter 2, table 2.5) for the urethane (-OCO-NH-) or amide (-CO-NH-) group. It was postulated that these individual -OCO- (or -CO-) and >N-H group contributions, which were determined from compounds that are polar but not strongly self-associated, represent the non-hydrogen bonded contributions of the amide group. Accordingly, we employ this method for polyamides and polyurethanes, since we believe we obtain good estimates of *both* the molar volume and the non-hydrogen bonded solubility parameter of such polymers.

An alternative approach was to calculate the non-hydrogen bonded solubility parameter for an analogous hypothetical molecule which does not contain the N-H proton. In effect we calculate the solubility parameter using a combination of the individual -OCO- (or -CO-) and >N- group contributions (F* and V* - Chapter 2, table 2.5) for the urethane or amide group. In this case the rationale is that the errors involved in eliminating the proton are reasonably small, especially when the repeat unit is relatively large and, again, both the -CO- and >N- group contributions were derived from essentially unassociated model compounds. It was pleasing to see that the two sets of non-hydrogen bonded solubility parameters are in close agreement (Chapter 2, table 4), *but it is important to recognize that the calculated molar volumes are significantly different.* The "proton extraction" method results in a serious underestimation of the magnitude of the molar volume (for example, calculation of the molar volume of the polyamides MPD4, MPD6 and MPD8 (which are isomorphous to the polyamides 2-I, 4-I and 6-I: Chapter 2, table 2.4) are undersetimated by 21, 17 and 14%, respectively).

This leads us to self-associated hydroxyl containing polymers (PVPh, phenoxy and PVOH etc.) where we do not have the luxury of being able to dissect the O-H group into two group contributions that are derived from essentially unassociated molecules. We can obtain an *initial* estimate of the non-hydrogen bonded solubility parameter from hypothetical analogues that are missing the hydroxyl proton by employing the ether group contributions, both F* and V*, for the hydroxyl group. Employing the **Miscibility Guide** we introduce into **Segment B** one of each of the following groups, $>CH_2$, $>CH-$, disustituted benzene ring and ether, as illustrated previously in figure 2.61 (page 138). The **bold** values give the results of the calculations. However, as mentioned above, we know that the estimated molar volume of PVPh calculated in this manner is subject to considerable error and underestimated by some 20%. Accordingly, we "revise" the value of the molar volume by substituting a value of 100 $cm^3 mole^{-1}$, calculated from the experimentally reported density of the polymer, into the appropriate box. At the same time, we can add 1 to the molecular weight to allow for the missing proton, and introduce 120 into its appropriate box. Finally, we ask the reader to accept for now that the *initial* estimated value of the non-hydrogen bonded solubility parameter for PVPh at 11.0 (cal. cm^{-3})$^{0.5}$ was somewhat overestimated and a more appropriate value is 10.6 (cal. cm^{-3})$^{0.5}$. This is within the range of estimated error of the calculations (Chapter 2, page 59) and has been determined from a large number of experimental studies of the phase behavior of different PVPh blends. Thus we imput a value of 10.6 into the pertinent box and strike **enter**. Now the segment file contains the information shown in figure 7.22.

Returning to the **Phase Calculator** program; selecting the --> arrow produces the **Association Information** window shown in figure 7.23, which now shows three boxes for the three necessary equilibrium constants and their corresponding enthalpies of hydrogen bond formation. As discussed in Chapter 5, page 279, the self-association equilibrium constants for PVPh, K_B and K_2, cannot be directly determined from infrared studies of the polymer, but may be determined from the infrared data obtained for phenol by Whetsel and Lady[9], after appropriate compensation for differences in molar volume.

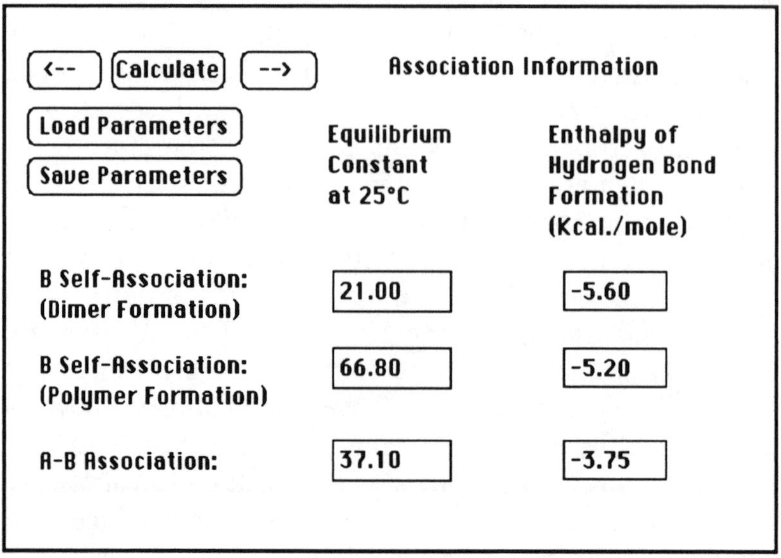

<--	Calculate	-->	Association Information		
Load Parameters		Equilibrium		Enthalpy of	
Save Parameters		Constant at 25°C		Hydrogen Bond Formation (Kcal./mole)	
B Self-Association: (Dimer Formation)		21.00		-5.60	
B Self-Association: (Polymer Formation)		66.80		-5.20	
A-B Association:		37.10		-3.75	

FIGURE 7.23

Values of 66.8 and 21.0 were determined for K_B and K_2 of PVPh (at 25°C), respectively. The equilibrium constants are assumed to obey a van't Hoff relationship over the temperature range of interest and Whetsel and Lady[9] also reported enthalpies of hydrogen-bond formation, $h_B=5.2$ and $h_2=5.6$ kcal. mole^{-1}, which are employed here.

We now require values of the inter-association equilibrium constant, K_A, and the related enthalpy of hydrogen bond formation and we will now very briefly summarize how we experimentally obtain such information. The use of FTIR spectroscopy to investigate the degree of molecular-level mixing in polymer blends possessing intermolecular hydrogen-bonding interactions is now firmly established and has been discussed fully in Chapter 5. The technique is particularly well suited to blends, such as the PVPh - poly(n-alkyl methacrylate) system considered here, in which one of the components contains a carbonyl group. The carbonyl stretching region of the FTIR spectra of these blends is characterized by two well separated bands which may be attributed to absorption by free and hydrogen-bonded carbonyl groups (figure 5.29, page 290). A quantitative analysis of the fraction of hydrogen bonded carbonyl groups is readily performed in the case of miscible PVPh - poly(ethyl methacrylate) (PVEE) and PVPh - poly(propyl methacrylate) blends at various temperatures (page 285).

Using the equations that define the stoichiometry of the system (page 286) in conjunction with the expression for the compositional variation of the theoretical fraction of hydrogen-bonded carbonyl groups, f_{HB}, (equation 5.46, page 289) the value of K_A (given values for K_B and K_2), at a given temperature, may be determined from a least-squares fit of f_{HB} to the experimental fractions of hydrogen-bonded carbonyl groups. A fine example for the miscible PVPh - PVEE blends at 150°C is given in the plot shown in figure 5.30, page 291. The results of K_A determinations for PVPh - PEMA and PVPh - PPMA blends in the temperature range between 150-200°C (above the T_g of PVPh) are summarized in table 5.4, page 292. A plot of ln K_A vs 1/T for this data is shown in the same chapter, figure 5.31. From the slope of the curves we obtain an average value of the enthapy of hydrogen-bond formation, h_A, of 3.8 kcal mole^{-1}.

Extrapolation of the van't Hoff plot to room temperature yields an average K_A value of 37.1 at 25°C. These values of K_A and h_A should be applicable to all PVPh blends with polymers containing methacrylates and it is these values that we enter into the appropriate boxes as shown in figure 7.23.

Again, we select the --> arrow to now give the **Calculation and Plot Options** window (figure 7.14) and first calculate the free energy and second derivatives at 150°C by selecting the button **Calculate** after choosing, in turn, the **Display Free Energy** and **Display 2nd Derivative** buttons (figure 7.9). The results for the PVPh - poly(n-alkyl methacrylate) blends are shown as multiple graphs in figures 7.24 and 7.25 following storage of the data from each calculation under the **File** menu by selecting **Save Data (TEXT)** and then employing the Cricket Graph® software.

The calculated results predict that at 150°C the PVPh bends with PMMA and PEMA are miscible.

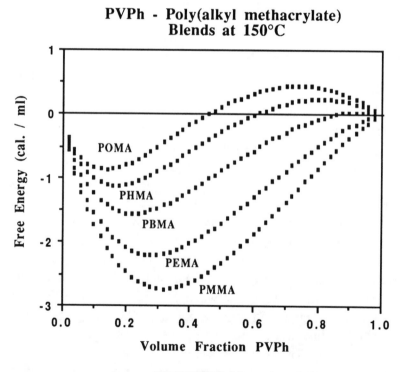

FIGURE 7.24

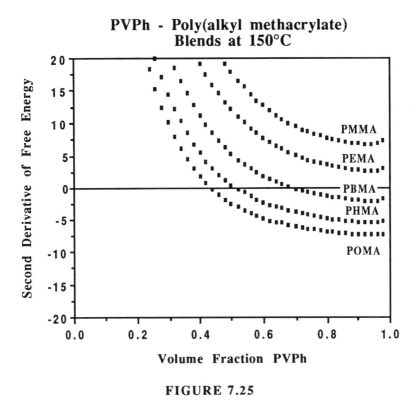

FIGURE 7.25

The free energy of mixing is negative and the second derivative is positive throughout the whole composition range - the definition of a miscible system. On the other hand, the PVPh - PHMA and POMA blend systems are predicted to be multiphased (immiscible). The free energy is not negative throughout the entire composition range and consequently the second derivative assumes negative values over a certain composition range. Finally, although by strict definition the PVPh - PBMA system is predicted to be immiscible at 150°C (the second derivative curve does assume a negative value over a very limited composition range at 150°C) it is, nonetheless, on the "edge" of miscibility. This suggests a system that might exhibit a LCST within an experimentally accessible temperature range. Returning to the **Calculation and Plot Options** window, let us now calculate theoretical binodal phase diagrams for the PVPh - poly(n-alkyl methacrylate) blends for the temperature range between -100°C and 250°C. The results are shown in figure 7.26.

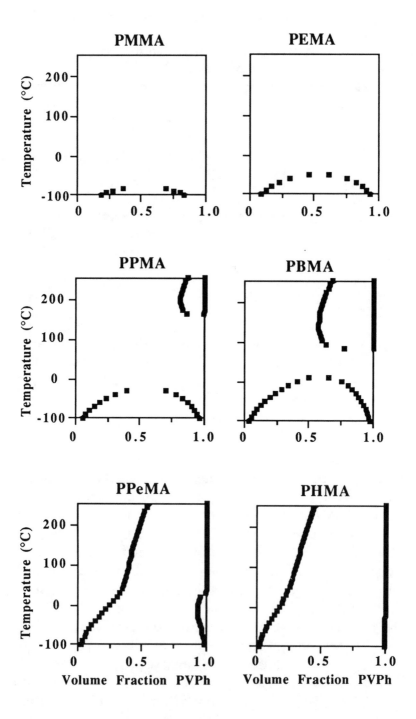

FIGURE 7.26

As indicated in the figure, blends of PVPh with PMMA and PEMA are predicted to be thermodynamically stable throughout this entire temperature range, except for a small region of immiscibility below −50°C. Results obtained for the PVPh - PPMA and PBMA systems predict the emergence of a closed immiscibility loop at high temperatures and high PVPh compositions, as well as an increase in temperature of the lowerphase boundary compared to that of the PMMA and PEMA blends. For PPeMA, PHMA and higher homologue blends, a larger two-phase region is predicted, resulting from the merger of the upper and lower phase boundaries, with the breadth of the two-phase region becoming progressively larger. The trend observed here--a decrease in miscibility with increasing size of the poly(n-alkyl methacrylate) pendant group--is identical to that predicted by the **Miscibility Guide** (Chapter 2, page 139).

These predictions of trends in the phase behavior of PVPh - poly(alkyl methacrylate) blends are in excellent agreement with experimental observations[8]. PMAA, PEMA and PPMA are indeed miscible with PVPh while PHMA, POMA and PDMA are not. . The most interesting blend in the series is, of course, the PVPh - PBMA system, because an LCST is predicted in the accessible range of temperature. The results of cloud point measurements were presented earlier in Chapter 1 (figure 1.14, page 35) and a comparison with the calculations shown in Chapter 6 (figure 6.13, page 371) indicates that our predictions are about 80°C off.

This is still pretty good. Not only have we correctly predicted the trend of phase behavior for mixtures of PVPh with a series of poly(alkyl methacrylates), we also identified the system that would have an LCST in the accessible range of temperature and got within 80°C of calculating the experimentally determined phase boundary. We believe that our results are even better than this, however, for two reasons. First, our simple (90° scattering) cloud point measurements can only be considered crude. Second, the T_g of PVPh ranges between 140°C and 190°C (depending on the molecular weight of the sample used) and once phase separation occurs a single phase system cannot be reestablished by lowering the temperature, due to the glassy nature of one of the phases, which our calculations indicate to be almost pure PVPh (see figure 7.26). Using FTIR as a probe of phase behavior we have indications that the binodal is about 50°C lower than these experimental cloud points[8]. Accordingly, we believe our calculations are surprizingly accurate, given the assumptions of the model and the effect of errors. We now turn our attention to this latter problem.

The Effect of Errors

We will use the PVPh - PBMA system to consider the effect of errors in the values of the solubility parameters, equilibrium constants and enthalpy of hydrogen bond formation on the calculation of phase diagrams, especially in this case where the χ and $\Delta G_H/RT$ contributions are finely balanced.

The most significant source of error in the estimation of χ is the value of the non-hydrogen solubility parameter of PVPh and it is here that we will commence. However, it must be remembered that χ is determined from the square of the difference of the solubility parameters of both polymers and errors in χ may be accentuated or fortuitously cancelled depending upon whether the solubility parameters are over- or underestimated.

Let us return to the **Segment Information** window via the **Calculate** menu, and introduce the segment information of PBMA and PVPh into **Segment A** and **Segment B**, respectively, from the segment files, but now alter the value of the solubility parameter of PVPh, in turn, from 11.0 to 10.8 to 10.6 and 10.4 (cal. cm^{-3})$^{0.5}$. Figure 7.27 illustrates the effects of these relatively small changes in the solubility parameter of PVPh on the spinodal calculations for PVPh - PBMA blends. Upon decreasing the solubility parameter of PVPh from 11.0 to 10.6 (cal. cm^{-3})$^{0.5}$, which is within the margin of error for calculated solubility parameters (see Chapter 2, page 60), an immiscible blend with a two-phase region extending throughout the whole temperature range transforms to one with separate upper and lower immiscibility regions. In other words, a decrease of only 0.4 (cal. cm^{-3})$^{0.5}$ is sufficient to yield a window of miscibility which extends throughout the entire composition range at ambient temperature.

PVPh - PBMA Blends: Effect of Errors in δ_{PVPh}

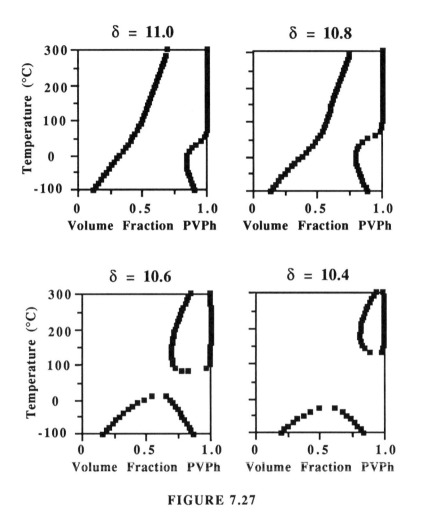

FIGURE 7.27

Since the estimation of the *non-hydrogen bonded solubility parameter* of PVPh, (or any strongly self-associated polymer for that matter), is subject to the most error in our scheme, it is perfectly reasonable to consider the interaction parameter χ, or the non-hydrogen bonded solubility parameter of PVPh from which it is calculated, as adjustable parameters. Thus, by adjusting the solubility parameter of PVPh in this manner, theoretical phase diagrams in excellent agreement with experimental observations may be obtained for all PVPh - poly(n-alkyl methacrylate) blends considered. What is probably most surprising is that our crude and simple method of determining the solubility parameter of PVPh from non-hydrogen bonded group contributions has yielded results that are close to that required to satisfactorily predict the gross phase behavior of a number of different polymer blend systems.

We have alluded that errors in the solubility parameter values will be particularly acute for the calculation of phase diagrams where the χ and $\Delta G_H/RT$ contributions are finely balanced such as those described above for the PVPh - PBMA blends.

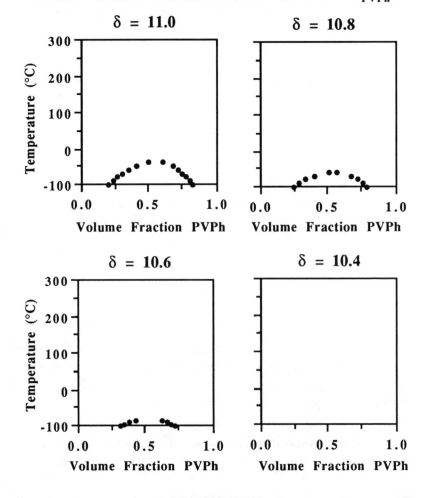

FIGURE 7.28

On the other hand, when either the χ or $\Delta G_H/RT$ contributions dominate, the effect of even quite large errors is rather small. For example, in PVPh-PMMA blends the $\Delta G_H/RT$ term dominates and the system is miscible. Figure 7.28 shows the effect of varying the solubility parameter over the range of 10.4 to 11.0 (cal. cm^{-3})$^{0.5}$ on the calculated phase diagrams. The upper critical solution temperature is predicted to increase somewhat over this range, but the overriding conclusion is all these calculations is that PVPh - PMMA blends are miscible in the accessible range of temperatures.

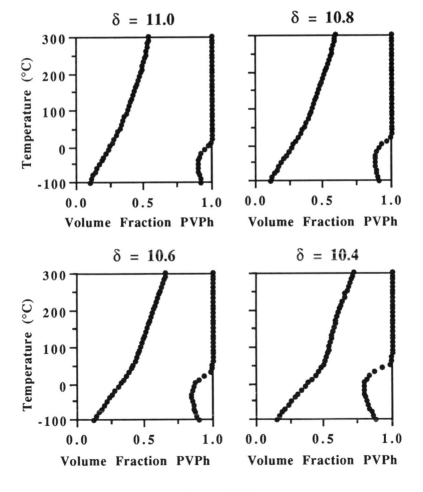

FIGURE 7.29

Similarly, in PVPh - PHMA blends the χ term is overwhelming, resulting in grossly phase separated materials, and as we can see from even a cursory glance of figure 7.29 changing the solubility parameter of PVPh over the range of 10.4 to 11.0 (cal. cm^{-3})$^{0.5}$ makes little difference in the predicted phase diagrams. This is why our model predicts *trends* in the phase behavior of a series of systematically changing blends very successfully, but in systems on the edge of miscibility errors have a significant effect on predicted positions of the phase boundaries.

Now let us consider the effect of errors in K_A. As we discussed a few pages ago, the value of K_A is calculated from a least squares fit of theoretical stoichiometric equations to infrared experimental data obtained from miscible PVPh - poly(n-alkyl methacrylate) blend samples, with the values of the self-association constants, K_2 and K_B, being held constant. What would be the effect of say a 20% error in K_A on the calculated phase diagrams of the PVPh - PBMA system ?

The experimental value of K_A was determined to be 37.1; the value employed to calculate the phase diagrams of the PVPh - poly(n-alkyl methacrylate) blend series shown in figure 7.26. Returning to the **Segment Information** window (figure 7.22), we restore the value of the solubility parameter of PVPh to 10.6 (cal. cm^{-3})$^{0.5}$; move forward to the **Association Information** window (figure 7.23) and replace the value of the **A-B Association** equilibrium with, in turn, values of 30, 40, 50 and 60 and **Calculate**. The results are illustrated in figure 7.30. While there are obvious differences in the calculated phase diagrams, it is apparent that reasonable errors (<10%) in the values of the equilibrium constants can be tolerated without seriously affecting the overall conclusions. In fact, we have performed numerous calculations where we have systematically varied all of the equilibrium constant values, and concluded that the calculated phase diagrams are not very sensitive to the absolute values of the equilibrium constants. We should not be surprised, however, since in free energy calculations it is the *natural logarithm of the equilibrium constant* that is paramount, so that errors tend to be minimized.

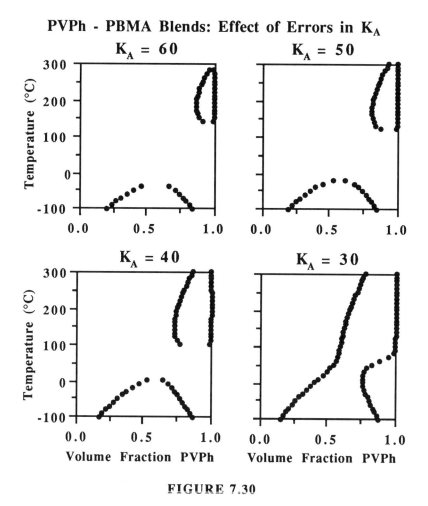

FIGURE 7.30

It is important to recognize that just one additional CH_2 group in the poly(n-alkyl methacrylate) repeat has a much greater repercussion upon the calculated phase diagram (figure 7.26) than does a 20% error in equilibrium constant values.

Finally, let us consider the effect of errors in h_A. Again, as we noted a few pages ago, the value of h_A is obtained experimentally from a van't Hoff plot of ln K_A versus inverse temperature. For the PVPh - poly(n-alkyl methacrylate) blends a value of 3.8 kcal. mol^{-1} was determined What would be the effect of say a 20% error in h_A on the calculated phase diagrams of the PVPh - PBMA system ?

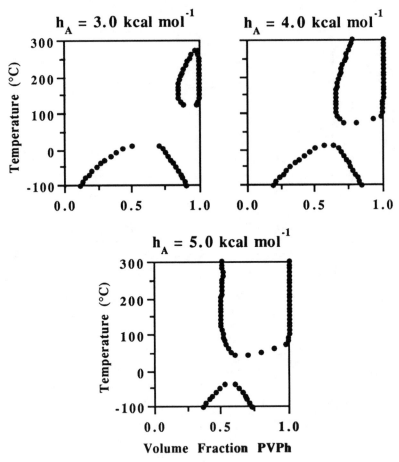

PVPh - PBMA Blends: Effect of Errors in h_A

FIGURE 7.31

Returning to the **Association Information** window (figure 7.23) we restore the value of the **A-B Association** equilibrium to 37.1 and replace the values of the enthalpy of hydrogen bond formation with, in turn, values of 5.0, 4.0 and 3.0 kcal. mol^{-1} and **Calculate**. The results are illustrated in figure 7.31. The effect of a 20% error in h_A on the calculated phase diagram is significant, but not earth shattering. We believe that we can determine h_A values to better than 10%, and thus in common with the values of the equilibrium constants, reasonable errors in h_A will not materially effect the major conclusions drawn from the calculation of the phase diagrams.

PVPh - Poly(alkyl acrylate) Blends

Let us now calculate the binodal phase diagrams for the analogous series of PVPh - poly(alkyl acrylate) blends. This is remarkably straightforward since we already have most of the necessary data stored in the computer. It is probably again most convenient to use the **Miscibility Guide** program to produce files of the segment information for a homologous series of poly(alkyl acrylates). Then all we have to do is introduce the segment information of the desired polyacrylate into **Segment A** of the **Segment Information** window, substitute the relevant values[6,10] for K_A (47.5) and h_A (- 4.0 kcal. mole[-1]) in the **Association Information** window run **Calculate**. Other parameters are the same as for the PVPh - poly(alkyl methacrylate) blends. The results of such an exercise are shown in figure 7.32, which indicate that blends of PVPh with poly(methyl acrylate) (PMA) (and its isomorphous equivalent poly(vinyl acetate) (PVAc)), poly(ethyl acrylate) (PEA), poly(n-propyl acrylate) (PPA) and poly(n-butyl acrylate) (PBA) are predicted to be thermodynamically stable throughout this entire temperature range except for a small region of immiscibility below –50°C. Results obtained for the PVPh - poly(n-pentyl acrylate) PPeA and poly(n-hexyl acrylate) (PHA) predict the emergence of a closed immiscibility loop at high temperatures and high PVPh compositions, as well as an increase in temperature of the lower phase boundary compared to that of the PMA, PEA, PPA and PBA blends. For PVPh blends with higher polyacrylate homologues, a larger two-phase region is predicted, resulting from the merger of the upper and lower phase boundaries, with the breadth of the two-phase region becoming progressively larger. Again, the trend calculated here - a decrease in miscibility with increasing size of the poly(n-alkyl acrylate) pendant group and immiscibility occurring after PPeA - is identical to that predicted by the **Miscibility Guide** (Chapter 2, figure 2.63) and in good agreement with experimental observations[6].

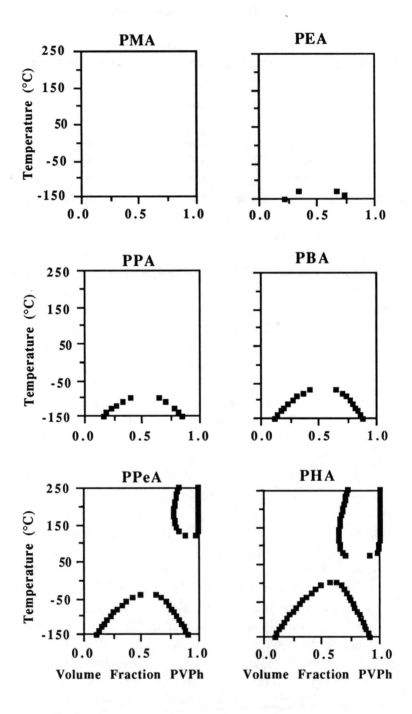

FIGURE 7.32

PVPh - Polyester Blends

Finally, we can repeat the process with a series of n-alkyl polyesters. We simply input the segment information of the desired polyester into **Segment A** of the **Segment Information** window and **Calculate**. The results are summarized in figure 7.33. The polyesters are defined as PnL which denote polylactones containing n methylene groups in the chemical repeat; i.e. $-(CH_2)_n-COO-$. Familar polylactones are also identified as poly(propiolactone) (PPL), poly(valerolactone) (PVL) and poly(caprolactone) (PCL). The results predict that amorphous blends of PVPh with polylactones containing between approximately 2 and 7 methylene groups should be thermodynamically stable throughout this entire temperature range except for a small region of immiscibility below –50°C. Calculations for the PVPh - P8L system predict the emergence of a closed immiscibility loop at high temperatures and high PVPh compositions and the system is on the margins of miscibility. For P10L and higher homologue blends, a larger two-phase region is predicted, resulting from the merger of the upper and lower phase boundaries, with the breadth of the two-phase region becoming progressively larger. At the time of writing, we know that PPL and PCL are miscible with PVPh in the amorphous state. We will have to wait to see if we have accurately predicted the phase behavior for the higher homologues.

PVPh - Polyether and Poly(vinyl pyridine) Blends

As we have discussed in Chapter 5, in specific cases, e.g. the polyurethane and polyamide - polyether blend systems, equilibrium constants describing both self-association and inter-association can be directly determined using infrared spectroscopy to obtain a quantitative analysis of the fraction of hydrogen bonded carbonyl groups in the pure self-associated polymer and blends, respectively.

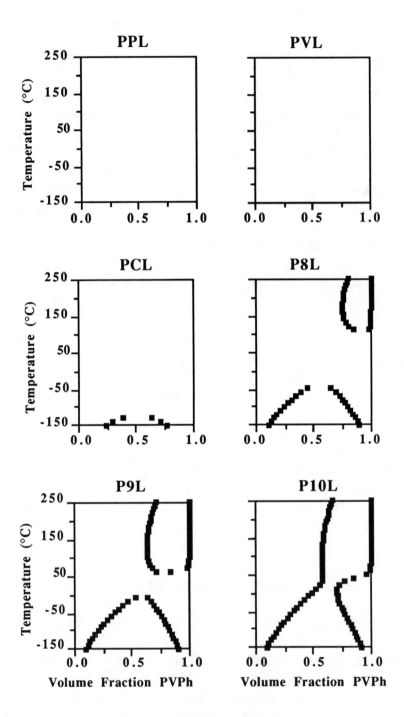

FIGURE 7.33

For strongly self-associating polymers containing hydroxyl, amine or similar groups, however, it is not presently feasible to measure the self-association equilibrium constants using samples in the solid state directly from the mid infrared region of the spectrum. Reiterating, K_B and K_2, for PVPh were obtained *indirectly* from the results of solution studies of the model compound phenol by scaling to the polymer repeat after correcting for molar volume differences, while the inter-association equilibrium constant for blends of PVPh with carbonyl containing polymers such as polyacrylates, polymethacrylates, polyesters and polyacetates etc., K_A, was measured *directly* from a quantitative analysis of the fraction of hydrogen bonded carbonyl groups. For the PVPh - polyether and poly(vinyl pyridine) (PVPy) blends considered now, both the equilibrium constants describing self-association and inter-association are scaled from literature values of low molecular weight analogues[11,12]. The results compare favorably with experimental observations and support the general concept of the transferability of equilibrium constants determined from appropriate model compounds to analogous polymeric blend systems (for polymers that are sufficiently flexible).

The self-association parameters of PVPh are independent of the second blend component and consequently those employed previously for blends of PVPh with polymethacrylates and the like are directly applicable to the PVPh - polyether system. Conversely, the inter-association parameters, K_A and h_A, for the PVPh - polyether (and PVPh - PVPy) blend systems are expected to differ significantly from those of the PVPh - polyacrylate (methacrylate etc.) systems due to the dissimilar nature of the respective hydrogen bonding interactions. As we have mentioned previously, K_A for the PVPh - polymethacrylate systems were obtained from a least-squares fit of the theoretical model to the experimentally determined fractions of hydrogen bonded carbonyl groups present in miscible blends as a function of composition. Unfortunately, this procedure is not applicable to the PVPh - polyether (or PVPy) blend systems.

Unlike the carbonyl stretching vibration, the C-O-C stretching vibration, for example, is not a localized mode and is conformationally sensitive, which precludes quantitative analysis of the fraction of hydrogen bonded ether groups in the blend.

Nevertheless, as in the case of the self-association parameters of PVPh, values of K_A and h_A for PVPh - polyether blends may be estimated from inter-association parameters determined from infrared studies of low molecular weight analogues. In particular, Powell and West[11] have investigated the thermodynamics of phenol - diethyl ether mixtures in carbon tetrachloride solutions using a near infrared spectrophotometric method and obtained a value of - 5.4 kcal. mole[-1] for h_A and 8.83 l mole[-1] for the concentration equilibrium constant, K_A^c . As we noted in chapter 5, equilibrium constants determined in CCl_4 may be subject to error due to interactions with this solvent. Unfortunately, there are no other data available, so we will use this work, but keep in mind the potential source of error. Conversion of the concentration-based equilibrium constant to the dimensionless parameters defined by the association model (Chapter 5) and correcting for molar volume differences leads to a value for K_A of 88.3. Similarly, values of $K_A = 598$ and $h_A = - 7.0$ kcal. mole[-1] for the inter-association parameters of PVPh - PVPy blends have been obtained from the literature[12].

Returning to the **Phase Calculator** program; select **New Association** via the **Calculate** menu and input the new values of K_A and h_A as illustrated in figure 7.34.

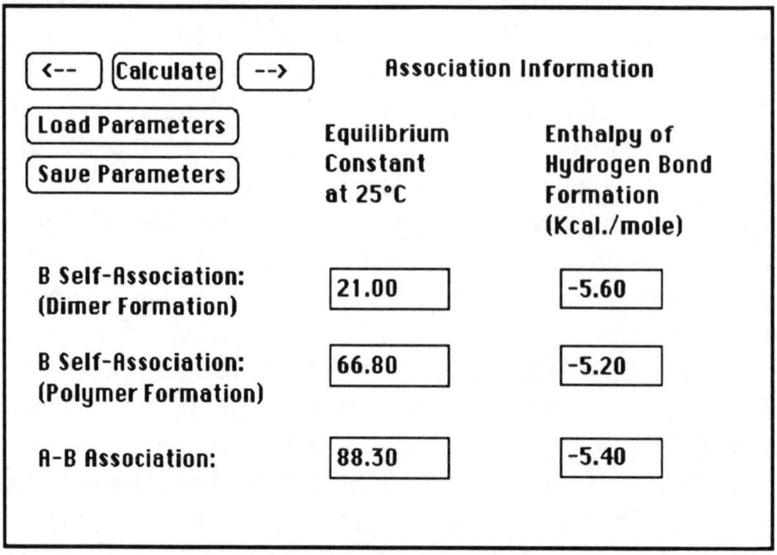

<-- Calculate -->	Association Information	
Load Parameters	Equilibrium Constant at 25°C	Enthalpy of Hydrogen Bond Formation (Kcal./mole)
Save Parameters		
B Self-Association: (Dimer Formation)	21.00	-5.60
B Self-Association: (Polymer Formation)	66.80	-5.20
A-B Association:	88.30	-5.40

FIGURE 7.34

Selecting the <-- arrow produces the **Segment Information** window shown earlier in figure 7.22 and we now introduce the segment information of desired polyether and PVPh into **Segment A** and **Segment B**, respectively, from the segment files. The results of the binodal phase diagram calculations are shown in figure 7.35 for PVPh blends with PVME, poly(vinyl ethyl ether) (PVEE), poly(vinyl butyl ether) (PVBE) and poly(vinyl hexyl ether) (PVHE).

A single phase region only was calculated for PVME and PVEE blends with PVPh, implying that are theoretically miscible throughout the entire temperature range considered.

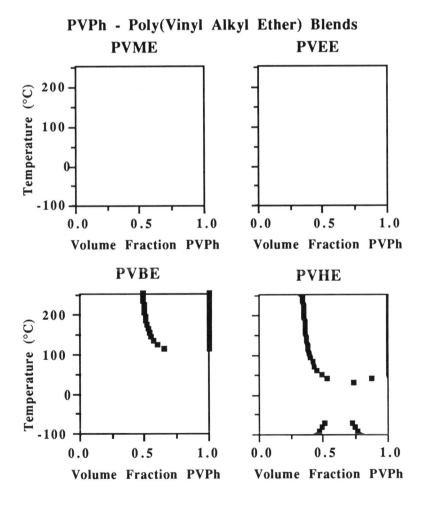

FIGURE 7.35

For blends of PVPh with PVBE and PVHE, a two phase region is calculated for blends rich in PVPh, with an LCST in the accessible range of temperature . Further addition of CH_2 groups to the side group in the poly(vinyl alkyl ether) repeat just serves to enlarge the two phase region. In summary, the results of these calculations indicate that PVPh blends should be miscible with PVME and PVEE; on the edge of miscibility with PVBE, but immiscible with PVHE and higher homologues in the amorphous state at ambient temperature. These predicted trends are again in good agreement with experimental observations.

In considering the ramifications of these results two factors are particularly important. First, as we increase the number of CH_2 groups in the repeat of the polyether the ratio of the molar volumes, $r = V_A / V_B$, increases, since V_B is constant. All other things being equal this is unfavorable to mixing, since it results in a decrease in the contribution from the $\Delta G_H/RT$ term (eq. 1). Second, the solubility parameter of the polyether decreases in magnitude with increasing number of CH_2 groups. Accordingly, the magnitude of χ increases as the difference between the PVPh and polyether solubility parameters widens. This is also unfavorable to mixing as the contribution from the $\chi\Phi_A\Phi_B$ term increases.

We will not spend too much time on PVPh blends with poly(2-vinyl pyridine) (PVPy) since the result is rather obvious. As we mentioned in the **Miscibility Guide** (Chapter 2, page 146) the interaction between the phenolic hydroxyl and pyridine nitrogen groups is much stronger than that occurring between PVPh and the ether oxygens or ester carbonyls. Here we have a polymer blend of two homopolymers in which there are very strong favorable intermolecular interactions but where the calculated non-hydrogen bonded solubility parameters are almost identical ($\delta = 10.6$ and 10.9 (cal. cm^{-3})$^{0.5}$, respectively). Substituting the new values of K_A and h_A (598 and - 7.0 kcal. mole^{-1} , respectively) in the **New Association** window (figure 7.34) and now introducing the segment information of PVPy into **Segment A**, the binodal calculation leads to the (terribly boring!) phase diagram shown in figure 7.36.

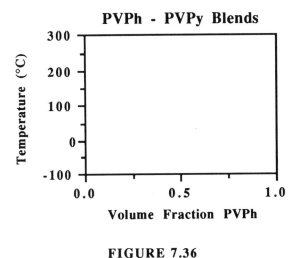

FIGURE 7.36

In a nutshell, PVPh - PVPy blends are predicted to be completely miscible throughout the entire temperature range considered, which is consistent with the experimentally observed strong polymer complex formed between these two polymers[14,15].

Poly(Methacrylic Acid) Blends

Introduction

The final set of homopolymer blends that we will consider are those based upon the carboxylic acid containing polymers such as poly(methacrylic) acid (PMAA). However, we must emphasize that the following results are *preliminary and subject to revision* because, at the time of writing, we do not have a significant body of experimental data pertaining to acid containing polymer blends that we can draw upon and make adjustments similar to those described for PVPh blends. Nevertheless, we will describe some calculations here as it allows us to illustrate the effect of self-association in the form of strongly hydrogen bonded cyclic dimers.

PMAA - Polyether Blends

Returning to the **Association Equilibrium Model** window via the **Calculate** menu, we need to select the model depicted by the icon on the right hand side of the window, as highlighted in figure 7.37. As discussed in Chapter 4, page 196, only one self-association equilibrium constant, K_B, which describes the exclusive formation of dimer structures of the type shown below, is usually necessary for carboxylic acid containing polymers.

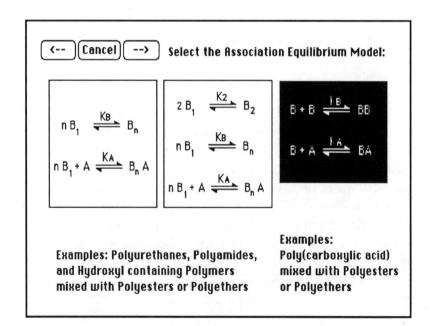

Carboxylic Acid Dimer

Moving to the **Segment Information** window shown in figure 7.38, and we now introduce the segment information of PEO and PMAA into **Segment A** and **Segment B**, respectively, from the segment files.

FIGURE 7.37

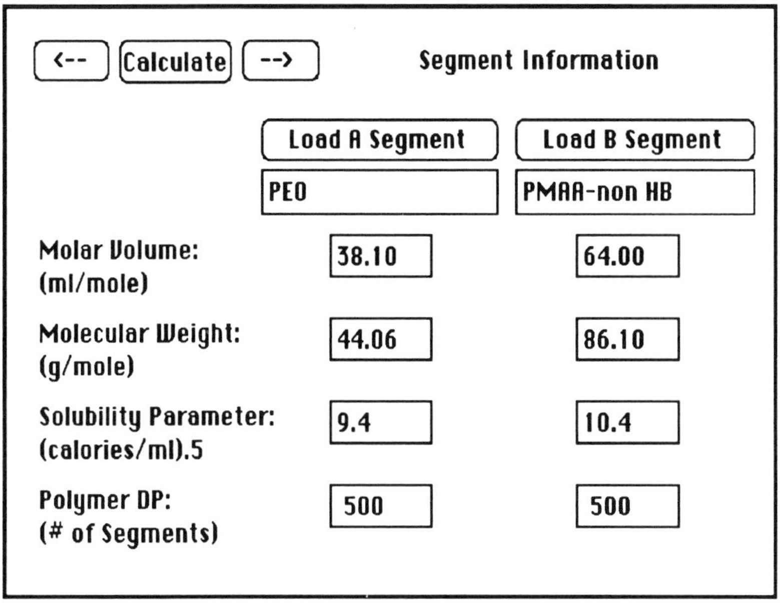

```
 ┌─────────────────────────────────────────────────────────────┐
 │  ┌──────┐ ┌─────────┐ ┌──────┐                               │
 │  │ <-- │ │Calculate│ │ --> │      Segment Information        │
 │  └──────┘ └─────────┘ └──────┘                               │
 │                                                               │
 │           ┌──────────────────┐  ┌──────────────────┐         │
 │           │ Load A Segment   │  │ Load B Segment   │         │
 │           └──────────────────┘  └──────────────────┘         │
 │           │ PEO              │  │ PMAA-non HB      │          │
 │                                                               │
 │  Molar Volume:       ┌─────────┐     ┌─────────┐             │
 │  (ml/mole)           │ 38.10   │     │ 64.00   │             │
 │                      └─────────┘     └─────────┘             │
 │                                                               │
 │  Molecular Weight:   ┌─────────┐     ┌─────────┐             │
 │  (g/mole)            │ 44.06   │     │ 86.10   │             │
 │                      └─────────┘     └─────────┘             │
 │                                                               │
 │  Solubility Parameter: ┌───────┐     ┌─────────┐             │
 │  (calories/ml).5       │ 9.4   │     │ 10.4    │             │
 │                        └───────┘     └─────────┘             │
 │                                                               │
 │  Polymer DP:         ┌─────────┐     ┌─────────┐             │
 │  (# of Segments)     │ 500     │     │ 500     │             │
 │                      └─────────┘     └─────────┘             │
 └─────────────────────────────────────────────────────────────┘
```

FIGURE 7.38

The segment information for PMAA, like that of PVPh, is "adjusted" from that produced from the **Miscibility Guide** for reasons that have been fully discussed in the PVPh case above (pages 406-408). An *initial* estimate of the non-hydrogen bonded solubility parameter of PMAA is obtained from the hypothetical analogue that is missing the carboxylic acid proton, by employing the ester group contributions, both F* and V*, for the carboxylic acid group (careful, this does not imply a carboxylate ion!). Employing the **Miscibility Guide** we introduce into **Segment B** one of each of the following groups, $>CH_2$, $>C<$, $-CH_3$ and $-COO-$, as illustrated in figure 7.39. The **bold** values give the results of the calculations. As we have discussed above, we know that the estimated molar volume of PMAA calculated in this manner is subject to considerable error and underestimated by at least 20%. Accordingly, as an initial estimate we use a value of 64 cm^3 mole^{-1} for the molar volume of PMAA. Now the segment file contains the information shown in figure 7.38. Moving to the **Association Information** window shown in figure 7.40, we now enter the self-association equilibrium constant for PMAA, K_B.

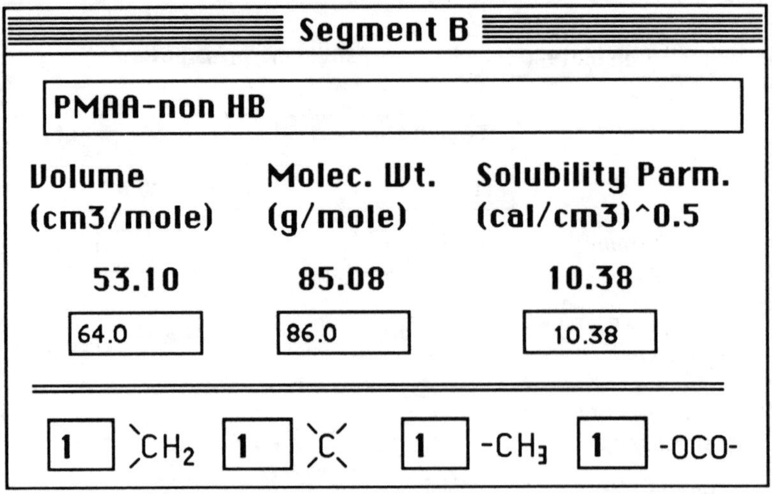

FIGURE 7.39

A seemingly very large value for K_B of 173,000 (at 25°C), which merely reflects the very strong predilection towards the formation of dimer structures, has been estimated from our temperatures studies of the infrared spectra of ethylene-co-methacrylic acid (EMAA) polymers, after appropriate scaling for molar volume differences[16,17]. In addition, from these studies we determined the enthalpy of hydrogen-bonded *dimer* formation to be h_B = - 14.4 kcal. mole[-1], (i.e. - 7.2 kcal. mole[-1] per hydrogen bond). Similarly, the equilibrium constant describing the inter-association of carboxylic acid and ether segments, K_A, may be directly obtained from infrared measurements of the fraction of free carbonyl groups in miscible blends of EMAA - polyether blends[17]. A value of approximately 325 was determined for K_A for the PMAA - PEO blends and the value of h_A was estimated to be about - 6.0 kcal. mole[-1]. The very large disparity between the values of K_A and K_B is quite striking, but one must remember that it is the natural logarithm of the equilibrium constants that is related to changes in free energy per hydrogen bond.

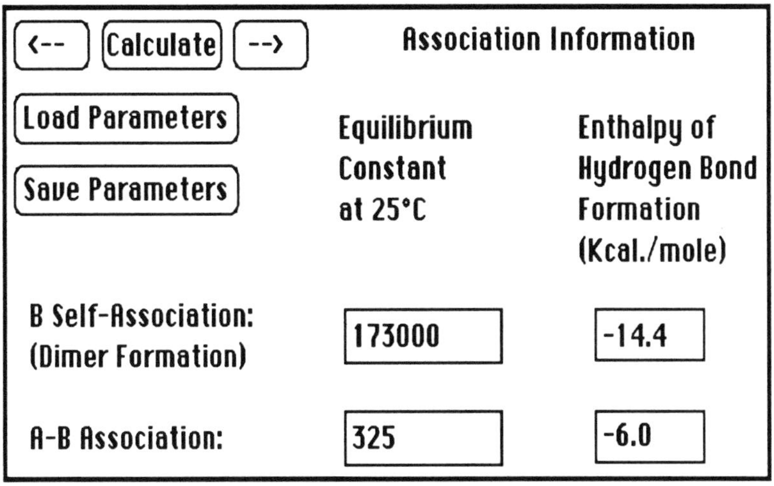

<--	Calculate	-->	Association Information

Load Parameters	Equilibrium Constant at 25°C	Enthalpy of Hydrogen Bond Formation (Kcal./mole)
Save Parameters		
B Self-Association: (Dimer Formation)	173000	-14.4
A-B Association:	325	-6.0

FIGURE 7.40

Moving to the **Calculation and Plot Options** window (figure 7.14) let us now calculate theoretical spinodal phase diagrams for the PMAA - polyether blends for the temperature range between -100°C and 250°C. The results are shown in figure 7.41. As indicated in the figure, blends of PMAA with PEO are predicted to be thermodynamically stable throughout this entire temperature range, except for a small region of immiscibility below –80°C. Results obtained for the PMAA - PTHF and PVME systems predict the emergence of a significant region of instability at temperatures below about 50°C and especially at high PMAA compositions. This region of instability becomes progressively larger as the number of methylene groups in the poly(vinyl n-alkyl ether) increases and PMAA - PVBE blends are predicted to be immiscible throughout the entire temperature range considered. These predictions have still to be tested.

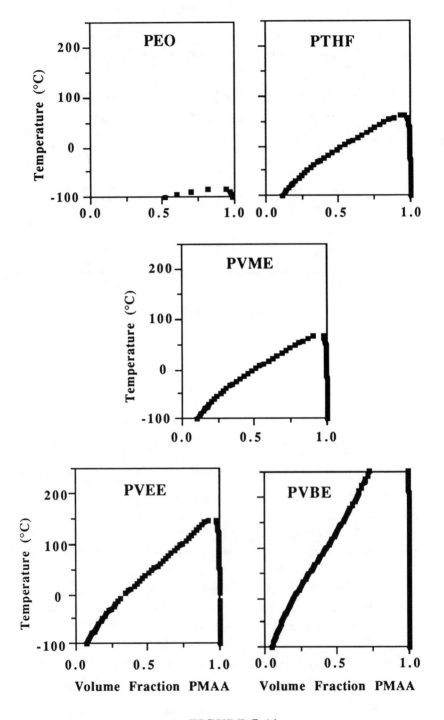

FIGURE 7.41

C. HOMOPOLYMER (Self-Associated)– COPOLYMER (Non Self-Associated) SYSTEMS

In the remainder of this chapter we will be considering the results of the calculation of three different types of miscibility "windows" or "maps" which are schematically represented in the **Calculation Type** window of the **Phase Calculator** program shown in figure 7.42. Miscibility windows are a convenient way of displaying the phase behavior of copolymers as a function of copolymer composition at a given temperature. In this particular section we will concentrate on the specific case of polymer blends where a self-associated homopolymer such as a polyamide, polyurethane, polyphenol or polyacid is mixed with a non self-associating copolymer such as styrene-co-methyl acrylate or ethylene-co-vinyl acetate. The applicable icon is highlighted at the top right hand side of figure 7.42.

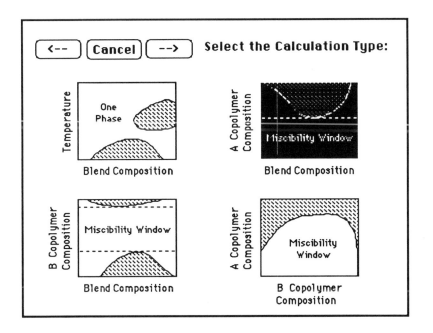

FIGURE 7.42

Poly(Vinyl Phenol) (PVPh) Blends

We should reiterate, as implied in the **Miscibility Guide**, that it is implicit in the way we have developed our model that we make no distinction between the chemical repeat of a homopolymer and the *average* chemical repeat of a random or alternating copolymer. Such is the case for a homologous series of polymethacrylate homopolymers which may be viewed as a "ethylene-co-methyl methacrylate copolymers". We will come to this later, but first we will consider a series of PVPh blends with different copolymers that are designed to illustrate some basic principles and the approach that we take when searching for potential miscible systems.

PVPh Blends with Styrene-co-Methyl Acrylate[18]

This first polymer blend system that we describe was deliberately chosen because the calculated solubility parameters of polystyrene (PS) and poly(methyl acrylate) (PMA) are almost identical (9.5 and 9.6 (cal. cm^{-3})$^{0.5}$, respectively), which infers that the solubility parameters of styrene-co-methyl acrylate (STMA) copolymers are essentially independent of composition. This is illustrated in figure 7.43 which shows a plot of STMA solubility parameters as a function of copolymer composition, together with that of the homopolymer PVPh. Since the magnitude of χ is determined by the difference in the solubility parameters ($\Delta\delta$) of STMA and PVPh and the value of the reference volume, V_B (a *constant* in this case being associated with the invariant PVPh chemical repeat), it follows that χ is also practically independent of the copolymer composition. Accordingly, as we dilute PMA with styrene, we are effectively only reducing the $\Delta G_H/RT$ term in equation 7.1. As we know that PVPh is miscible with PMA at 25°C (figure 7.32), an interesting question arises, "can we calculate the amount of styrene that can be incorporated into PMA before the system becomes immiscible?" Obviously we can, or we would not have posed the question! And it is a remarkably simple procedure, as we hope to demonstrate.

Returning to the **Phase Calculator** program, one first selects the appropriate model for PVPh blends from the

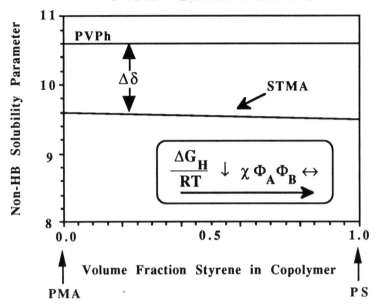

FIGURE 7.43

Association Equilibrium Model window (figure 7.21), followed by the **Segment Information** window, where the relevent information for PMA and PVPh segments are copied from the segment files (figure 7.44). Selecting the --> arrow leads to a new window called the **Copolymer Diluent Information** window (figure 7.45) where the segmental information of the diluent, in this case the polystyrene chemical repeat, may be introduced from the segment files into the **A Diluent** box. (Note that one cannot access the **B Diluent** box as it is the unassociated segment that is to be diluted).

The values of the equilibrium constants, K_2, K_B and K_A are identical to those shown previously for PVPh - polyacrylate type blends illustrated previously in the **Association Information** window (figure 7.23). Now we are ready to calculate the miscibility window and by selecting the --> arrow the **Calculation and Plot Option** window shown in figure 7.46 is displayed.

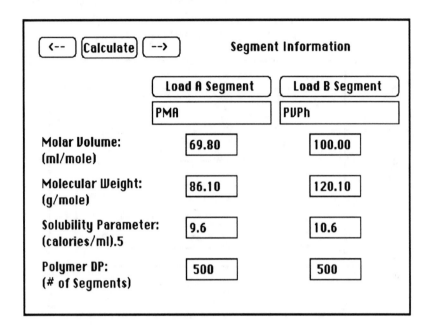

FIGURE 7.44

	Load A Diluent	Load B Diluent
	St	

Molar Volume:
(ml/mole) — `93.90`

Molecular Weight:
(g/mole) — `104.15`

Solubility Parameter:
(calories/ml).5 — `9.5`

<-- | Calculate | --> | Copolymer Diluent Information

FIGURE 7.45

Under the label **Copolymer A Composition** one can choose the units of copolymer composition and enter appropriate values for the **Starting, Ending and Stepping** (the latter being the interval in copolymer composition over which the calculations are performed). In essence, one is calculating the free energy of mixing (equation 7.1) as a function of blend composition and testing whether or not the particular blend is miscible at a specific temperature (points related to the **Spinodal** or **Binodal** may be chosen). Initially, a blend of PVPh with PMA is considered. The process is then repeated, in succession, for STMA copolymers of varying composition (using appropriately calculated parameters for the average chemical repeat). The resulting miscibility window is presented in figure 7.47.

The results of this calculation predict that PMA and STMA copolymers containing up to about 80 weight % styrene should be miscible with PVPh at 25°C, assuming equilibrium conditions. In other words, at concentrations above 80% styrene the contribution from the $\Delta G_H / RT$ term (equation 7.1) is not sufficient to overwhelm the unfavorable $\chi \Phi_A \Phi_B$ term. Experimental results performed in our laboratories substantiate the general form of the calculated miscibility window[18].

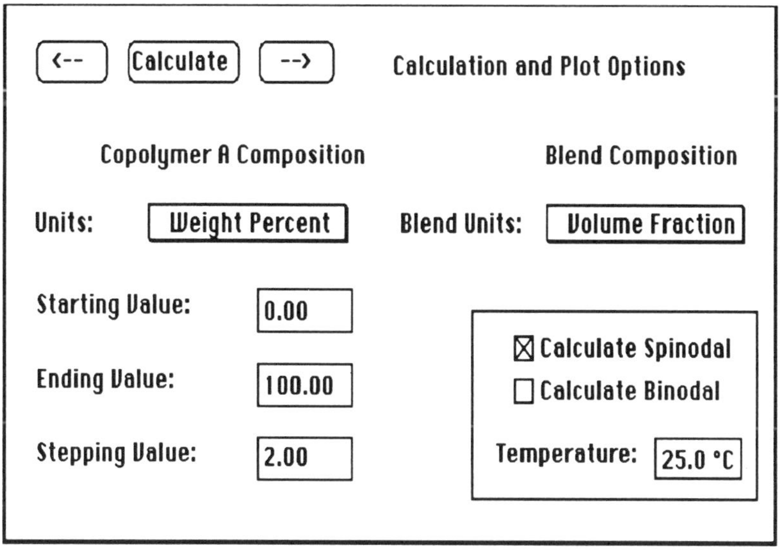

| $\leftarrow$ | Calculate | $\rightarrow$ | Calculation and Plot Options |

Copolymer A Composition   **Blend Composition**

Units: **Weight Percent**   Blend Units: **Volume Fraction**

Starting Value: 0.00

Ending Value: 100.00   ⊠ Calculate Spinodal
☐ Calculate Binodal

Stepping Value: 2.00   Temperature: 25.0 °C

FIGURE 7.46

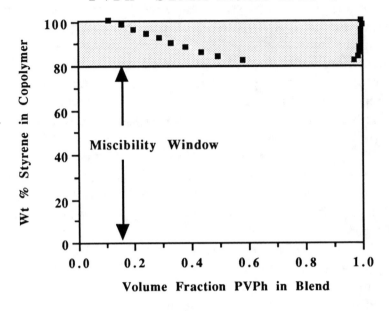

FIGURE 7.47

Figure 7.48 shows a comparison of our experimental results to the miscibility window calculated *at 150°C* (this is readliy performed by returning to the **Calculation and Plot Options** window (figure 7.46) and simply substituting 150°C for 25°C in the appropriate box and recalculating.). We choose 150°C to compare theory with experiment because at this temperature we are above the glass transition temperatures of both pure polymers, so that it is reasonable to assume that equilibrium conditions can be attained. The reader will note, not unexpectedly, that the range of miscibility is more restricted at 150°C compared to that calculated for 25°C. PVPh blends with STMA copolymers containing up to about 65% styrene are now predicted to miscible at 150°C. The black filled circles in figure 7.48 represent blends that have been shown to be immiscible on the basis of experimental FTIR spectroscopic and thermal analysis data, while the corresponding unfilled circles represent single phase systems[18]. Agreement between theoretical prediction and experimental data is good and this was one of the results that we found most encouraging.

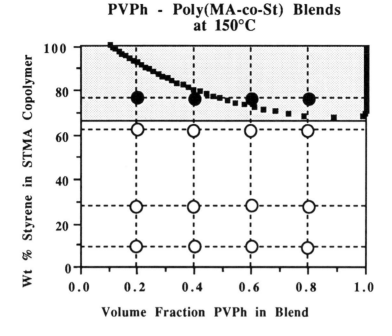

FIGURE 7.48

PVPh Blends with Ethylene-co-Methyl Acrylate (EMA) and Ethylene-co-Vinyl Acetate (EVA)[19]

We have shown that the value of χ (as define by us) for PVPh - STMA blends is essentially constant with copolymer composition. Let us now consider what happens to the miscibility windows when χ varies with copolymer composition. The first example concerns PVPh - EMA (or EVA) blends. (Incidentally, we should mention that the calculated *segment information* for methyl acrylate and vinyl acetate is identical since they are composed of the same chemical groups; i.e. $>CH_2$, $>CH-$, $-COO-$ and $-CH_3$. There are subtle differences in the values of the inter-association equilibrium constants, K_A, but such differences have only a minor effect upon the calculated miscibility window.)

Figure 7.49 shows a plot of the EMA (or EVA) solubility parameter as a function of copolymer composition, together with that of the homopolymer PVPh.

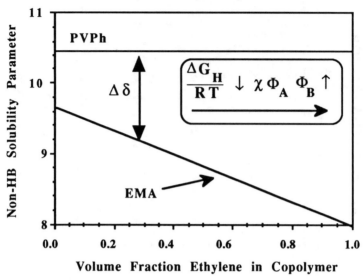

PVPh - EMA (EVA) BLENDS

Volume Fraction Ethylene in Copolymer

FIGURE 7.49

As stated above, the magnitude of χ as a function of copolymer composition is determined by the difference in the solubility parameters ($\Delta\delta$) of the EMA (or EVA) and PVPh . Accordingly, it follows in this case that χ increases with increasing concentration of ethylene in the copolymer. In other words, as we dilute PMA (or PVAc) with ethylene, we are effectively *reduce* the $\Delta G_H/RT$ contribution and *increase* the $\chi\Phi_A\Phi_B$ contribution of equation 7.1. Both these factors are unfavorable in terms of forming miscible mixtures.

Returning to the **Phase Calculator** program, it is a simple matter to replace the segmental information of the diluent styrene (figure 7.45) with that pertaining to ethylene and recalculate. This results in the miscibility window shown in figure 7.50. Comparing figures 7.47 and 7.50, it is evident that the predicted miscibility window for PVPh - EMA blends is, as expected, more restricted than that of the corresponding STMA copolymers (although the use of the more industrially common units of weight % tends to accentuate the differences somewhat), ranging in composition from 0 to approximately 40 weight % ethylene.

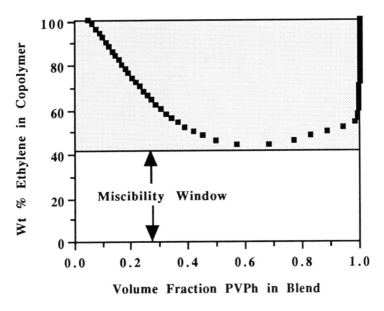

PVPh - EMA (EVA) Blends at 25°C

FIGURE 7.50

This agrees well with experiment as studies performed in our laboratories[10] have demonstrated that PVPh is miscible with PMA and EMA copolymers containing 26 and 41% ethylene, but immiscible with EMA copolymers containing 60 and 70% ethylene.

The calculated miscibility window for PVPh - EVA copolymer blends is very similar to that shown for the analogous EMA blends (figure 7.50) and we will leave the reader to perform the calculations. Note that the only changes necessary in the **Phase Calculator** program is to change the values of K_A and h_A to 47.5 and 4.0 kcal. mol^{-1}, respectively, in the **Association Information** window to reflect the difference for the PVPh - acetate interaction[19].

Again, experimental results obtained in our laboratories have shown that EVA copolymers containing 70 and 100% vinyl acetate, (EVA[70] and PVAc, respectively) are miscible, while those containing 45 and 25% vinyl acetate (EVA[45] and EVA[25]) are multiphased, which agrees nicely with the theoretical miscibility window.

PVPh Blends with Styrene-co-Vinyl Pyridine (STVPy)

Another example of a PVPh blend where the magnitude of χ increases with increasing dilution over most of the range of copolymer composition concerns mixtures with STVPy (figure 7.51). As we have mentioned previously, PVPh and PVPy have, within error, identical calculated non-hydrogen bonded solubility parameters (10.6 and 10.8 (cal. cm^{-3})$^{0.5}$, respectively), which infers a very small value of χ. Compound this with a very strong intermolecular interaction between the PVPh and PVPy segments and it is not surprizing that a polymer complex, which precipitates out of a common solvent, is formed. Hence, dilution of PVPy by copolymerization with styrene leads to the same tendancy as EMA copolymers, i.e. an increasing value of χ, but now this is offset somewhat by the increased interaction strength of inter-association that is reflected in the relative magnitude of K_A.

PVPh - STVPy BLENDS

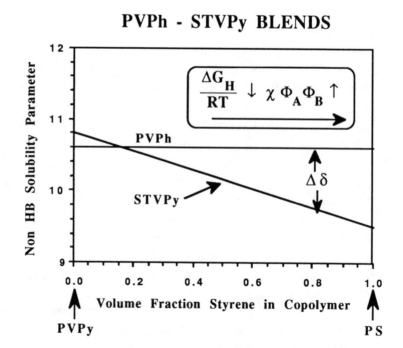

FIGURE 7.51

Returning to the **Phase Calculator** program and before recalculating, we need to substitute the segment information of PVPy into **A Segment** (figure 7.44) and that of PS into **A Diluent** (figure 7.45). In addition, we must change the values of K_A to 598 and h_A to - 7.0 kcal. mole[-1] in the **Association Information** window (figure 7.34). The resultant miscibility window is shown in figure 7.52 which predicts that PVPh is miscible with PVPy and STVPy copolymers containing up to about 83 mole % styrene. There is only very limited experimental data concerning PVPh - STVPy blends, but Meftahi and Frechet[15] have shown that. copolymers containing 50 mole % styrene are miscible with PVPh, while those containing 80 mole % are not, in reasonable agreement with these predictions (particularly considering that equilibrium constants were obtained from phenol[9] and phenol/pyridine mixtures[12].

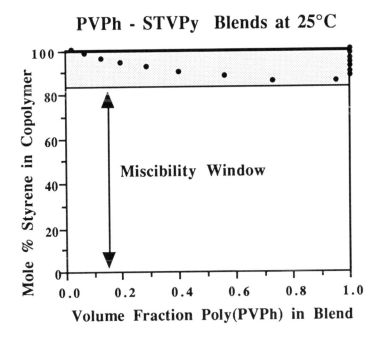

FIGURE 7.52

PVPh Blends with a Homologous Series of Poly(n-alkyl Methacrylates) (PAMA) or an Equivalent Ethylene-co-Methyl Methacrylate (EMMA) Copolymer

We have discussed in some detail in the previous section the calculation of temperature / blend composition phase diagrams of PVPh blends with a systematic series of poly(n-alkyl methacrylates) (PAMA) and the results of these calculations are presented in figure 7.26. For those just "scouting" for miscible systems, however, it is convenient to view this type of information in summary form using miscibility windows at a particular temperature. A plot of PAMA solubility parameters as a function of the number of methylenes in the side group has been presented in the Chapter 2, figure 2.63, and it shows the variation of the solubility parameter difference ($\Delta\delta$) between PAMA homopolymers and PVPh. This is another case where χ increases with increasing concentration of the diluent. In other words, as we dilute PMMA with methylenes in the side groups, we effectively *reduce* the $\Delta G_H/RT$ and *increase* the $\chi\Phi_A\Phi_B$ terms of equation 7.1 in a similar manner to that depicted in figure 7.47. Both these trends are unfavorable to mixing.

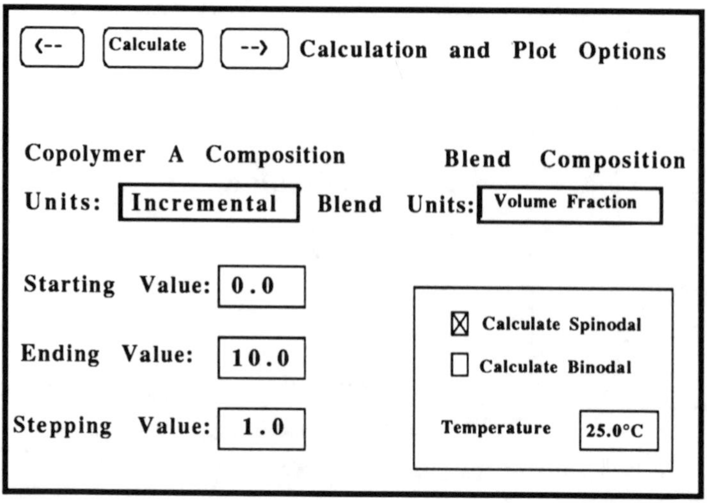

(--	Calculate	-->	Calculation and Plot Options

Copolymer A Composition   Blend Composition

Units: | Incremental | Blend Units: | Volume Fraction |

Starting Value: | 0.0 |

Ending Value: | 10.0 | ☒ Calculate Spinodal

☐ Calculate Binodal

Stepping Value: | 1.0 | Temperature | 25.0°C |

FIGURE 7.53

Returning to the **Phase Calculator** program, replace the segmental information of **A Segment** (figure 7.44) with that of PMMA and the **A Diluent** (figure 7.45) with methylene. Now move forward to the **Calculation and Plot Option** window (figure 7.53), choose **Incremental** from the **Units** box below the **Copolymer A Composition** label and using appropriate values for the **Starting, Ending** and **Stepping,** recalculate. This results in the miscibility window shown in figure 7.54.

The conclusions that one can draw from the miscibility window of the PVPh - PAMA blends calculated at 25°C shown in figure 7.54 are pleasingly consistent with those we obtained previously from the conventional phase diagrams shown in figure 7.26. Assuming equilibrium conditions can be attained, PMMA, PEMA, PPMA and PBMA are all predicted to be miscible at ambient temperature, while the polymers containing greater than 4 methylenes in the side group are predicted to be immiscible, in good agreement with experimental observations.

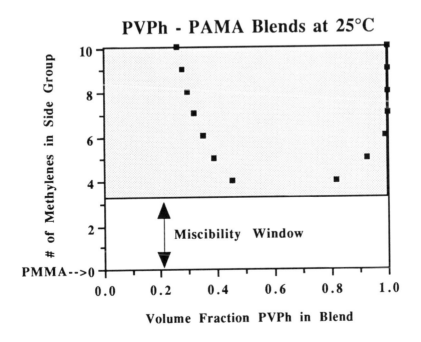

FIGURE 7.54

Since we are dealing primarily with blends containing one unassociated copolymer in this section, it is important to recognize that we make no distinction between, for example, PVPh blends with the "homopolymer" poly(n-propyl methacrylate) (PPMA) and its isomorphous equivalent, an *average* repeat of an equimolar ethylene-co-methyl methacrylate (EMMA) random "copolymer".

PPMA   EMMA (1:1 molar)

Thus we can calculate an equivalent miscibility window to that shown in figure 7.54 for PVPh - EMMA copolymer blends by simply returning to the **Calculation and Plot Option** window (figure 7.53), choose **Weight Percent** from the **Units** box below the **Copolymer A Composition** label and using appropriate values for the **Starting, Ending** and **Stepping**, recalculate. The results of the calculation yield the miscibility window shown in figure 7.55, which predicts that EMMA random copolymers containing up to approximately 30 weight % ethylene should be miscible. Note also that we have identified on the right hand side of figure 7.55 the PAMA "homopolymers" as if they were EMMA copolymers of equivalent ethylene concentration (weight %). While the reader may find this interchangability of homo- and random copolymers somewhat surprizing, experimental evidence to date supports this concept. This evidence will be considered in more detail below, where we compare experimental results obtained from various copolymer blends involving the alkyl methacrylates and EMMA copolymers to theoretically predicted miscibility maps.

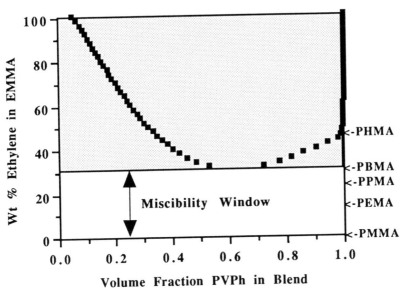

FIGURE 7.55

Amorphous Polyurethane (APU) Blends

APU Blends with Homologous Series of Poly(n-Alkyl Ethers) and Poly(Vinyl n-Alkyl Ethers)

At the beginning of this chapter we discussed the calculation of the free energy of mixing and corresponding conventional temperature / blend composition phase diagrams for blends of an amorphous polyurethane (APU - for schematic representation see page 134) with PEO and PVME. The results of these calculations were presented in figures 7.15 and 7.16. In addition, we have presented (Chapter 2, figure 2.60) a plot of the relationship of the solubility parameter of APU to that of homologous series of poly(n-alkyl ethers). The solubility parameter difference ($\Delta\delta$) increases with increasing number of methylene groups and this is another case where χ increases with increasing concentration of the diluent.

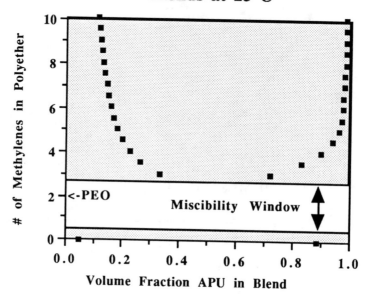

FIGURE 7.56

Again, for scouting purposes, it is convenent to calculate the appropriate miscibility window. Returning to the **Phase Calculator** program, first choose the correct **Association Equilibrium Model** (figure 7.2); move to the **Segmental Information** window and replace the segmental information of **A Segment** with that of an ether oxygen; the **B Segment** (figure 7.44) with APU and the **A Diluent** (figure 7.45) with methylene. Now move forward to the **Association Information** window and introduce the appropriate equilibrium and enthalpy values ($K_B = 205$; $K_A = 9.8$; $h_A = h_B = 5.0$ kcal. mole^{-1} - figure 7.4). Finally, move to the **Calculation and Plot Option** window (figure 7.53), choose **Incremental** from the **Units** box below the **Copolymer A Composition** label and using appropriate values for the **Starting, Ending** and **Stepping,** recalculate. This results in the miscibility window shown in figure 7.56.

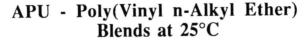

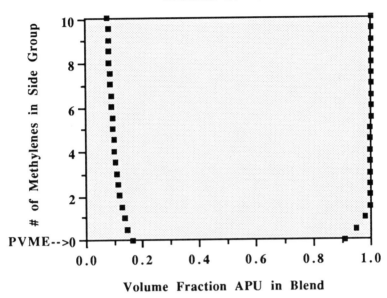

FIGURE 7.57

From the calculated miscibility window of the APU - poly(n-alkyl ethers) blends shown in figure 7.56 we can see that only poly(methylene oxide) (polyformaldehyde) and poly(ethylene oxide) (PEO) are predicted to be miscible at 25°C. This is consistent with experimental studies performed in our laboratories[2]. The corresponding miscibility window for the APU - poly(vinyl n-alkyl ethers) blends, obtained by simply replacing the ether in the **Segmental Information** window with the PVME segment and recalculating, is given in figure 7.57 and predicts that PVME and all higher homologues are immiscible with APU at 25°C, which is again bourn out by experiment.

Poly(Methacrylic Acid) (PMMA) Blends

PMAA Blends with Butadiene-co-Vinyl Pyridine (BDVPy)

Just for fun, before completing this section, let's speculate and calculate a miscibility window for a self-associated homopolymer - unassociated copolymer blend system for which we have only very limited information - one for which we have no experimental data with which to compare the results and where we might get "nailed" if someone bothers to test the predictions! We know that PMAA has a tremendous affinity for, and forms insoluble polymer complexes with, PVPy. An interseting question might be, "how much butadiene (BD) can we incorporate into the PVPy polymer chain before a blend with PMAA becomes immiscible ?" We choose to consider BD because its solubility parameter is low (8.1 (cal. cm^{-3})$^{0.5}$) and incorporation into PVPy significantly reduces the average solubility parameter.

Returning to the **Phase Calculator** program, first choose the acid dimer icon in the **Association Equilibrium Model** (figure 7.37); move to the **Segmental Information** window and replace the segmental information of **A Segment** with that of PVPy; the **B Segment** (figure 7.44) with PMAA and the **A Diluent** (figure 7.45) with butadiene. Now move forward to the **Association Information** window (figure 7.40) and introduce the appropriate equilibrium constants (K_B = 173000 and K_A = 3000; the latter, attributed to the very strong acid - pyridine interaction and which is an order of magnitude greater than that of the analogous acid - ether interaction, was estimated from the ratio of K_A to K_B that we determined in our studies of ethylene-co-methacrylic acid copolymer blends with PVPy.) Finally, move to the **Calculation and Plot Option** window (figure 7.51), choose **Weight Fraction** from the **Units** box below the **Copolymer A Composition** label and using appropriate values for the **Starting**, **Ending** and **Stepping**, recalculate. This results in the miscibility window shown in figure 7.56.

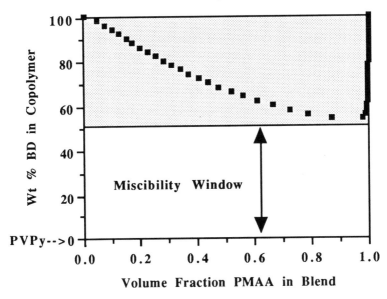

FIGURE 7.58

The results suggest that PMAA should be miscible with BDVPy copolymers containing up to about 50% BD. This is not a particularly earth shattering result, but as we will see later, if we take it a step further and simulateously dilute the PMAA polymer with say ethylene, things get very interesting.

D. HOMOPOLYMER (Non Self-Associated)– COPOLYMER (Self-Associated) SYSTEMS

Preamble

In this section we will be considering what happens when we reverse the procedure and dilute the polymer that is strongly self-associated. In Chapter 5 we have shown infrared spectra (figure 5.23, page 278) and discussed the effect of co-polymerization of styrene with vinyl phenol using the infrared data from styrene-co-vinyl phenol (STVPh) polymers.

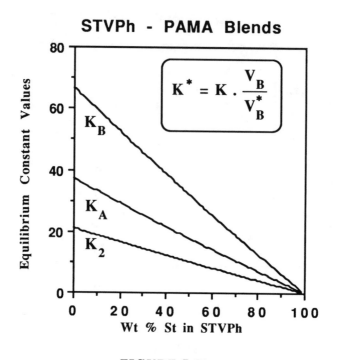

FIGURE 7.59

Copolymerization with an "inert" monomer such as ethylene, styrene, butadiene etc. effectively dilutes the system. Using PVPh - poly(alkyl methacrylate) blends as an example, we can readily determine the equilibrium constants, K_2 and K_B, for hypothetical STVPh copolymers of varying composition by computing an average chemical repeat that contains one VPh unit and scaling according to the ratio of the molar volumes (page 296). As we have mentioned before, the magnitudes of K_2 and K_B vary with the composition of the STVPh copolymer, but the ratio of these two equilibrium constants to K_A is theoretically predicted to be constant. This is because all the equilibrium constants are defined with respect to the molar volume of the self-associating *interacting unit* used to define the lattice cell size. Figure 7.59 shows graphically the magnitude of K_2, K_B and K_A as a function of concentration of styrene in STVPh copolymers. The adjustment for the different size of the chemical repeat units enters through the factor r = V_A / V_B. The value of χ for the STVPh blends may be determined as before using the relationship $\chi = (V_B / RT) [\delta_A - \overline{\delta_B}]^2$, where the molar volume of the STVPh repeat is employed as the reference volume, V_B. It is important to recognize that the magnitude of χ *as defined here* is determined by two factors: (i) the size of the reference volume V_B, which increases with increasing styrene content in the STVPh copolymer, and (ii) the difference in the solubility parameter of STVPh, $\overline{\delta_B}$, and its value relative to δ_A, the solubility parameter of the unassociated homopolymer. [Naturally, as the size of the reference volume B increases, the degree of polymerization of the self-associating copolymer *decreases* and this must be accounted for.]

Vinyl Phenol Copolymer Blends

Styrene-co-Vinyl Phenol (STVPh) Blends with Poly(Alkyl Methacrylates) (PAMA)[17,18]

One of the more intriguing questions posed over the last decade by those working in the general field of polymer blends is, "how many favorable intermolecular interactions are required in order to achieve miscibility ?" To give a relevent example, poly(methyl methacrylate) (PMMA) is *immiscible* with polystyrene (PS), but *miscible* with poly(vinyl phenol) (PVPh). An obvious question comes to mind, "how many vinyl phenol (VPh) units would we need to incorporate into PS to render it miscible with PMMA ?" Chen and Morawetz have reported fluorescence studies of blends of polymethacrylates with styrene copolymers containing hydrogen bond donors[20]. A number of findings of this work are germane here. Surprizingly, it was determined that only 2% of VPh in PS was required to form a miscible blend with poly(butyl methacrylate) (PBMA), while even less (approximately 1%) was necessary in the case of PMMA and poly(ethyl methacrylate) (PEMA). The reader may recall that the qualitative **Miscibility Guide** predicts this rather well (see chapter 2, figure 2.65), but that was after the fact. Since we have all the data necessary to calculate miscibility windows for the STVPh - PAMA system it is now a trivial task to test whether or not our quantiative theoretical approach predicts this behavior.

However, before calculating the miscibility windows for STVPh - PAMA blends, let us spend a little time reviewing the simple underlying principles that point the direction torwards miscible blends of this type. The calculated solubility parameters of PVPh and PS are 10.6 and 9.5 (cal. cm^{-3})$^{0.5}$, respectively, which sets the limits of the solubility parameter range of STVPh copolymers. This is illustrated in figure 7.60 which shows a plot of the STVPh solubility parameter as a function of copolymer composition together with those of the homopolymers, PMMA and poly(n-hexyl methacrylate) PHMA. As we have mentioned above, the magnitude of χ is determined by the difference in the solubility parameters ($\Delta\delta$) of STVPh and PAMA and the value of the reference volume, V_B (a *variable* in this case because it is related to the STVPh *average* chemical repeat). The value of χ as a function of STVPh copolymer composition for blends with PMMA, PBMA and PHMA is illustrated in figure 7.61. Thus, as we dilute PVPh with styrene we are still effectively reducing the $\Delta G_H/RT$ term in equation 7.1, but the magnitude of the $\chi\Phi_A\Phi_B$ term varies not only with $\Delta\delta$, but also with V_B.

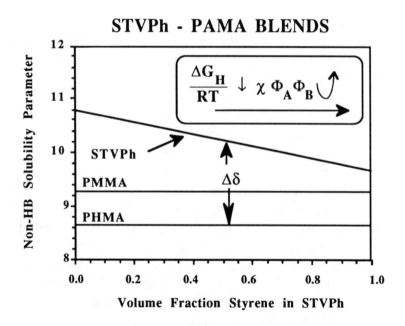

FIGURE 7.60

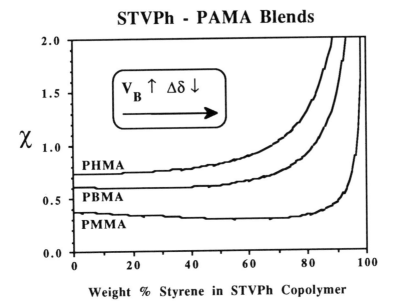

FIGURE 7.61

This is because we choose to define an average repeat unit for the self-associating copolymer, for computational convenience. This "artificial" increase in χ is, of course, balanced by the corresponding decrease in the degree of polymerization of the B copolymer chain, defined in terms of the same reference volume, and the program also adjusts this automatically using the initially defined degree of polymerization of segment B as a starting point.

Let us return to the **Phase Calculator** program and select the specific program describing blends where a self-associated copolymer such as a STVPh, ethylene-co-methacrylic acid etc. is mixed with an essentially unassociated homopolymer such as PMMA, PEO, PVAc etc. The applicable icon is highlighted at the bottom left hand side of figure 7.62. Now select the appropriate model for PVPh blends from the **Association Equilibrium Model** window (figure 7.21), followed by the **Segment Information** window (figure 7.44), where the relevent information for the particular poly(alkyl methacrylate) and PVPh segments are read from the segment files.

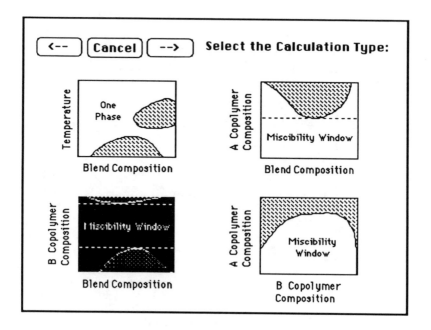

FIGURE 7.62

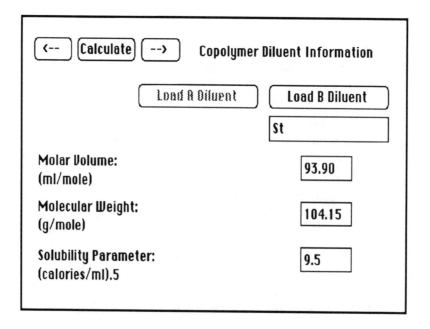

FIGURE 7.63

Next, select the **Copolymer Diluent Information** window (figure 7.63) where the segmental information of the diluent, the polystyrene chemical repeat, is introduced from the segment files into the **B Diluent** box. (Note that this time one cannot access the **A Diluent** box as it is the self-associated segment that is to be diluted.) The values of the equilibrium constants, K_2, K_B and K_A are simply those given previously for PVPh - polyester type blends (figure 7.23) illustrated in the **Association Information** window (values for STVPh copolymers as a function of composition are calculated automatically within the computer program). Now we are ready to calculate the miscibility window by selecting the **Calculation and Plot Option** window (similar to that shown in figure 7.46). One can choose a **Spinodal** or **Binodal** calculation by entering appropriate values for **Starting, Ending and Stepping**, and under the label **Copolymer B Composition**, the units of copolymer composition.

Let us first consider the results of the calculations for the STVPh blends with PMMA and PBMA which are shown in figures 7.64 and 7.65, respectively. The results of these calculations predict that PMMA and PBMA should be miscible with STVPh copolymers containing up to approximately 98 and 96 weight % styrene[21].

This outstanding agreement with the experimental studies of Chen and Morawetz[20] is most gratifying and encouraged us to extend our theoretical and experimental studies to other STVPh - PAMA blends. Figures 7.66 through 7.69 show calculated miscibility windows for STVPh blends with poly(n-hexyl methacrylate) (PHMA), poly(n-octyl methacrylate) (POMA), poly(n-decyl methacrylate) (PDMA) and poly(n-dodecyl methacrylate) (PDoDMA). As we discussed above, PVPh is immiscible with these particular poly(alkyl methacrylates). STVPh copolymers are miscible with the first three of these over certain composition ranges, but the fairly wide miscibility window obtained for STVPh - PHMA blends narrows as the methacrylate series is ascended, so that STVPh - PDoDMA blends (and higher homologues) are predicted to be immiscible throughout the entire composition range.

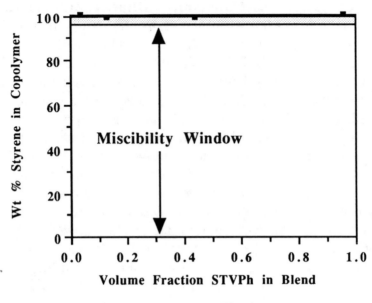

FIGURE 7.64

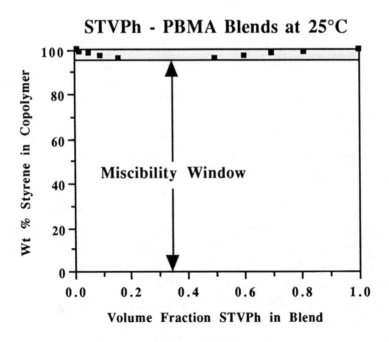

FIGURE 7.65

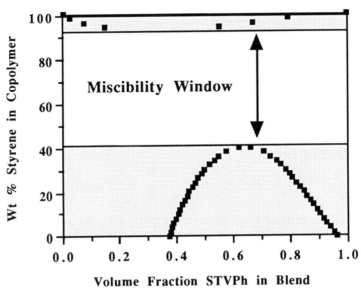

FIGURE 7.66

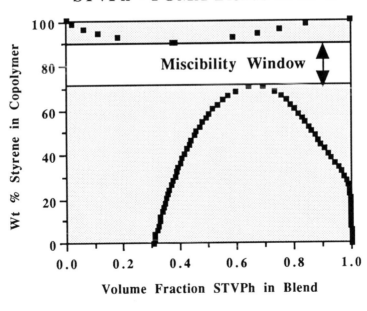

FIGURE 7.67

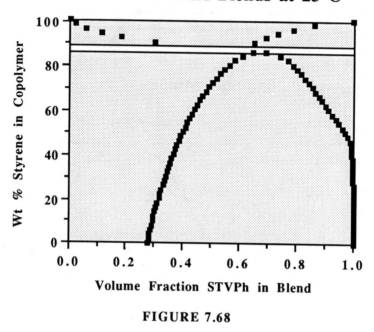

FIGURE 7.68

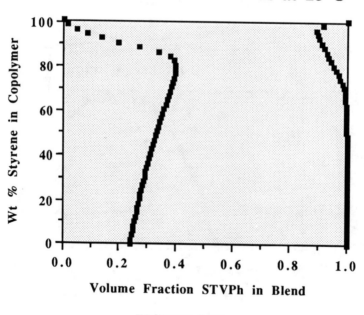

FIGURE 7.69

We have recently synthesized six random copolymers of STVPh containing 75, 43, 25, 15, 8 and 2 weight % vinyl phenol and experimental studies were performed on blends of these copolymers with PBMA, PHMA, PDMA and PDoDMA[20]. Analyses of the fraction of hydrogen bonded carbonyl and hydroxyl groups were obtained by FTIR spectroscopy and the data compared to the theoretical predicted values for miscible systems.PVPh and the STVPh copolymers containing 75, 43, 15, 25 and 8% vinyl phenol were found to be miscible with PBMA, while the corresponding copolymer containing 2% vinyl phenol was found to be immiscible. For the PHMA blends, PVPh and the two STVPh copolymers containing 75 and 2% vinyl phenol were found to be immiscible with PHMA, while the corresponding copolymers containing 43, 25, 15 and 8% vinyl phenol were found to be miscible. In the case of POMA blends, only the STVPh copolymer containing 15% vinyl phenol was found to be miscible and we have found no miscible blend of STVPh with PDMA. These experimental results are in excellent agreement with the theoretical predictions (see figures 7.66 through 7.69) and lend support to the general validity of our association model approach.

Butadiene-co-Vinyl Phenol (BDVPh) Blends with Poly(Alkyl Methacrylates) (PAMA)

Encouraged by our success in predicting miscibility windows for STVPh - PAMA blends, it is time to abandon caution to the wind and yet again go out on a limb! Let's say, for example, that you wished to make a miscible blend which included as one component PDoDMA (God knows why you would want to, but we're academics and fully prepared to ignore issues of practical utility). We have seen that all compositions of STVPh copolymers are *immiscible* with PDoDMA. Now, PDoDMA has a low solubility parameter value, 8.3 (cal. cm^{-3})$^{0.5}$, compared to PVPh (10.6), PS (9.5) or any STVPh copolymer that ranges between these latter two values. Accordingly, $\Delta\delta$ is always rather large, which in turn implies a large unfavorable $\chi\Phi_A\Phi_B$ term (equation 7.1), which cannot be overwhelmed by the favorable $\Delta G_H/RT$ contribution.

However, what if we copolymerize vinyl phenol with butadiene instead of styrene, which effectively increases the range of copolymer solubility parameters from 10.6 to 8.1 (cal. cm^{-3})$^{0.5}$?

Now, as we show in figure 7.70, we can come close to and even cross the solubility parameter line of PDoDMA as we dilute PVPh with butadiene. It therefore follows that there will be a range of BDVPh copolymer compositions in blends with PDoDMA where the $\Delta G_H/RT$ contribution will overcome that of the $\chi\Phi_A\Phi_B$ term resulting in a window of miscibility. Returning to the **Phase Calculator** program, replace the segmental information of the **B Diluent** (figure 7.63) with butadiene and recalculate in order the miscibility windows for the series of BDVPh - PAMA blends.

The results are summarized in the miscibility windows shown in figure 7.71. BDVPh copolymers containing from zero to greater than 95 weight % butadiene are predicted to be miscible with PMMA, PEMA, PPMA and PBMA, at ambient temperature. But it is the theoretical miscibility windows for BDVPh blends with PHMA, POMA, PDMA and PDoDMA that are most interesting and should be compared to the corresponding STVPh blend (figures 7.66 - 7.69).

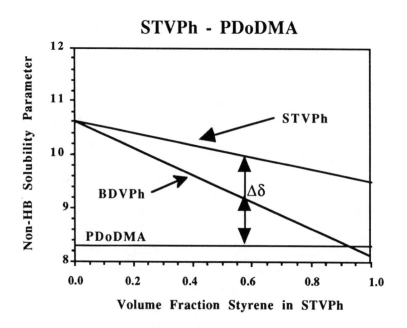

FIGURE 7.70

BDVPh - PAMA Blends at 25°C

PMMA

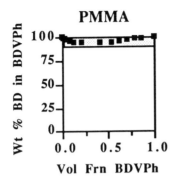

PBMA

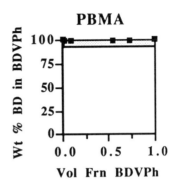

PHMA

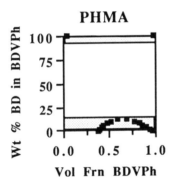

POMA

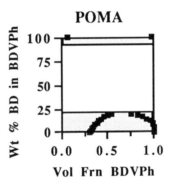

PDMA

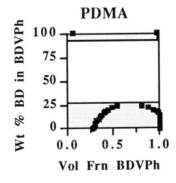

PDoDMA

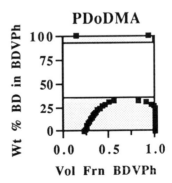

FIGURE 7.71

In marked contrast to the latter, we now predict that there are significant ranges of BDVPh copolymer compositions that are miscible with PDoDMA with windows extending from approximately 15, 20, 25 and 30 for PHMA, POMA, PDMA and PDoDMA, respectively, through to greater than 98% butadiene. To our knowledge, nobody has synthesized BDVPh copolymers, but at the time of writing we are attempting to do just this and test these predictions. [Our early success has given us a possibly foolish confidence in this model, time will tell!]

Polyamide "Copolymer" Blends

Polyisophthalamide (PIPA) Blends with Poly(Ethylene Oxide) (PEO)

We have previously considered (pages 399 to 404) the calculation of conventional temperature / blend composition phase diagrams for mixtures of aromatic / aliphatic polyamides (the nylon n-I series or the corresponding isomorphous MPDn homologues) with PEO. The results of these calculations for MPD5 (3-I), MPD6 (4-I), MPD7 (5-I) and MPD8 (6-I) blends with PEO are presented in figures 7.20. As we have mentioned before, however, for scouting purposes it is more convenient to calculate a miscibility window. The reader will not be surprized by now that we simply view blends of a homologous series of aromatic / aliphatic polyamides with PEO, as being entirely equivalent to a polymer blend system consisting of methylene-co-aromatic amide "copolymers" mixed with a homopolymer, PEO.

Without further ado, let us return to the **Phase Calculator** program; select the appropriate model for nylon blends from the **Association Equilibrium Model** window (figure 7.2); move to the **Segment Information** window (figure 7.44) and replace the segmental information of **A Segment** with that of PEO, the **B Segment** with that of the PIPA chemical repeat (see Chapter 2 page 130) and the **B Diluent** (figure 7.63) with methylene.

Now move forward to the **Association Information** window (figure 7.4) and substitute the values of 168 and 16.8 for the values of K_B and K_A and also put $h_B = h_A = 3.2$ kcal.

mole^{-1}, the value of K_B corresponding to the aromatic amide chemical repeat (the hypothetical nylon MPD2 which contains no methylene groups - see table 7.1). Next move to the **Calculation and Plot Option** window, choose **Incremental** from the **Units** box below the **Copolymer B Composition** label and using appropriate values for the **Starting, Ending** and **Stepping**, recalculate.

This leads to the miscibility window calculated at 70°C (above the T_m of PEO) shown in figure 7.72. In a nutshell, the results predict that aromatic / alipahatic polyamides containing greater than 3 methylene groups in the *specific* repeat (i.e. corresponding to nylon 6-I or MPD8) should be miscible with PEO at 70°C. Experimentally, we have found that nylon 8-I and 10-I are miscible, while nylons 6-I, 4-I and 2-I are immiscible - acceptable agreement. The upper limit of the window is greater than 10 methylene groups and this is difficult to test experimentally.

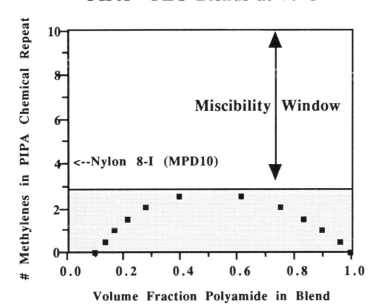

PIPA - PEO Blends at 70°C

FIGURE 7.72

In fact, it now it is better to consider a hypothetical methylene-co-aromatic amide (MPIPA) copolymer and ask the question, "how many aromatic amide groups must we incorporate, on average, into an otherwise polymethylene polymer backbone in order to render mixtures with PEO a single phase?"

This is a simple calculation requiring only that the **Calculation and Plot Option** window be changed to **Weight Percent** in the **Units** box below the **Copolymer B Composition** label and using appropriate values for the **Starting, Ending** and **Stepping**. A miscibility window shown in figure 7.73 is thus calculated, which predicts that MPIPA random copolymers containing from about 30 to 90 weight % methylene should be miscible. For comparison purposes, note also that we have identified on the right hand side of figure 7.73, the nylon 8-I or MPD10 "homopolymers" as if they were MPIPA copolymers of equivalent ethylene concentration (weight %). Again, this serves to emphasize the interchangability of homo- and random copolymers that is a fundamental component of the model we have constructed.

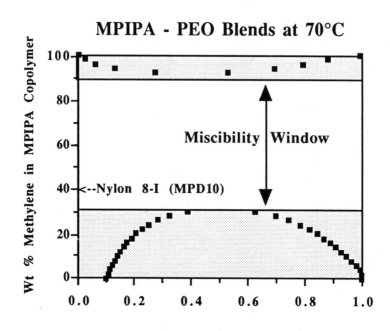

FIGURE 7.73

Methacrylic Acid Copolymer Blends

Butadiene-co-Methacrylic Acid (BDMAA) Blends with
Poly(Vinyl Pyridine) (PVPy)

In the previous main section (page 452), we left the safe haven of experimental data and calculated a miscibility window for a hypothetical blend system consisting of a *self-associated homopolymer*, polymethacrylic acid (PMAA) and an *unassociated copolymer*, butadiene-co-vinyl pyridine (BDVPy), and asked question, "how much butadiene (BD) can we incorporate into the PVPy polymer chain before a blend with PMAA becomes immiscible ?" The answer was predicted to be about 50% (figure 7.58). We thought it might now be interesting to reverse the question and ask, "how much butadiene (BD) can we incorporate into the PMAA polymer chain before a blend with PVPy becomes immiscible ?" We do not have any data on these blends, but we are presenting the results of our calculations to make a specific point.

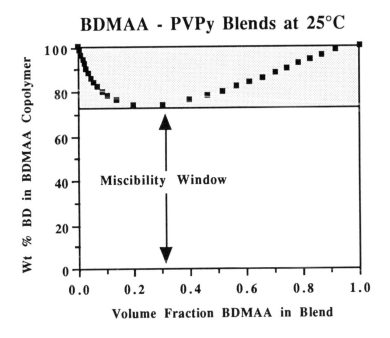

FIGURE 7.74

In copolymer / homopolymer blends there are some differences in the predicted miscibility windows between a system where we dilute the non self-associating component with "inert" diluent and a system where we dilute the self-associating component. Using the identical procedure to that described on page 452, with the exceptions that we now select the appropriate calculation procedure for a self-associated copolymer with an unassociated homopolymer (the bottom left hand icon in **Selection Calculation Type** window - figure 7.62) and introduce the polybutadiene segment information into **B Diluent** (figure 7.63), we simply recalculate. The result is shown in figure 7.74 and predicts that BDMAA copolymers containing up to about 75 weight % butadiene should be miscible with PVPy at ambient temperature.

It may be of interest to the whimsical reader to note that for the three segments PMAA, PVPy and PBD, diluting PMAA with butadiene leads to an wider miscibility window than if PVPy is copolymerized with the same diluent (c.f. figures 7.58 and 7.74).

E.   COPOLYMER (Non Self-Associated)– COPOLYMER (Self-Associated) SYSTEMS

Preamble

In this last section we will be considering the case of blends composed of two copolymers, one consisting of units that can self-associate copolymerized with "inert" (non hydrogen bonding) units, and the second consisting of units that do not self-associate, but with functional groups such as ethers or esters that can take part in inter-association, again copolymerized with inert diluent. Here we calculate miscibility windows at a particular temperature and vary the composition of both copolymers. For example, we might wish to simultaneously consider the effect of copolymerizing vinyl phenol with butadiene and methyl acrylate with ethylene. This is not as crazy as it sounds, and illustrates an important principle, as we hope to show!

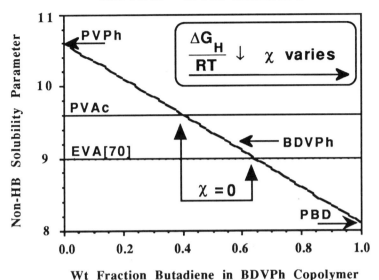

FIGURE 7.75

The calculated solubility parameters of PVPh and PBD are 10.6 and 8.1 (cal. cm^{-3})$^{0.5}$, respectively, which sets the limits of the solubility parameter range of BDVPh copolymers. This is illustrated in figure 7.75. A plot of the BDVPh solubility parameters as a function of copolymer composition is given together with, for simplicity, just two examples of EVA copolymers, EVA[100] (equivalent to PVAc) and EVA[70]. Note that the lines depicting the BDVPh and the two EVA copolymer solubility parameters would cross at different points, depending upon the composition of both copolymers. At these specific cross-over points, of course, χ is by definition zero, which presumably guarantees miscibility in the absence of a large unfavorable free volume contribution. The actual magnitude of χ as a function of both copolymer compositions is determined by the difference in the solubility parameters ($\Delta\delta$) of BDVPh and EVA and the value of the reference volume, V_B (a *variable* in the way we have defined it, because it is related to the *average* BDVPh chemical repeat). Values of χ as a function of BDVPh copolymer composition for blends with PVAc and EVA[70] is illustrated in figure 7.76.

BDVPh - EVA BLENDS

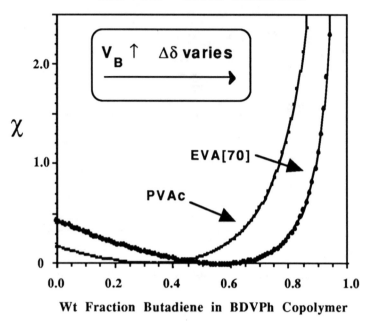

FIGURE 7.76

Thus, as we dilute PVPh with butadiene and PMA with ethylene, we still are effectively reducing the $\Delta G_H/RT$ term in equation 7.1. The crucial point we are making here, however, is that χ is now a more complex function of composition and we should anticipate a wide variety of miscibility windows and maps. The particular system we have used as an illustration has yet to be studied, so we will now turn our attention to some copolymer - copolymer blends where data is available.

Vinyl Phenol Copolymer Blends

Styrene-co-Vinyl Phenol (STVPh) Blends with Ethylene-co-Methyl acrylate (EMA) and Ethylene-co-Vinyl acetate (EVA)[10]

We will first consider the results of the calculations for the STVPh blends with EMA (and EVA), since we can compare the results to experimental data recently obtained in our laboratories.

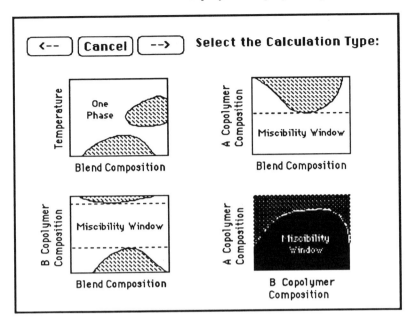

FIGURE 7.77

Returning to the **Phase Calculator** program; we first select the icon for the calculation of blends of two copolymers, where a self-associated copolymer B (STVPh, BDVPh, EMAA etc.) is mixed with copolymer A that can engage in inter-association interactions only (EMA, STMA, STVPy etc.) The applicable icon is highlighted at the bottom right hand side of figure 7.77.

Next, we select the appropriate model from the **Association Equilibrium Model** window (figure 7.21), followed by the **Segment Information** window (figure 7.44), where the relevent information for homopolymers PVPh and PMA segments are read from the segment files. Now select the **Copolymer Diluent Information** window (figure 7.78) where the segmental information of the *two* diluents, ethylene and styrene, are introduced from the segment files into the **A Diluent** and **B Diluent** boxes, respectively. (Note that both boxes may be accessed in this case).

The values of the equilibrium constants, K_2, K_B and K_A are again simply those given previously for PVPh - poly(alkyl methylacrylate) type blends and these are introduced into the **Association Information** window (figure 7.23).

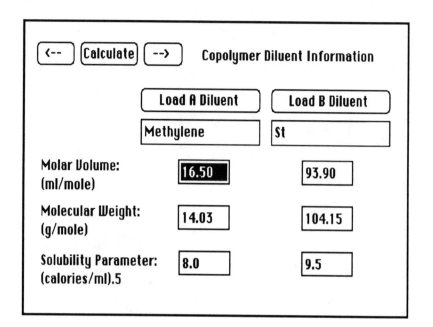

FIGURE 7.78

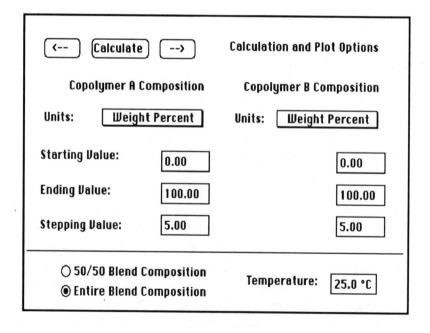

FIGURE 7.79

Now we are ready to calculate the miscibility window by selecting the **Calculation and Plot Option** window, which now appears as shown in figure 7.79. Appropriate values for **Starting, Ending and Stepping** for both **Copolymer A Composition** and **Copolymer B Composition** are now required, together with the **Temperature** at which the calculations will be performed. In addition, one may select either the radio button **50/50 Blend Composition** or **Entire Blend Composition** and this warrents further explanation.

The copolymer - copolymer miscibility window is calculated in the following manner. First, the second derivative of the free energy of mixing of a blend of the two homopolymers, PVPh ("copolymer" B ≡ STVPh[100]) and PMA ("copolymer" A ≡ EMA[100]), is calculated over the entire *blend* composition range at the desired temperature. This is equivalent to drawing a line at a temperature **T** in the schematic temperature / blend composition phase diagram presented in figure 7.80.

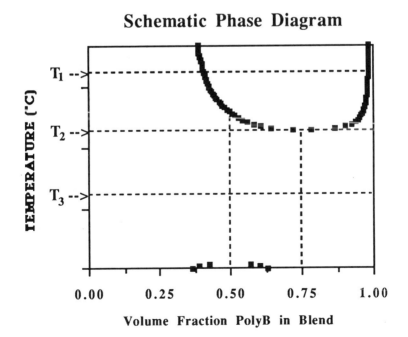

Schematic Phase Diagram

FIGURE 7.80

If the line at **T** crosses a two phase region of the spinodal phase diagram as in T_1 then the blend is defined as immiscible. Conversely, if the line at **T** does not cross a two phase region of the spinodal phase diagram as in T_3 then the blend is defined as miscible. A line at T_2 (the LCST in this example) is still defined as immiscible, however, using the selection **Entire Blend Composition**, as it cuts the two phase region somewhere over the blend composition range (in this case at about 75% PolyB). Nonetheless, several authors have employed 50/50 weight % blends to prepare experimental miscibility windows and we include an option where the second derivative of the free energy is tested only at **50/50 Blend Compositions**. The results are equivalent if the phase diagram is symmetrical about the 50/50 composition, but it is interesting to note that many of the calculated phase diagrams for hydrogen bonded systems are skewed. The reader may note that line at T_2 the blend is immiscible under the selection **Entire Blend Composition**, but miscible, albeit just so, under the selection **50/50 Blend Composition.**

This process is now repeated for PVPh and an EMA copolymer containing 98% methyl acrylate (EMA[98]) and then at 2% composition intervals (defined by the **Interval**) down to EMA[2]. Now the whole proceedure is repeated for STVPh[98] etc. etc., until the matrix corresponding to the copolymer - copolymer miscibility window is complete. The result of this large number of computations is shown for the STVPh - EMA blends at 25°C in figure 7.81.

The bottom left hand corner of the miscibility window (0,0) corresponds to a blend of PVPh and PMA which is predicted to be miscible. The top right hand corner (100,100) corresponds to a blend of *essentially*, (but not actually), PS and PE which is predicted to be immiscible. (The closest calculation to pure PS and PE is a blend of x % VPh in STVPh and y % ethylene in EMA, where x and y are defined by the **Interval** chosen). The top left hand corner (100,0) coincides with a blend of PVPh and essentially PE while that of the corresponding bottom right hand corner (0,100) reflects a blend of PVAc and essentially PS. Elsewhere in the matrix are all the other blends of the two STVPh and EMA copolymers of different copolymer compositions.

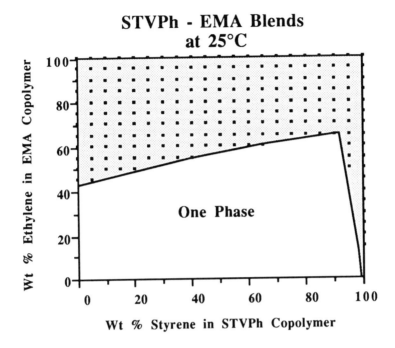

STVPh - EMA Blends at 25°C

Wt % Styrene in STVPh Copolymer

Wt % Ethylene in EMA Copolymer

One Phase

FIGURE 7.81

The "wedge" shaped one phase region shown in figure 7.81 predicts that as we dilute PVPh with up to about 90% styrene, the range of miscibility with EMA copolymers steadily increases from about 42 to 65 % ethylene. Further increasing the styrene content from about 90 to 98%, reverses the trend and miscibility is only obtained for EMA copolymers containing progressively less and less ethylene. To reiterate, this whole process is a reflection of the balance of the $\chi\Phi_A\Phi_B$ and $\Delta G_H/RT$ terms of equation 7.1.

As mentioned previously, we have synthesized a series of STVPh copolymers in our laboratories. In addition, we were fortunate to obtain a series of well characterized EMA copolymers from Dr. J. Harrell of the E. I. du Pont de Nemours & Company. Thus we were in a position to be able to put the theoretical miscibility window (figure 7.81) to the test[10]. At the time of writing the experimental study has just been completed and the results are truly encouraging. These are summarized in figure 7.82.

Miscibility Map for STVPh - EMA Blends
Comparison of Theory with Experiment

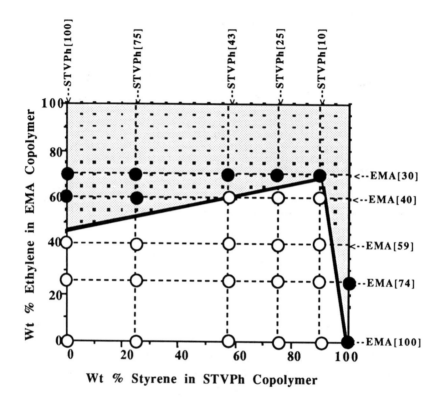

FIGURE   7.82

The black and unfilled filled circles represent, repectively, experimental data which show that blends of particular copolymers are grossly phase separated (immiscible) or single phase (miscible) materials. All in all, the agreement between theory and experiment is outstanding.

As we have mentioned previously, the calculated *segment information* for poly(methyl acrylate) and poly(vinyl acetate) is identical since they are composed of the same chemical groups. There are subtle differences in the values of the inter-association equilibrium constants, K_A, (58.7 versus 47.0 at 25°C for PVAc and PMA, respectively), but such differences have only a minor effect upon the calculated miscibility map. Thus the miscibility map shown in figure 7.81 is, to all intent and purposes, identical to that calculated for STVPh - EVA blends.

A comparison of the predicted miscibility map of STVPh - EVA blends to experimental results obtained in our laboratories is shown in figure 7.83. Again, the black and unfilled filled circles represent, repectively, experimental data which show that blends of particular copolymers are grossly phase separated (immiscible) or single phase (miscible) materials. The experimental result for the STVPh[10] - EVA[25] blend is denoted by a cross hatched circle signifying that the blend is "highly mixed", but we are not sure if it is indeed single phase. The agreement between theory and experiment for STVPh - EVA blends is remarkably good and again we believe this substantiates the general validity of our association model.

Miscibility Map for STVPh - EVA Blends
Comparison of Theory with Experiment

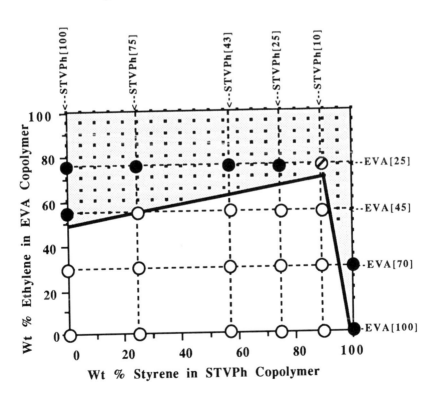

FIGURE 7.83

Styrene-co-Vinyl Phenol (STVPh) Blends with Styrene-co-Methyl Acrylate (STMA)[18]

Miscibility windows for PVPh blends with STMA copolymers (figures 7.47 and 7.48) have been previously described. Next we consider the case of a copolymer - copolymer miscibility window where we simultaneously consider the effect of diluting both poly(vinyl phenol) and poly(methyl acrylate) with styrene.

Returning to the **Phase Calculator** program we need to change the following: select the **Segment Information** window (figure 7.44), introduce the data for PVPh and PMA from the segment files, move to the **Copolymer Diluent Information** window (figure 7.78) and input the segmental information for styrene into both the **A Diluent** and **B Diluent** boxes and **Calculate**. This leads to figure 7.84.

The bottom left hand corner of the miscibility window (0,0) corresponds to a blend of PVPh and PMA which is predicted to be miscible.

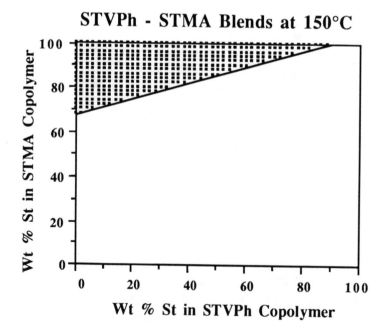

FIGURE 7.84

The top right hand corner (100,100) corresponds to PS mixed with itself which, of course, is also predicted to be miscible. The top left hand corner (100,0) coincides with a blend of PVPh and essentially PS while that of the corresponding bottom right hand corner (0,100) reflects a blend of PMA and essentially PS. Elsewhere in the matrix are all the other blends of the two STVPh and STMA copolymers of different copolymer compositions.

We again see a "wedge" shaped one phase region indicating that as we dilute PVPh with styrene, miscible mixtures are obtained for STMA copolymers ranging in composition from about 68% to 100% styrene. Using the STVPh and STMA copolymers synthesized in our laboratories we were again able to compare the theoretical miscibility window (figure 7.84) to experimental data[18].

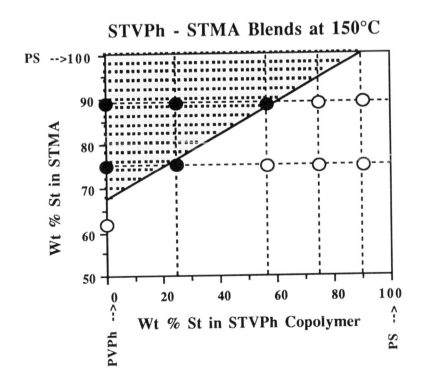

FIGURE 7.85

This is summarized in the scale expanded miscibility map shown in figure 7.85. The black and unfilled filled circles once more represent experimental FTIR data, which show that blends of particular copolymers are grossly phase separated (immiscible) or single phase (miscible) materials, respectively. Yet again, the agreement between theory and experiment is very good.

Styrene-co-Vinyl Phenol (STVPh) Blends with a Homologous Series of Poly(n-alkyl Methacrylates) (PAMA)[22]

We have previously described the results of calculations of conventional temperature / blend composition phase diagrams of PVPh blends with individual members of a series of PAMA's (figure 7.26). Composition / composition miscibility windows for PVPh blends with a homologous series of PAMA's (figure 7.54), or the equivalent EMMA copolymers (figure 7.55), have also been described. Next we considered miscibility windows calculated for STVPh copolymer blends with different PAMA polymers (figures 7.64 - 7.69). To complete the picture, we can now consider the case of a copolymer - copolymer miscibility window where we simultaneously consider the effect of diluting vinyl phenol with styrene and methyl methacrylate with methylene.

Returning to the **Phase Calculator** program we need to make only minor modifications; select the **Segment Information** window (figure 7.44) and introduce the data for PVPh and PMMA from the segment files; move to the **Copolymer Diluent Information** window (figure 7.78) and input the segmental information of the *two* diluents, methylene and styrene, into the **A Diluent** and **B Diluent** boxes, respectively. Moving to the **Calculation and Plot Option** window (figure 7.79) select **Incremental** for the units of **Copolymer A Composition**, with appropriate values for **Starting, Ending and Stepping**.

We obtain the miscibility window shown in figure 7.86 which was calculated by incrementally adding methylene groups to PMMA on the y-axis, while changing the weight % concentration of styrene in a STVPh copolymer on the x-axis.

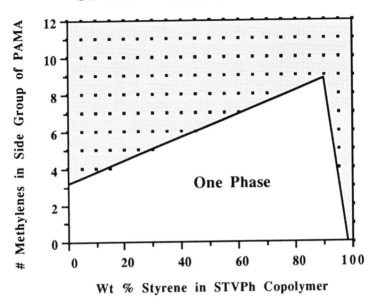

STVPh - PAMA Blends at 25°C

One Phase

Methylenes in Side Group of PAMA

Wt % Styrene in STVPh Copolymer

FIGURE 7.86

We again see a "wedge" shaped one phase region indicating that as we dilute PVPh with up to about 90% styrene, the range of miscibility with PAMA steadily increases from PBMA to poly(n-nonyl methacrylate). Further increasing the styrene content from about 90 to 98%, reverses the trend and miscibility is only predicted for PAMA's with smaller and smaller numbers of methylenes in the side group.

Using our STVPh copolymers, together with a series of previously synthesized amorphous poly(n-alkyl methacrylates), we are again in a position to experimentally test the accuracy of this theoretical miscibility window. The results are displayed schematically in figure 7.87[22]. The black filled circles represent experimental data which show that a particular blend is immiscible, such as the PHMA blends with PVPh, STVPh[75] and STVPh[2]. The unfilled circles represent experimental data for blends that have been established to be miscible, such as the PHMA blends with STVPh[43], STVPh[25] and STVPh[8]. The reader may note that the the two copolymers, STVPh[25] and STVPh[8], were found to be immiscible with POMA, but these lie at the extreme edge of the miscibility window for this system. We decided to "push our luck" and intentionally synthesized a copolymer of intermediate composition, STVPh[16], and found to our absolute delight that STVPh[16] is indeed miscible with POMA. We find this truly remarkable and most encouraging. At the risk of sounding immodest, we are forced to observe that the overall agreement between theory and experiment is astonishingly good and we believe this strongly supports the general validity of the model. (Incidentally, the reader may wish to compare figure 7.87 with that obtained for the same system using the qualitative **Miscibility Guide** - chapter 2, figure 2.66).

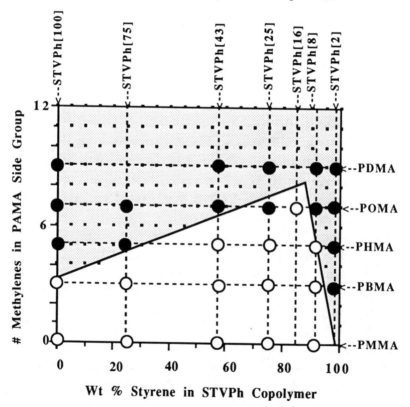

FIGURE 7.87

Butadiene-co-Vinyl Phenol (BDVPh) Blends with Ethylene-co-Methyl acrylate (EMA) and a Homologous Series of Poly(n-alkyl Methacrylates) (PAMA)

Since we appear to have been remarkably successful in theoretically predicting the copolymer - copolymer miscibility maps of STVPh blends with EMA (figure 7.82), EVA (figure 7.83), STMA (figure 7.85) and PAMA (figure 7.87) perhaps we can afford to return to the world of speculation. We encourage the reader to "play with" the computer and ask questions like, , "what happens if we substitute the diluent butadiene in place of styrene?" We do not have immediate access to random BDVPh copolymers and it is not a trivial task to synthesize and characterize these polymers. It is, however, a trivial task to perform calculations of miscibility windows to judge whether or not it is worthwhile embarking on such an experimental project, *and that is where we believe the model will prove to be very useful.* We will now consider some of these calculations as an illustrative exercise and see if the theoretical miscibility maps reveal anything of interest.

Returning to the **Phase Calculator** program, all we have to do is substitute the segment information information of butadiene for styrene in the **B Diluent** box of the **Copolymer Diluent Information** window (figure 7.78) and recalculate. The results are displayed for the BDVPh blends with EMA and PAMA in figures 7.88 and 7.89, respectively.

First, let us compare the miscibility windows of STVPh and BDVPh blends with EMA copolymers, shown respectively in figures 7.81 and 7.88. In the latter case the one phase (miscible) region is considerably larger and predicts that a blend of an ethylene copolymer containing 2% or less of methyl acrylate (or vinyl acetate) should be miscible with a butadiene copolymer containing 2% or less of vinyl phenol. If this doesn't make our industrial friends sit up and take notice then our worst fears are realized and this book will only serve as a handy projectile to throw at unproductive graduate students, because here we are predicting the possibility of designing a rubber modified plastic with a "compatible" interface by introducing a relatively small number of specific interaction sites.

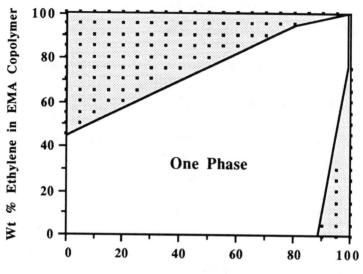

BDVPh - EMA (EVA) Blends at 25°C

Wt % Ethylene in EMA Copolymer

One Phase

Wt % Butadiene in BDVPh Copolymer

FIGURE 7.88

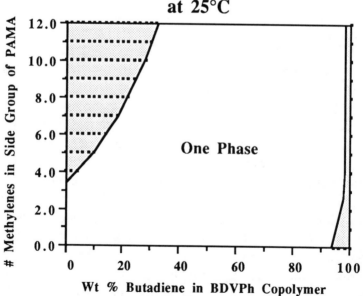

BDVPh - PAMA Blends at 25°C

Methylenes in Side Group of PAMA

One Phase

Wt % Butadiene in BDVPh Copolymer

FIGURE 7.89

Remember, this is a question of balance and involves both the $\chi\Phi_A\Phi_B$ and $\Delta G_H/RT$ terms of equation 7.1. If one now compares the miscibility windows of STVPh and BDVPh blends with the PAMA polymers shown respectively in figures 7.84 and 7.87, the one phase (miscible) region is again considerably larger in the latter case. Whereas, for example, PDMA is immiscible with all compositions of STVPh copolymers, it is predicted to be miscible with BDVPh copolymers containing from about 40 to 95 % VPh. This is a prediction that we intend to experimentally test in the near future. Incidentally, if one calculates the copolymer - copolymer miscibility window for BDVPh blends with hypothetical methylene-co-methyl methacrylate (MMMA) copolymers one obtains a result that resembles that of the BDVPh - EMA blends (figure 7.89) and the comments made above are also pertinent to these blends.

Methacrylic Acid Copolymer Blends

Ethylene-co-Methacrylic Acid (EMAA) Blends with Styrene-co-Vinyl Pyridine (STVPy) and Butadiene-co-Vinyl Pyridine (BDVPy)

These are our final two examples. Ethylene copolymers containing different amounts of methacrylic acid are precursors to the formation of ionomers and are commercially available. We thought it might be interesting to compare the calculated miscibility windows for EMAA copolymers with styrene or butadiene copolymers containing a segment known to exhibit a strong association to the acid group, e.g. vinyl pyridine. Without further ado, let us return to the **Phase Calculator** program, choose the acid dimer icon in the **Association Equilibrium Model** (figure 7.37); move to the **Segment Information** window (figure 7.44) and replace the segmental information of **A Segment** with that of PVPy; the **B Segment** with PMAA and the **A** and **B Diluents** with styrene (or butadiene) and ethylene, respectively (figure 7.78).

Now move forward to the **Association Information** window (figure 7.40) and introduce the appropriate equilibrium constants ($K_B = 173000$ and $K_A = 3000$) and recalculate. This results in the miscibility windows shown in figures 7.90 and 7.91.

It is unlikely that miscible EMMA - STVPy blends are worth pursuing because EMAA copolymers containing large amounts of MAA are required. However, a cursory glance at the EMAA - BDVPy miscibility window suggests again that only very few vinyl pyridine units need to be incorporated into polybutadiene to make it miscible with an ethylene copolymer containing very few methacrylic acid groups. The statement we made at the conclusion of chapter 2 is still valid and bears repetition - *"it is axiomatic in our scheme that the closer the match of the two non-hydrogen bonded solubility parameters and the greater the relative strength of the potential intermolecular interactions present between the polymeric components of the blend, the greater the probability of miscibility"*.

That's enough–you're on your own!

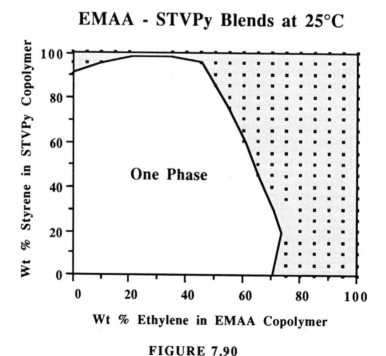

FIGURE 7.90

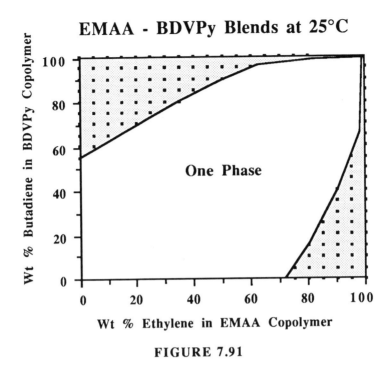

EMAA - BDVPy Blends at 25°C

Wt % Butadiene in BDVPy Copolymer (y-axis)

One Phase

Wt % Ethylene in EMAA Copolymer (x-axis)

FIGURE 7.91

F. REFERENCES

1. (a) Coleman, M. M., Skrovanek, D. J., Hu, J. and Painter, P. C., *Macromolecules*, 1988, **21**, 59.
    (b) Painter, P. C., Park, Y. and Coleman, M. M., *ibid.*, 1988, **21**, 66.

2. Coleman, M. M., Hu, J., Park, Y. and Painter, P. C., *Polymer*, 1988, **29**, 1659.

3. Hu, J., Painter, P. C., Coleman, M. M. and Krizan, T. D., *J. Polym Sci, Phys Ed.*, 1990, **28**, 149.

4. Bhagwagar, D. E., Painter, P. C., Coleman, M. M. and Krizan, T. D., *J. Polym Sci, Phys Ed.*, submitted.

5. Moskala, E. J., Howe, S. E., Painter, P. C. and Coleman, M. M. *Macromolecules*, 1984, **17**, 1671.

6. Coleman, M. M., Lichkus, A. M. and Painter, P. C. *Macromolecules*, 1989, **22**, 586.

7. Serman, C. J., Xu, Y., Painter, P. C. and Coleman, M. M., *Macromolecules*, 1989, **22**, 2015.

8. Serman, C. J., Painter, P. C. and Coleman, M. M., *Polymer,* in press.
9. Whetsel, K. B. and Lady, J. H., *"Spectrometry of Fuels"*, H. Friedle, Ed., Plenum, London, 1970.
10. Coleman, M. M., Xu, Y., Painter, P. C. and J. R. Harrell, *Makromol. Chem. Macromol. Symp.,* two papers accepted.
11. Powell, D. L. and West, R., *Spectrochim. Acta,* 1964, **20**, 983.
12. Gramstad, T., *Acta Chem. Soc.,* 1962, **16**, 807.
13. Serman, C. J., Xu, Y., Painter, P. C. and Coleman, M. M., *Polymer,* in press.
14. Lee, J. Y., Moskala, E. J., Painter, P. C. and Coleman, M. M., *Appl. Spectrosc.,* 1986, **40**, 991.
15. Meftahi, M. V. and Frechet, J. M. J., *Polymer,* 1988, **29**, 477.
16. Lee, J. Y., Painter, P. C. and Coleman, M. M., *Macromolecules,* 1988, **21**, 346.
17. Coleman, M. M., Lee, J. Y., Serman, C. J., Wang, Z. and Painter, P.C., *Polymer,* 1989, **30**, 1298.
18. Coleman, M. M., Zhang, H. X. and Painter, P. C., to be published.
19. Xu, Y., Painter, P. C. and Coleman, M. M., *Makromol. Chem. Macromol. Symp.,* accepted.
20. Chen, C-T. and Morawetz, H. *Macromolecules,* 1989, **22**, 159.
21. Serman, C. J., Xu, Y., Painter, P. C. and Coleman, M. M., *Macromolecules,* 1989, **22**, 2015.
22. Xu, Y., Painter, P. C. and Coleman, M. M., *Polymer,* in press.

Absorption Coefficients
  effect of frequency on, 266, 271
  effect of temperature on, 274
  relationship to dipole moment, 272
Absorptivity ratios
  determination of, 274
  values of, 275
Association Equilibrium Constants
  chain stiffness, effect of, 341
  definition of, 172, 176
  measurement of, 275
  molecular weight, effect of, 341
  thermodynamic relationships, 173
  transferability of, 214, 293
Association Models, 176, 309
  chemical potentials, 339
  cyclic dimer association, 377
  enthalpy and entropy of mixing, 360
  free energy, 325
  free volume effects, 368
  melting point depression, 354
  open chain association, 314
  phase behavior, 344
  spinodal equations, 344
  spinodal, sample calculations, 349
  two equilibrium constant model for linear
    self-association, 379

Binary Mixtures of Small Molecules, 1
  butyl alcohol in water, 28
  ideal mixtures, 2
  m-toluidine and glycerol, 27

χ Values
  estimation by solubility parameters, 54
Competing Equilibria, 185
  equilibrium constants, 186

lattice model, 212
  relationship to fraction of free groups,
    187-190
Copolymer Blends
  see *Polymer Blends*
  transferability of equilibrium constants,
    293
Cosolvency, 37, 70
Critical Values
  effect of inert diluent on, 73-75
  of χ, 22, 25, 50, 73
  of solubility parameter difference, 52, 73
  upper limits for different specific
    interactions, 152

Enthalpy of Hydrogen Bond Formation
  for phenolic hydroxy-ester interaction,
    292
  for polyamides, 266
  for self-association in poly(vinyl
    phenol), 285
  from van't Hoff plot, 284

Group Molar Constants
  by addition or subtraction ,68
  unassociated groups, 59
  weakly associated groups, 59

Hydrogen Bond
  characterization of, 162
  definition of, 159
  effect on vibrational spectra, 222
  in polymers, 164, 199
  in small molecules, 314
  nature of, 157
  relative strength, 160

Hydrogen Bonding in Polymers
distribution in mixture, 331
equilibrium constants, 199
free volume effects, 368
inter-association, description of, 167
linear association, 329
reference states, 334
self-association, description of, 164
stoichiometry, 199

Infrared Spectra of:
amorphous polyamide, 267
amorphous polyurethane
poly(ethylene oxide-co-propylene
oxide) blends, 251
poly(ethylene oxide) blends, 252
poly(vinyl methyl ether) blends, 252
nylon-11, 242
poly(ethylene-co-methacrylic acid)
poly(vinyl methyl ether) blend, 239
polyimide model compound, 262
poly(styrene-co-vinyl phenol), 276
poly(vinyl phenol) blends with:
poly(ethyl methacrylate), 290
poly(ethylene-co-vinyl acetate), 241
poly(vinyl pyridine), 260
semi-crystalline polyurethane, 258
Infrared Spectroscopy
Beer-Lambert law, 237
curve resolving, 237-252
equilibrium constants; determination of,
275-297
FTIR spectrometers, 233
interferogram, 235
Michaelson interferometer, 234
quantitative analysis, 237
sample preparation, 236
second derivative spectra, 2434
Inter-association, 168
amide-ether, 173
urethane-ether, hydroxyl-ester,
acid-pyridine, 167
Intermolecular Interactions, 3
"chemical" interactions, 5, 8
cohesive energy density, 7
hydrogen bonding in polymers, 164
interaction strength, 3
moderate specific interactions, 107, 127
"physical" interactions, 5, 6, 52
polar forces, 85
strong dipolar forces, 9
very weak specific interactions, 76
weak specific interactions, 93
weak to moderate specific interactions,
102

Mapping phase diagrams, 297
MPD6-PEO system, 304
single phase, 299

single phase versus two phase, 303
Miscibility Guide
calculation choices, 80
constant critical value option, 120
copolymer-copolymer option, 89
critical windows, 81
input of segment information, 77
specific repeat definition, 129, 134
Miscibility Windows and Maps
amorphous polyurethane
aliphatic polyethers, 136, 450
poly(ethylene oxide-co-propylene
oxide), 135
poly(vinyl alkyl ethers), 451
polyarylate-aromatic polyesters, 88
poly(bis-phenol A carbonate)
aliphatic polyesters, 104
poly(butadiene-co-methacrylic acid)
poly(vinyl pyridine), 469
poly(butadiene-co-vinyl phenol)
poly(n-alkyl methacrylates), 465, 485
poly(ethylene-co-methyl acrylate), 485
poly(chloromethyl methacrylate)
poly(p-methyl styrene-co-
acrylonitrile), 118
poly(cyclohexyl methacrylate)
poly(styrene-co-acrylonitrile), 118
poly(ethylene-co-methacrylic acid)
poly(butadiene-co-vinyl pyridine), 488
poly(styrene-co-vinyl pyridine), 487
poly(ethylene oxide)
polyisophthalamides, 131
poly(ethylene-co-methacrylic acid),
149
poly(styrene-co-acrylic acid), 151
poly(hydroxy ether of bis-phenol A)
aliphatic polyesters, 128
polyisophthalamides
poly(ethylene oxide), 467
poly(methacrylic acid)
poly(butadiene-co-vinyl pyridine), 453
poly(methyl methacrylate)
aliphatic polyethers, 87
poly(p-methyl styrene-co-
acrylonitrile), 118
poly(styrene-co-acrylonitrile), 115
poly(styrene-co-vinyl phenol), 142
poly(n-alkyl methacrylates)
poly(ethylene oxide), 86
poly(styrene-co-acrylonitrile), 116
poly(styrene-co-vinyl phenol), 143
polystyrene
poly(butadiene-co-acrylonitrile blends),
80, 83
poly(styrene-co-acrylonitrile)
poly(methyl methacrylate-co-t-
butyl methacrylate), 124
poly(methyl methacrylate-co-
cyclohexyl methacrylate), 123
poly(methyl methacrylate-co-N-
phenyl itaconimide), 126

Miscibility Windows and Maps (cont.)
poly(styrene-co-acrylonitrile)
  poly(methyl methacrylate-co-
    phenyl methacrylate), 121, 122
  poly(styrene-co-maleic anhydride),
    98-102
  poly(styrene-co-N-phenyl itaconimide),
    106
poly(styrene-co-methyl methacrylate)
  poly(styrene-co-N-phenyl
    itaconimide), 93
poly(styrene-co-vinyl phenol)
  poly(alkyl methacrylates), 460-462,
    481, 483
  poly(ethylene-co-methyl acrylate),
    144, 477, 478
  poly(ethylene-co-vinyl acetate), 477
  poly(styrene-co-methyl acrylate),
    479, 480
poly(tetramethylene-bis-phenol A
    carbonate)
  aliphatic polyesters, 104
poly(vinyl chloride)
  aliphatic polyesters, 108, 109
  poly(butadiene-co-acrylonitrile), 97
  poly(n-alkyl methacrylates), 111
  poly(ethylene-co-vinyl acetate), 111
  poly(styrene-co-acrylonitrile), 95
poly(vinyl methyl ether)
  polyisophthalamides, 132, 133
poly(vinyl phenol)
  poly(alkyl acrylates), 138
  poly(alkyl methacrylates), 139, 447
  poly(ethylene-co-methyl methacrylate),
    447
  poly(ethylene-co-vinyl acetate),
    140, 443
  poly(styrene-co-methyl acrylate), 443
  poly(styrene-co-vinyl pyridine), 445
  poly(vinyl n-alkyl ethers), 146
poly(vinyl pyridine)
  poly(styrene-co-vinyl phenol), 147

Phase Behavior
binodal, 22
closed immiscibility loops, 26
conditions for phase separation, 19
cosolvency, 34, 40
free volume, 26, 30
lever rule, 297
lower critical solution temperatures, 26,
    298
"repulsion" effect, 37
repulsion model, 14, 37
spinodal, 22, 344
systems that hydrogen bond, 35
temperature dependence, 348
upper critical solution temperatures,
    22-26
Phase Calculator Program

association equilibrium model, 381, 405,
    430
association information, 386, 408, 426,
    433
calculation and plot options, 439, 446,
    474
  degree of H-Bonding, 387
  free energy, 391
  phase behavior, 395
  second derivative of the free energy,
    391
calculation type, 383, 435, 458, 473
copolymer diluent information, 438,
    458, 474
segment information, 385, 406, 431,
    435
Phase Diagrams
amorphous polyurethane
  polyethers, 396-399
aromatic-aliphatic polyamides
  poly(ethylene oxide), 403
effect of errors on, 414
poly(methacrylic acid)
  polyethers, 434
poly(vinyl phenol)
  aliphatic polyesters, 424
  poly(alkyl acrylates), 422
  poly(alkyl methacrylates), 412
  poly(n-butyl methacrylate), 415, 419,
    420
  poly(n-hexyl methacrylate), 417
  poly(n-methyl methacrylate), 416
  poly(vinyl methyl ethers), 427
  poly(vinyl pyridine), 429
Polymer Blends
amorphous polyurethane
  polyethers, 133, 382-399, 449-451
aromatic-aliphatic polyamides, 37
aromatic polyamides
  poly(ethylene oxide), 129, 399-404
    466-468
copolymer-homopolymer, 37, 44, 70
homopolymer systems, 37
melting point depression, 354
polyarylate
  aromatic polyesters, 87
  poly(butylene terephthalate), 83
poly(alkyl and chloroalkyl
    methacrylates)
  poly(p-methylstyrene-co-acrylonitrile),
    113
poly(bis-phenol A carbonate)
  aliphatic polyesters, 102
poly(butadiene-co-methacrylic acid)
  poly(vinyl pyridine), 469
poly(butadiene-co-vinyl phenol)
  poly(ethylene-co-methyl acrylate), 484
  poly(ethylene-co-vinyl acetate), 471
  poly(n-alkyl methacrylates), 463
polyethylene
  poly(propylene glycol, 28

Polymer Blends (cont.)
  poly(ethylene-co-methacrylic acid)
    polyethers, 148
    poly(butadiene-co-vinyl pyridine), 486
    poly(styrene-co-vinyl pyridine), 486
  poly(ethylene oxide)
    poly(styrene-co-acrylic acid), 150
  polyisophthalamides-polyethers, 129
  poly(methacrylic acid)
    poly(butadiene-co-vinyl pyridine), 452
  poly(methyl methacrylate)
    aliphatic polyethers, 84
    poly(ethylene oxide), 55, 84
    poly(p-methylstyrene-co-
      acrylonitrile), 113
    poly(styrene-co-acrylonitrile), 113
    poly(styrene-co-vinyl phenol), 72
  poly(n-alkyl methacrylates)
    poly(ethylene oxide), 86
    poly(styrene-co-vinyl phenol), 143
  polystyrene
    polybutadiene, 26
    poly(butadiene-co-acrylonitrile), 77
    poly(methyl methacrylate), 18
    poly(phenylene oxide), 84
    poly(vinyl methyl ether), 18, 30, 36,
      84
  poly(styrene-co-acrylonitrile)
    poly(alkyl and chloroalkyl
      methacrylates), 113
    poly(methyl methacrylate-co-t-butyl
      methacrylate), 113
    poly(methyl methacrylate-co-
      cyclohexyl methacrylate), 113
    poly(methyl methacrylate-co-N-phenyl
      itaconimide), 125
    poly(methyl methacrylate-co-phenyl
      methacrylate), 113
    poly(styrene-co-maleic anhydride), 98
    poly(styrene-co-N-phenyl itaconimide),
      105
  poly(styrene-co-methyl methacrylate)
    poly(styrene-co-N-phenyl itaconimide),
      89
  poly(styrene-co-vinyl phenol)
    poly(n-alkyl methacrylates), 455, 481
    poly(n-butyl methacrylate), 297
    poly(ethylene-co-methyl acrylate), 472
    poly(styrene-co-methyl acrylate), 478
  poly(tetramethyl-bis-phenol A carbonate)
    aliphatic polyesters, 103
  poly(vinyl acetate)
    poly(ethylene oxide), 84
  poly(vinyl chloride)
    aliphatic polyesters, 108
    poly(butadiene-co-acrylonitrile), 96
    poly(ε-caprolactone), 102
    poly(ethylene-co-vinyl acetate), 111
    poly(n-alkyl acrylates), 110
    poly(n-alkyl methacrylates), 110
    poly(styrene-co-acrylonitrile), 93
  poly(vinyl acetate), 112
  poly(vinyl phenol)
    aliphatic polyesters, 136, 423
    poly(alkyl acrylates), 137, 421
    poly(alkyl methacrylates), 139,
      405-420, 446-449
    poly(ethyl methacrylate), 289
    poly(ethylene-co-vinyl acetate),
      139, 441
    poly(methyl methacrylate), 416
    poly(n-butyl methacrylate), 35,
      136, 297, 372, 415, 419, 420
    poly(hexyl methacrylate), 417
    poly(n-propyl methacrylate), 289
    poly(styrene-co-methyl acrylate), 441
    poly(styrene-co-vinyl pyridine), 444
    poly(vinyl acetate), 139
    poly(vinyl alkyl ethers), 145, 423
    poly(vinyl pyridine), 146, 423
  poly(vinyl pyridine)
    poly(styrene-co-vinyl phenol), 146
  phenoxy-poly(ε-caprolactone, 127
Polymer Solutions
  lattice model, 15
  polyisobutylene in nonpolar solvents,
    30
  polystyrene-acetone system, 29
  polystyrene fractions in cyclohexane,
    24

Quantitative infrared analysis, 237
  fraction of hydrogen bonded groups, 273
  inter-association equilibrium constants,
    285
  self-association equilibrium constants,
    273

Regular Solutions, 10
  geometric mean assumption ,12
  regular solution theory, 11

Self-association, 164
  definition of equilibrium constant, 176
  definition of the "interacting unit", 204
  in amides, urethanes, hydroxyls, 165
  in ureas, acids, urazoles, 166
  lattice model, 202
  number average degree of association,
    185
  relationship to fraction of free groups,
    183
  two equilibrium constant model, 191
Solubility Parameters
  definitions, 7, 12, 14
  discrepancies in calculations, 55
  errors involved in calculations, 60
  experimental vs calculated, 65
  for strongly associated polymers, 64
  non-hydrogen bonded, 68

Stoichiometry of Hydrogen Bonding, 180
  carboxylic acids and urazoles, 196
  competing equilibria, 185
  effect of chain flexibility, 200
  simple self-association, 180
  temperature dependence of equilibrium
    constants, 190

Thermodynamic Properties
  associated solutions, 320

Vibrational Spectroscopy
  anharmonic diatomic oscillator, 229
  harmonic diatomic oscillator, 226
  hydrogen bonding systems, 253
    absorption coefficients, 266
    acid salts, 263
    band widths, 266

classification of, 256
correlation between frequency and
    A - - B bond length, 264
effect in O-H and N-H stretching region,
    258, 266
free and hydrogen bonded groups,
    266-272
frequency shifts, 262
infrared spectra of, 258-263
hydrogen bonding systems, 253
model describing intensity
    variations, 268
potential energy diagram for
    A-H - - B system, 254
"satellite" bands, 259
normal modes of vibration, 223
origin of spectrum, 222
selection rules, 227

THE **MG&PC** SOFTWARE

(MISCIBILITY GUIDE & PHASE CALCULATOR)

Software support for "Specific Interactions and the Miscibility of Polymer Blends: Practical Guides for Predicting & Designing Miscible Polymer Mixtures."

Only purchasers of this book are entitled to purchase *The MG&PC Software*. As proof of purchase, **please return this form** when ordering your copy of the software.

☐   Apple Macintosh version (628238-A)
☐   IBM or compatible version (628238-B)

Price: **$350.00**

Prepayment is required for personal orders and orders from outside the U.S. Postage and $3.00 handling charge added to billed and credit card orders. PA orders: add 6% sales tax. CANADA: add 7% GST plus US $2.00 processing fee. GST Reg. #RI24I70630.

(Please print)
ORGANIZATION _____

NAME _____

ADDRESS _____

CITY _____ STATE _____ ZIP _____

CREDIT CARD ORDERS: ☐ VISA   ☐ Mastercard   ☐ American Express

CARD NO. _____ EXP. DATE _____

SIGNATURE _____

Please detach and return with order to:
TECHNOMIC Publishing Company, Inc.
851 New Holland Ave., Box 3535, Lancaster, PA 17604, U.S.A.